AF673266

Plant Maintenance and Reliability Engineering

N. V. S. Raju
Professor of Mechanical Engineering and Principal,
Jawaharlal Nehru Technological University Hyderabad (JNTUH)
University College of Engineering Wanaparthy,
Wanaparthy (D), Telangana, India.

4-4-309/316, Giriraj Lane,
Sultan Bazar, Hyderabad- 500 001,
India.

Plant Maintenance and Reliability Engineering ***by*** **N. V. S. Raju**

Published by:

4-4-309/316, Giriraj Lane,
Sultan Bazar, Hyderabad-500 001, India.
Phone: 040 - 23445688
e-mail: info@bspbooks.net
www.bspbooks.net

ISBN: 978-93-49562-73-8 (Hardback)

Dedicated to
the Feets of Sadguru Sainath
and
All those Gurus
(Including my mother & father)
who brought me up

JAWAHARLAL NEHRU TECHNOLOGICAL UNIVERSITY
HYDERABAD
KUKATPALLY, HYDERABAD, ANDHRA PRADESH.

FOREWORD

Plant Maintenance and Reliability Engineering is an important subject for the students of Mechanical Engineering and its related branches. This subject is originally derived from the concepts of industrial engineering and production management. However, it has experienced many revisions as it is applied to solve industrial and even problems of daily life. Particularly, with the focus shifting to reliability studies, the subject offers a new dimension to effective and efficient decision to enhance the life of equipment.

The topics covered in this book are integral part of the syllabi in almost all universities in India. The subject is taught either as a complete paper or partially under various papers such as operations management, production management, industrial engineering, reliability engineering, maintenance management, total productive maintenance, etc. *Plant Maintenance and Reliability Engineering* is an important area owing to its practical applicability in daily life.

In my view, there are two types of learning—first, the *chalk and talk* method practised in a conventional classroom (learning like 'Arjunas'); second, the *distance learning* mode, in which case there is no help from the teacher and the learner has recourse only to study material (learning like 'Ekalavyas'). So a book may be considered to be the best if it can cater to and benefit both Arjunas and Eklavyas. This book authored by Dr. N.V.S. Raju is one among such good books. Dr. Raju's rich experience in teaching as well as in industry is embedded in every page of this book to the benefit of even the layman.

The contents in this book are a perfect blend of sufficient theory and a large number of illustrations with practical examples, which not only create but also sustain the interest of the reader. Since there is limited literature available in this field, I truly believe that this comprehensive book is a welcome addition to the texts available in the market. Finally, since it covers the entire field of Maintenance Management and Reliability Engineering and related disciplines, I strongly recommend this book to not only the students of B.Tech. and M.Tech., but also the practitioners in the industry.

Dr. T. Kishen Kumar Reddy

Dr. T. Kishen Kumar Reddy
Vice Chancellor
Professor of Mechanical Engineering
Jawaharlal Nehru Technological University, Hyderabad
Formerly Advisor, AICTE, New Delhi, India (Apr '02 to Apr '05)
Director, Academic Audit, JNTUH
Dt. 20 December 2010

Contents

Preface

Easier said than done

I experienced the truth in the above proverb practically in my career several times. During my research days, I had a lot of difficulty collecting literature related to my areas of interests and focus. There were no comprehensive books available in the area and I had to collect most materials from a vast number of books. Further, when I joined industry (M/s Hyderabad Allwyn Ltd, Patancheru), I was part of the Production Department. Later, I was transferred to the Plant Engineering Department. This switch necessitated further understanding of the plant engineering field on my part. Here again, the dearth of a systematic and comprehensive book in the subject area resurfaced as a major challenge. Subsequently, when I started teaching and training, I again felt the need for a consolidated book that would be useful to all concerned—students, researchers, practitioners, and teachers.

With my knowledge and experience, I thought I could write this book quite easily. I believed writing a book posed no challenge. It was while writing that I experienced the real pains and realised that it is easier said than done. However, now when I present this book in its present form, I truly believe that the pains of writing this book have been worthwhile and an inexplicably fulfilling experience.

About the field

Maintenance Management has gained importance in the recent times with the streaming technological developments. The need for enhanced knowledge to deal with the advanced plant operations has been not only felt but also acknowledged and, thus, an initiation has been taken up to bring forth the emerging issues faced by plant engineers on a day-to-day basis and the timely resolution and prevention of such issues.

Enhanced reliability of systems is essential for progressive operation or environment. This field is now attracting the attention of everyone in the industry. Although reliability engineering is common interest of all the streams of engineering and management, it is more appropriate to be associated with plant maintenance.

About the book

This book is intended to majorly serve four segments of people related to this field, namely, students, researchers, practitioners, and teachers. The basic aim of the book is to provide theoretical and practical knowledge accompanied with the latest technological concepts and research in the field. This book is unique in its kind due to its comprehensive and up-to-date offering on the subject. The book is organised into chapters, touching most aspects of plant maintenance and reliability engineering. The text is supplemented by a number of pedagogical features to ensure easy understanding of the concepts discussed. Suitable illustrations, figures, and case studies are provided to clarify and amplify the concept. Several numerical problems are included in each chapter, for both illustrative and practice purposes. Extensive exercises are included to test the readers' understanding of the topics discussed. In addition, under every concept some practical activities are suggested in the form of Bustles. These are intended to not only enhance the understanding of the concept but also enable proper application under relevant conditions. These bustles can also be used as main or mini projects. This practical exposure can boost the readers' confidence working in the field of plant engineering. A list of references is provided at the end of the book for those who are interested in further reading.

Contents and structure

Chapter 1 introduces the basic concepts of maintenance and plant engineering. The objectives of maintenance department and the functions that are carried out by the maintenance department of an organisation are thoroughly discussed. Some light is thrown on emerging issues in maintenance/plant engineering area, such as tero-technology or total productivity maintenance, etc.

Chapter 2 discusses the importance of planning in maintenance and plant engineering and explains the functions of planning maintenance and plant engineering and management (PEM). The planning functions of maintenance are identified and executed. Various systems such as centralised and decentralised systems for organising maintenance are discussed. This chapter focuses on bringing out and establishing interface between PEM and other departments planning and organising the maintenance activities.

Chapter 3 emphasises on various methods or policies of maintenance from which an organisation may select one or more to meet its requirements. The policies are designed depending on prevailing factors in a given organisation. The details on various maintenance methods, policies, and strategies are clearly discussed in this chapter. The systematic procedures for carrying out maintenance and setting a maintenance strategy are also discussed in this chapter.

Chapter 4 focuses on the importance of maintenance scheduling and the methods of scheduling. Various scheduling procedures for breakdown maintenance and preventive maintenance are discussed here. The sequence of the jobs can be determined depending on the type of the organisations and their requirements and availability of resources. The scheduling formats of a product industry are also discussed for preparing maintenance master schedule.

Chapter 5 emphasises on the classification and controlling of the spare parts and solving the problems related to spare parts inventory. This chapter discusses three inter-related topics of spare parts management, namely, selective inventory control techniques, codification systems, and standardisation of the spare parts. Discussions are extended to the maintenance system optimisation and other relevant applications. The codification and classification systems are then explained including parts classification, Opitz classification, multi-class, and general classifications. The standardisation, its significance, methods, levels, advantages, and barriers are described with a special reference to the maintenance department. A list of some applications of computers in maintenance department is provided at the end of the chapter.

Chapter 6 lays emphasis on the concept of facility and determines the situation when the decision of facility planning arises. It is greatly swayed by the type of the industry that they support particularly, in the decisions such as product design, plant layout, buildings and grounds, location planning, size of the facility, expansion, diversification, organisation development, etc. All the factors influencing the facility planning and considerations in facility location planning are explained in detail. This chapter will give a better understanding of selection of a location and site that can maximise the facilities, design a layout plan for a plant, and design the material flow line.

Chapter 7 deals with the failure pattern of given equipment. It helps distinguish if the equipment is repairable or replaceable (non-repairable). It focuses on assessing replacement age based on the costs. The replacement age of a given machine is discussed in this chapter. It identifies and determines suitable replacement policy, if the equipment is of replaceable type. It helps understand and interpret the failure trend of the equipment if it is repairable type.

Chapter 8 focuses on the approaches of failure analysis—failure mode and effect analysis (FMEA) and failure mode, effects, and criticality analysis (FMECA). The elements of FMECA, the basic information required for FMECA and the basic analysis procedure of FMEA/FMECA are discussed. The different analysis approaches such as hardware approach and functional approach are also covered in this chapter. The requirement or utility of block diagram in such failure analysis has been explained. Statistical process control uses sampling and other statistical methods to monitor the quality of an ongoing process. Probability sampling methods and non-probability sampling methods are frequently used for cost and time advantages. A few probability sampling methods available are discussed. Design FMEA and process FMEA are discussed and a brief comparative study has been presented in this chapter.

Chapter 9 focuses on the various maintenance philosophies. Before adopting a particular philosophy, it must be thoroughly examined. The implementation of diagnostic maintenance by using fault diagnosis techniques is discussed. The philosophies such as failure-based, time-based, condition-based, reliability-centred, etc., are discussed. The condition-based maintenance techniques and approaches are discussed in this chapter.

Chapter 10 defines and explains various philosophies of maintenance. Various philosophies of the maintenance management, their central themes, principles and their evolution are extensively discussed. The significance of condition monitoring, the principles, the method, and the techniques of CBM are detailed with appropriate support. This chapter helps understand diagnosing methods to find the fault in a machine. It also explains various destructive and non-destructive tests to estimate the machine condition. The advantages of CBM are also discussed here.

Chapter 11 focuses on the concept of trend and its computation. It analyses the failure patterns of equipment. Behaviour of machines, machine life cycle (MLC), and bathtub curve are used to explain different types of failure patterns. Various types of failures based on the volume, the mode of failure are discussed. The trends and patterns of failures with time of repairable systems are described using graphical methods such as cumulative plot test, eye ball analysis, and analytical trend tests such as Laplace test, Mil-hdbk-189 test are elucidated. Test for presence of correlation using serial correlation coefficient of correlation test can also help in tracking the trend.

Chapter 12 emphasises on identifying, collecting relevant data, and analysing a problem related to reliability of the equipment. A clear cut path is established to predict the repairable equipment behaviour in this chapter. It conducts trend analysis and examines various reliability-centred maintenance models for fitting the data. A goodness of fit test is performed which confirms the distribution of failure patterns. A suitable maintenance policy/strategy and schedules can be suggested based on the analysis. This chapter will help design and assign the reliability characteristics.

Chapter 13 focuses on reliability, availability, and maintainability requirement as design goals of specifications. Various combinations of the components such as series system, parallel system, and combined (series and parallel) system are discussed that will help enhance or design the overall reliability of the systems as needed. The advantages of modular design or modularisation are explored and then quality and reliability are distinguished. Different types of availability such as intrinsic availability (A_i), achieved availability (A_a), operational availability (A_o) and the factors affecting availability and hence the reliability are discussed. The effect of redundancy and group redundancy on reliability is narrated through the numerical problems. The evaluation of overall equipment effectiveness (OEE), maintenance performance, and efficiency are also depicted through numerical problems.

Chapter 14 focuses on total productive maintenance (TPM) that aims at maximising the equipment effectiveness like tero-technology. The eight pillars of total productive maintenance and equipment effectiveness in maintenance are discussed. This chapter also covers the TPM template, expected contributions by application of TPM, and the barriers of adopting the TPM in the organisation. It also explains the concept of zero defects, zero maintenance, and zero accidents. This chapter will also help understand and implement Kaizen with right outlook.

Chapter 15 delves on the importance of safety in industry and the role of safety manager during discharging the duties of safety management entrusted with three-fold responsibilities, i.e., before, during, and after the accident. It analyses the causes and prevention of the occurrences of accidents. Various legislative provisions relevant to industrial safety are discussed in detail. The topics discussed in this chapter should be of utmost importance to the employees of plant engineering due to the risk-associated task they involve in on a daily basis.

N.V.S. Raju

ACKNOWLEDGEMENT

At the outset, I express my deep sense of gratitude for encouragement and support extended by

- Dr. D.N. Reddy, Vice Chancellor, JNTU Hyderabad; Former Principal and Professor of Mechanical Engineering, O.U., Hyderabad
- Dr. K. Narayana Rao, Professor and Head, Department of Mechanical Engineering, O.U., Hyderabad; Former Member Secretary, AICTE, New Delhi
- Dr. K.R.M. Rao, Former Dean and Professor (Retd.) of Electrical Engineering, O.U., Hyderabad
- Dr. T. Kishen Kumar Reddy, Director, Academic Audit, JNTU, Hyderabad
- Dr. A. Vinaya Babu, Professor of CSE, Chairman, Board of Studies for CSE, JNTU; Director, Admissions, JNTU, Hyderabad
- Dr. C.B. Krishna Murthy, Professor (Retd.) in Mechanical Engineering, JNTU (Former Principal of JPNEC-Mahaboobnagar and NIET-Deshmukhi, Nalgonda District)
- Dr. (Retd. Col.) K. Prabhakar Rao, Principal, Raja Mahendra Engineering College, Hyderabad
- Dr. A. Govardhan, Principal, JNTUH College of Engineering, Nachupally (Kondagattu), Karimnagar
- Dr. B.S. Prem Kumar, Vice-Principal, JNTUH College of Engineering, Nachupally (Kondagattu), Karimnagar

I particularly thank Mr. S.R. Akkina, Mr. T. Ramalingachary, Mr. A.S. Ananad, Mr. R.S. Rao, Mr. Sk. A. Basha, and all my former colleagues at M/s Hyderabad Limited, Watch Division, Patancheru, for the inputs and cooperation.

My special thanks to Mr. N. Venkateshwarlu, Reader, and other friends at SOET, IGNOU, New Delhi.

I specially thank all my present and former colleagues and students in general and Dr. D.V. Ravi Shankar, Principal, NIET; Mr. N.L. Narayana, Vice-Principal, PIET; Mr. K. Vasantha Kumar, APME, JNTUHCEN; Mr. A. Suresh Rao, Professor of CSE, Vagdevi CEW; Mr. Somasundaram Kalaga (USA); Dr. S. Venkatesh Associate Professor, OU; Dr. M. Manzoor Hussain and Dr. K. Eashwar Prasad, Professor of ME, JNTUHCEH in particular.

While authoring this book, I went through numerous books on plant engineering, maintenance management, reliability engineering and met many professors/practitioners. I thank all those who directly or indirectly helped me while authoring this book. In case, I, inadvertently missed out mentioning anyone who helped me in the process, my heartfelt apologies to one and all.

Last but not least, I thank my wife Mrs. Prasanna and children Kumari Srija Nataraj, Kumari Himaja Padmapriya, and Kumari Sanja Karunamayi for their help, cooperation, and patience, and parents Smt. and Sri. N.V. Murali Manohar Raju and brothers Dr. NVSN Raju (USA) and Mr. NV Srinivasa Raju (Nippon Paints) and sisters for their encouragement. Finally, I thank the entire Kasu and Chennamadhavuni families for their good wishes.

Any suggestion or feedback will be highly appreciated and will go a long way in the improvement of subsequent editions.

N.V.S. Raju

CHAPTER 1

Maintenance Management —A Conceptual Framework

PREAMBLE

This chapter focuses on the basic concepts of Maintenance and Plant Engineering. The objectives of maintenance department and the functions that are carried out by maintenance department of an organisation are thoroughly discussed. The inter-relationships and possible interactions of plant engineering with various departments of a given organisation are portrayed. Various management functions such as planning, coordinating, directing applying to plant engineering are depicted. Some light is thrown on emerging issues in maintenance/plant engineering area, such as tero-technology or total productivity maintenance (TPM), etc.

CHAPTER OUTLINE

1.1 INTRODUCTION

The concept of maintenance is as old as Stone Age. When man learnt to manufacture tools for hunting and fighting, he also learnt how to preserve, sharpen and keep them ready to use at any moment he wanted. He understood the need for upkeeping (maintenance) of his tools and assets, and found ways and means to safeguard and enhance their life. This function has been slowly gaining importance from Stone Age to Modern Age. Yet, in many industries today, maintenance function is given a low priority. Perhaps this is because it was considered to be a third rate job till the Middle Age, a rather unfortunate perception which continues to this age. Maintenance, being an absolute necessity, has not been given due importance over centuries and not much thought has been given to its significance. There are studies to support this. The results of the studies revealed that the equipment utilisation is found to be as low as 30% in few cases while 50%–60% in many cases. However, in the present-day environment, particularly when limited capital is available to spend for capital equipment, the best method to increase the productivity is by enhancing the useful life of the assets and their maximum utilisation. Therefore, maintenance function has become a challenging job. In India, organisations like National Productivity Council (NPC), in addition to a few universities, e.g., Indira Gandhi National Open University (IGNOU), are playing a vital role in disseminating the knowledge about the importance and significance of maintenance management in an organisation.

1.1.1 Maintenance Management

Maintenance management, in a broader sense, is termed as Plant Engineering and Management (PEM). This is because its activities are related to engineering and management. Generally, maintenance is regarded as a repair function. However, in reality it is a combination of four basic activities, retain, preserve, prolong, restore (known as 2Ps and 2Rs of maintenance) carried out to prolong the acceptable condition of a machine or equipment. However, a much broader concept has been assigned to PEM by adding the activities like enhancing reliability, availability, and maintainability (R-A-M) of machine, equipment, assets, or tools. And these functions have taken broader shapes in the modern world with names like "Physical Assets Management", "Tero-technology", and "Total Productive Maintenance".

1.2 AWARENESS ON MAINTENANCE AND PLANT ENGINEERING AND MANAGEMENT

Plant Engineering and Management can be compared with a doctor's job if we assume our body as a machine. When a refrigerator or a television in our house fails to work, we immediately call a repairman to get it rectified. This activity of the repairman in an organisation will be equal to that of maintenance or plant engineering.

Generally, in small organisations, plant engineering activities are carried out by production personnel (self-maintenance). In a few firms, this is carried out by an external agency or any appropriate outside parties on call as and when required (contractual maintenance). However, medium to larger industries usually go for in-house maintenance with a separate department for the following reasons:

- Maintenance and repair of equipment is not so easy and specialised personnel are required to handle the situation.
- It is not economical to call or contract the services of outside parties.

- Breakdown maintenance becomes too difficult sometimes due to non-availability of repairmen in the needed time.
- When maintenance is given to outside parties, it leads to too much dependability which is not desirable.
- Loss of production due to lack of maintenance cannot be compensated by contractual maintenance because outsider may not stand responsible.
- Shifting, day-to-day preventive maintenance activities, and other practical difficulties cannot be entrusted to outside parties.

The function of maintenance is not just repairing the equipment when it fails, but it is more concerned in preventing it from failing. Further, the maintenance has to find ways and means to enhance the useful and productive life of the equipment. However, outside parties may not do any justice to these two activities. Therefore, most firms often maintain a separate department of maintenance or plant engineering to look after the act of enhancing the life or effective and efficient usage of machinery and equipment.

Let us now define maintenance* or plant engineering.

A plant, in industrial jargon, is a place where men, materials, machinery and equipment, etc., are brought together to convert the inputs to desired outputs through well-defined process or method. Plant engineering is a study that looks after the smooth operation of a function without any interruption by keeping the machinery available to the maximum extent possible. *Therefore, the maintenance or plant engineering is the activity mix of erecting and installing the machinery, maintaining the good condition, enhancing the useful life, preventing the failures, and restoring the condition of the machinery/equipment in case of any failures.*

Thus, the plant maintenance consists of the following three major types of activities:

- **One-time Activities**: Erection, installation, and commissioning of equipment or machinery.
- **Ongoing or Continuous Activities**: Preventing the failures of the equipment, maintaining good condition of machine, and enhancing the machine life.
- **Intermittent or Limited Activities**: Restoring the condition of the equipment in case of failure or replacing the failed or poorly performing parts.

The technical activities mentioned above also include some analytical and managerial aspects such as planning, scheduling the preventive maintenance, analysing the failures, and spare parts management.

In fact, every machine is thoroughly inspected or tested by the seller before it is sold and by the buyer before it is put into use. Though the user operates the machine as per the specifications and directions by the original equipment manufacturer (OEM), even in the idealistic conditions also, there is no 100% guarantee that the machine will never fail. No machine is immune to failures. Moreover, as the machine is put into use, the wear and tear develops on running. These worn-out parts lead to the failure or breakdown of the machine and hence to stoppage or shutdown for repair or replacement of parts. At certain times, any delay in taking action as required may prove very costly or crucial. The role of Plant Engineering and Management comes into play at this moment to restore the condition or to control the machine from further deterioration or failure. As a matter of fact, the PEM enters the picture much before the failure occurs, i.e., to prevent the failure.

---

*Strictly speaking, Plant Engineering and Management is the broader concept of maintenance. But in this book, both these terms are more or less synonymously used.

Strictly speaking, the function of maintenance or PEM begins much before the activity of preventive maintenance. The activity starts with the erection and installation of machinery. In some industries, the unloading of machinery when they are procured is detached from purchase or stores management and is assigned to the PEM owing to the reason that the maintenance personnel are well-versed with the knowledge of machine care and method of unloading, etc. In the case of sophisticated machinery, proper handling and unloading becomes utmost necessity.

1.3 OBJECTIVES OF MAINTENANCE AND PLANT ENGINEERING

1.3.1 Need for an Objective

The maintenance manager needs an objective for interpreting the goals of the organisation into terms of departmental aims both in the long-term (policy/strategy) and short-term (task) bases. To achieve the design life of the equipment, safety of the operators, assets and the production output of the plant, a suitable policy is required to be framed. Thus, the maintenance objective should be in accordance and compatible with the objective of the organisation. At the same time, the use of energy and consumption of certain consumables and spare parts is to be optimised. For this the maintenance manager must fully understand his resources and requirements of his department as well as that of the organisation. These are exhibited in Figure 1.1.

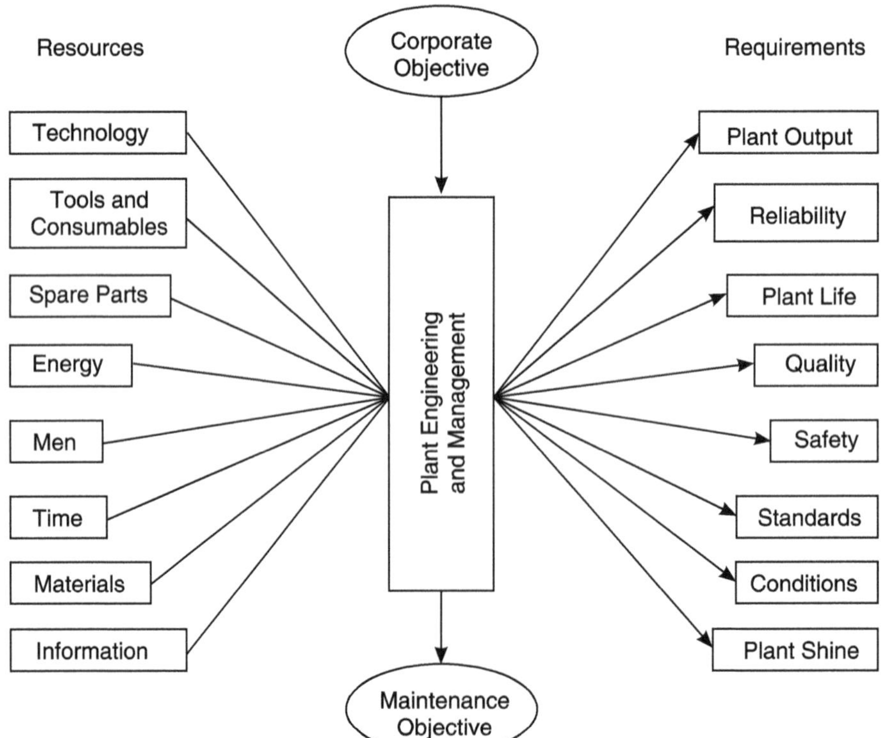

Figure 1.1. Resources and requirements for reflecting the corporate objectives into maintenance objectives

1.3.2 Objectives of Maintenance and Plant Engineering

From the above discussion, we get a clear picture about plant engineering and its need for objectives and goals. Now let us discuss the commonly adopted objectives of Plant Engineering and Management (PEM) in an organisation.

- To maximise the uninterrupted available time of the assets or machinery or equipment or services so that they can be used for the intended purpose they were procured.
- To preserve the value of assets by reducing the rate of deterioration by maintaining them to work in good condition. For example, lubricating, removing rust, dust, and dirt periodically and applying grease, etc.
- To facilitate maximisation of output of production by maintaining utilisation of machinery without affecting its deterioration.
- To perform the maintenance activities in the most economical way, such as restoring the condition of machine at the earliest possible time when it fails or is about to fail, planning and scheduling preventive maintenance without interrupting the production schedules.
- To update the machine and/or reconditioning with the latest technological or engineering features and/or management philosophies/styles.

In a nutshell, the objective of plant engineering is to concern itself with the provision of uninterrupted production process at minimum cost and maximum reliability and availability.

BUSTLE - 1.1

List a few organisations in your vicinity or those you are familiar with in which there is

(i) a separate plant engineering department in the factory itself.

(ii) contract of the services from outside agencies.

(iii) system of hiring the services of the outside parties as and when needed.

Interview them about the problems that they are facing due to the prevailing system. Analyse their merits and demerits. Suggest some remedies for the problems.

Test Your Understanding 1.1

1. Why do large industries prefer to establish their own in-house maintenance department? Give examples of the existence of plant engineering department in large scale industries. Justify their existence.
2. What are the expected problems when large industries undertake contractual maintenance? Explain with one or two illustrative examples.

1.4 STATE OF PLANT

Often people speak and take an account of the terms such as off time and on time or productive hours and non-productive hours depending on the running of the plant. With reference to the Plant Engineering and Management and Maintenance functions, the plant is usually found in one or more of the following states.

1.4.1 On Production

In this state, the production continues and there is no loss of output. This is the most ideal state that any production engineer or manager always desires to have. To retain the same state, the responsibility is on the maintenance engineer to maintain the equipment or machinery in perfect working condition or operable state. The maintenance engineer or the plant manager will monitor the equipment without disturbing the production. This is done online, i.e., while the machine is running and hence called online monitoring. Some small activities exist, such as lubricating, cleaning, inspection, and setting at contacting parts, etc., that will prevent the probability of failure of a machine and enhance the life. Generally, these activities can be performed while the machine is running by monitoring at suitable intervals. Such activities take less time rather few minutes (five to ten) only and hence will not interrupt the production.

In some occasions, production state is further classified into two sub-states:

(a) **Production State with On-line Maintenance**: A little disturbance in the production occurs rather than any loss of production.

(b) **Production State with Off-line Maintenance**: No maintenance activities are involved while production is going on, i.e., maintenance is carried out when production is not planned or engaged (such as lunch time or small gaps whenever the machine is free or not engaged).

1.4.2 Dead State

Dead state is a state of plant when the machine is not required for production for a given period of time. In other words, since the machine is not taken up for production, this time is usually said to be dead time. This is the most suitable time for Plant Engineering and Management to take up maintenance such as preventive maintenance or scheduled maintenance. Commonly, the night, Sundays or holidays or play days, etc., are considered to be the dead time of the machines as they are not operated during these times. The maintenance manager can plan and schedule the preventive maintenance during this time without any production loss. However, planning and performing should not exceed the dead time otherwise it may disturb the production.

1.4.3 Production Outage

This state of a plant arises due to shortage of stocks or product changeover or shift changeover or machine changeover, etc. In such state, the production is not extracted from the machine and remains idle for shorter duration. This shorter duration also can be made use of by the Plant Engineering and Management for effective maintenance.

1.4.4 Down State or Under Shutdown

This is a state of a plant, which comes into existence when the production is stopped and the equipment is shut down for maintenance by pre-planned agreement. The moment the machine is shut down, it will be handed over to the department of Plant Engineering and Management. Unlike the above two states (dead state and outage), this state is the stoppage of machine intentionally and purposefully for maintenance and a considerable loss of production may take place and attributed to maintenance department. However, this is an anticipated loss of production and hence will be inclusive of the production plan with due consideration.

1.4.5 Under Emergency Maintenance State

When the machine fails unexpectedly, it will be referred to the plant engineer to repair and put it in use at the earliest. This state occurs unanticipated and may happen due to any reason, such as erratic operation, unsafe conditions, environmental conditions, failure of a machine member, etc. This maintenance is also known as *unplanned maintenance.*

1.4.6 Waiting for Maintenance

This state usually occurs to the machine when it fails and gets into a state of emergency maintenance. This state occurs when a machine fails unexpectedly and one or more of the following conditions prevail:

- No spares/materials are available to replace
- No repairmen
- Proper tools are not available
- No information (in time)
- Suitable environment or conditions for repair is not hailing
- Non-availability of any resource required for restoring the machine operating condition
- Any other reason that demands postponement of the repair work

Considerable production loss is incurred in this state and the extent of the loss is unpredictable. This loss can be minimised to some extent by adopting a good inventory control strategy and policy.

These states are summarised through a model for plant states with reference to plant engineering objective and function as shown in Figure 1.2.

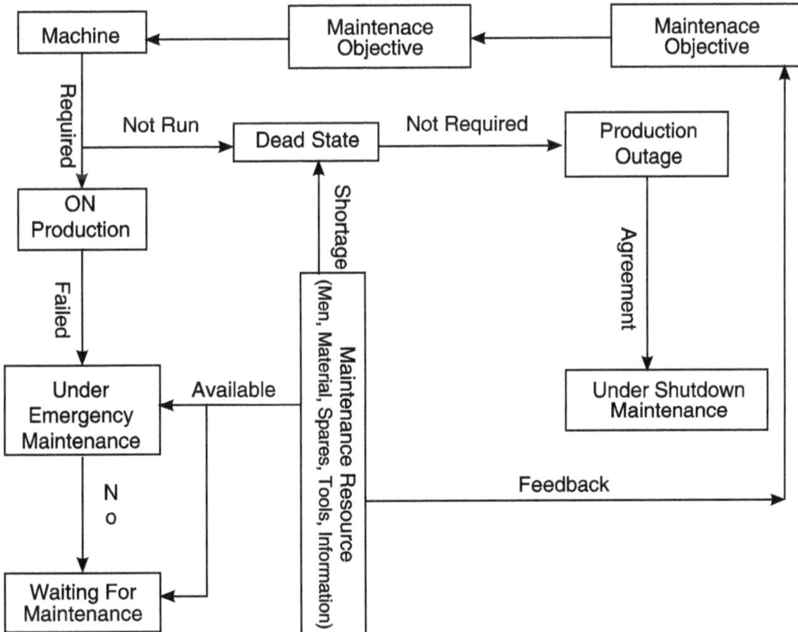

Figure 1.2 Model for plant states

Summarily, a plant or machine can be stated to be in one of the two states, namely, up state or down state. The up state includes the states when the machine is available for production, i.e., on production state, production outage while all the other above-mentioned states are down states during which the machine is not available for production.

BUSTLE - 1.2

Consider the equipment such as a car or bike, household refrigerator, personal computer or laptop, television, washing machine, tape recorder, or CD/DVD player, etc., and identify the states they encounter in one month and put the duration of these states in the above model. Based on your findings comment on the condition and functioning of the equipment.

____

____

____

____

Test Your Understanding 1.2

1. Explain various possible states of a machine, which it may experience in its life. Support with suitable examples.
2. What do you understand by the dead state of a machine? What are the adverse effects of this status? How can a plant engineer exploit this state to contribute to the productivity?

1.5. FUNCTIONS AND RESPONSIBILITIES OF PLANT ENGINEERING

Generally, in small to medium scale industries, a separate plant engineering department does not exist. In most cases, the production manager looks after the functions of plant engineering or adopts contractual maintenance. However, if a separate department of plant engineering and maintenance exists (generally in a large scale industry), it holds responsibility for the following functions:

1.5.1. Installation, Erection, and Commissioning

The function of installing, erection, and commissioning is one-time activity of plant engineering department (PED), i.e., at the beginning of the establishment of the organisation or whenever a new machine is procured. However, it revokes when the management takes a decision to shift a machine from one place to the other within the organisation or when the total firm is shifted to any other place or in the case of expansion, etc.

1.5.2. Preventive Function

Preventive function is the major responsibility of plant engineering department. The Maintenance or PE department prepares short-term as well as long-term schedules to check each and every machine at regular intervals throughout the year. The PE crews will carry out the inspections, overhauls, and activities like lubrication along with checking of the operating parameters. The worn out parts (if any) will be replaced if required, so as to prevent further deterioration and not to lead to major failures or

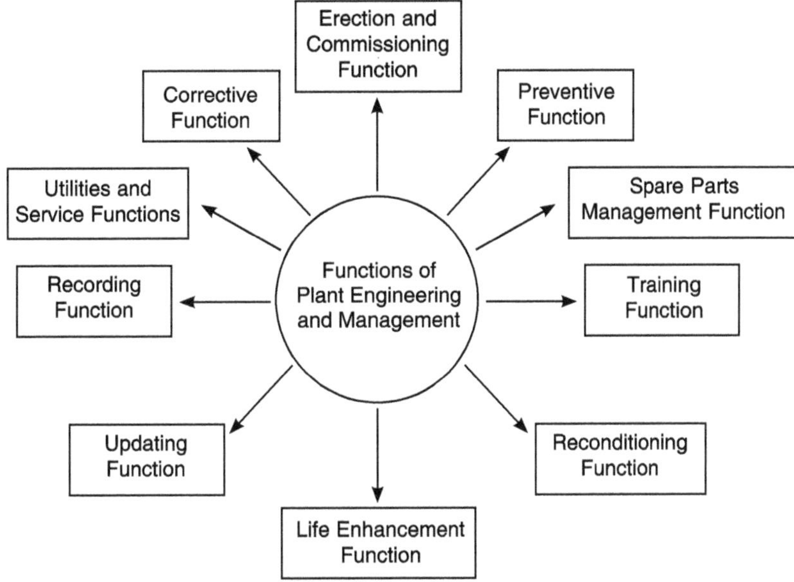

Figure 1.3 Functions of Plant Engineering and Management

catastrophes. The basic objective of this function is to forestall or try to prevent the occurrence of a failure. There are various strategies and approaches to accomplish this objective such as planned maintenance, shutdown maintenance, productive maintenance, predictive maintenance, preventive maintenance, maintenance prevention, which will be explained in detail in Chapter 3.

1.5.3. Corrective Function

Corrective function is one of the basic functions of plant engineering. When a machine fails to perform the intended work in a production department, the production department will immediately inform the PED through a maintenance service requisition. The maintenance crew, notified for the purpose, will then be assigned to repair the machine/equipment and restore its condition. The order of priority or sequence of the jobs of the maintenance will be decided by the head or officer in-charge or foreman of the Maintenance/PED. The priority of taking up the jobs is usually decided on the basis of First Come First Serve (FCFS), Earliest Due Date (EDD), Service In Random Order (SIRO), Any Pre-emptive or Non-pre-emptive preferences or Johnson's and Jackson's rules of sequencing or using other operations research techniques such as assignment problem, etc.

Depending on the nature of the organisation, type of production, and other prevailing conditions or situations, this function is observed in different forms such as breakdown maintenance (BDM), corrective maintenance (CM), operate to failure and corrective maintenance (OFCM), emergency maintenance (EM), etc. These functions will be discussed in Chapter 3 in detail.

1.5.4. Updating Function

Though the existing machines run at full efficiency, the latest technological developments can influence the productivity. Often, these new technological developments are introduced to enhance productivity in the form of ease of operation, increased output, wastage reduction, less time consumption, etc.

Owing to these reasons, the top management will insist the plant engineering department to update the technological features of the machinery. This function is often referred to as design out maintenance, which not only improves the productivity but also keeps the organisation alive or enhances its life. Further, if this function is ignored or neglected, the equipment becomes obsolete early and hence its resale value diminishes.

For example, you have purchased a personal computer with say, 40 GB Hard Disk Drive, which is the highest configuration during the time of your purchase. Suppose two or three years hence, a new configuration, 80 GB Hard Disk Drive, is prevalent in the market. It is considered to be the latest configuration that is in use. In such case, it can be suggested that 40 GB drive be replaced by 80 GB immediately, otherwise it will become completely outdated after a couple of years. It cannot even be sold in the second sale and just have to be discarded. Thus, updating the configurations or specifications with reference to the changing technology is very essential to prevent declination which may result in the collapse of the equipment, its products and even the organisation in the course of time.

1.5.5 Recording Function

"Men may come, men may go; But I go forever", Robert Frost, the famous poet describes the nature of a river. These words hold true for the maintenance department. It continues as some people may join or leave the organisation. The new entrants, particularly, if they are inexperienced or fresh to the field of engineering/technology/management, to facilitate them, the department has to maintain the operating and maintenance manuals, machine specification and details of its procurement/installation/commissioning, instructions of safety, repair history, preventive maintenance history, etc. The time and cost details of day-to-day repairs and preventive maintenance (PM) activities will be recorded in their logbooks or registers. Based on this information in the logbooks, a weekly/monthly report will be generated and subsequently the reliability, availability, maintainability calculations can be performed. The maintenance department will also maintain a record of the procurement of spares and inventory stock registers of inward/outward movements, etc.

The manager's function is not all-embracing unless there is a proper feedback system. Many experts on management speak about authorities and responsibilities of a manager and the proportion in which they should be delegated. But they speak least about accountability. Every manager has to be accountable for every job he/she does. In fact, this function is more significant in maintenance department. The plant manager also has to take accountability for his/her success as well as failure. This can be possible only if there is a good reporting system.

However, a good reporting system can be established in an enterprise with

- simplified organisation and organisational structure
- well-defined programmes and policies without any ambiguity
- well-designed methods of communication
- smooth relations among the managers horizontally and vertically
- good record-keeping
- faith in one another

Good recording system provides better control over the PE activities and events. Records give scope for introspection to the managers and the management by which a quality in work culture can be inculcated. A good recording system will bring healthy atmosphere, unambiguous communications, smooth run of programmes, cordial relationships, and accurate information system.

The record of maintenance jobs are usually figured by keeping track of the materials used and the labour hours expanded on any one job. Most generally, the following types of records are found in plant engineering.

(a) **Material Costs Record**: Material costs are recorded by the use of storeroom requisitions. A mechanic needing a new bearing to repair a machine takes the bearing from the storeroom with a material requisition. All such materials used are accumulated in the storeroom and a record of its use is maintained. Finally, based on this record, total material costs are calculated.

(b) **Labour Costs Record**: Labour hours spent on any one job can be recorded in various ways. The simplest way is to maintain a work report of each mechanic in a special card submitted at the end of each day indicating how much time was spent on each job assigned to him/her. The mechanic's time spent on a maintenance job is translated into money as the actual rupees paid to that mechanic. Further, this data can be used to estimate the average wage of the department. However, the average wage includes the overheads also, such as storeroom upkeep, power used in the shop, utility charges (compressed air, gases, etc.), safety equipment, etc.

(c) **Equipment Records**: Records of equipment are a necessary part of every good maintenance programme. Such records include all the pertinent data concerning the equipment itself, such as serial number, supplier and initial cost. Such information is also necessary while ordering parts or seeking information about such equipment from the supplier. A history card also contains a record of repair work performed, the schedule for inspections, and the costs of inspection and repair. Repair cost information is especially valuable, as it can determine the replacement age, i.e., the age when running costs are higher than normal preventive maintenance cost.

Some companies are now making excellent use of data processing equipment and digital computers to assist in maintaining maintenance records. One aircraft company utilises punched cards to feed data into the system and can get almost immediate answers to any of the following queries:

- Who received the service?
- When and where did they receive the service?
- What kind of service was performed?
- How much time did the maintenance man spend?
- What was accomplished?
- How much did it cost?
- Where were the excessive maintenance costs being incurred?

Also, this recording system is of tremendous assistance in making decisions on what to do when, to forecast equipment-replacement schedules and so forth.

1.5.6 Training Function

The department has its responsibility to update the technology, knowledge, skill, efficiency, and effectiveness of its crew as well as the machine operators. It also conducts a time to time expert lectures or workshops to make the machine operators aware about the cleaning and upkeeping of their machines in addition to upgrading themselves with the latest technical knowledge.

The main objectives of training function with reference to plant engineering are

- updating the technology
- optimisation techniques for effective utilisation of resources
- grievance handling with reference to the work difficulties
- interactive sessions with the employees about their experiences while working
- waste reduction, cost reduction, innovations, etc., in the field
- feedback about the troubles faced by the PE personnels.

The training programmes in maintenance provide the following advantages:

- The training programmes help in sharing the views of the problems of the workers with others that bring forth a common solution to the similar problems in their areas.
- Though the crew is thorough and skilled, the training programmes will further help in upgrading them with the technological information for troubleshooting.
- The human behavioural sciences indicate that the joy doubles while troubles are halved if shared with others.
- They pave an easy way for performance appraising particularly, during promotions and awarding merit increments, etc.
- Records of these discussions will be helpful in the future.
- The training programmes help develop a proper work culture in people and produce effective output.
- The training programmes create enthusiasm and reduce fatigue.

1.5.7 Spare Part Management Function

In most of the industries, the inventory of maintenance and repair such as oils, lubricants, grease, machine spares (bearings, bushes, belts, chains, electrical equipment, etc.) are taken care of by the PED and a separate store is maintained in coordination with the production planning and control (PPC), materials planning and control (MPC), and purchase department.

The management of spares has many dimensions such as the availability of spares in the market, the type of market (such as government market, open market, foreign market, etc.) in which they are available, time or season of availability, size, standards, quality aspect, traceability, replaceability, compatibility, etc.

1.5.8 Utilities and Service Function

Utilities mean those aspects which supplement the production. The utilities include telephones, fax machines, compressed air, water circulation, electric power and/or generator, air conditioners, lighting (both regular and emergency), fire extinguishers, and other safety measures (in many cases, fuels such as petrol, diesel, liquefied gases of gaseous fuels, etc., are also included). These utilities will be under the maintenance department which is expected to provide service in all these aspects along with the machine servicing.

1.5.9 Machine Life Enhancement Function

The main motto of PED is to increase the length of the service life of the machinery and equipment. In the interest of this, the department will coordinate with the production departments to upkeep the machinery. The maintenance crew will also perform the activities like removal of rust, cleaning, lubricating, applying grease, etc., to prolong the useful life of the machinery.

For example, a CNC machine will be maintained under specified atmospheric conditions by providing air conditioning. Thus, maintaining correct conditions such as pressure, humidity, noise level, dust level, vibration level, and temperature will enhance the machine life as well as human lives.

1.5.10 Reconditioning Function

As the machines grow old, they tend to create lot of problems due to wear and tear. In such cases, the companies will be under dilemma to replace the equipment owing to their huge costs. Moreover, the decision making to discard the old machines which incurred huge investments is very difficult in many cases. Hence, the maintenance managers will think of renovating or reconditioning the old machine. In this function, the old machine and its base or basic parts will remain the same but some parts may be changed. Thus, the condition of the machine will be restored generally by replacing the worn-out parts and accessories, and, in rare cases, a few of the basic parts may be replaced.

For example, a scooter after using it for about 10 to 15 years, keeping all other parts unaltered, the cylinder can be re-bored to next size and a suitable piston can be replaced to use for another three to five years.

1.5.11 Maintenance Planning Function

In fact, maintenance planning is the most important function that sequences, schedules, and coordinates all the above functions. This function starts with forecasting and procurement of the material requirement, prediction of the failures, monitoring machine health, and assigning time and men to perform the planned activities. Owing to its significance, an elaborate discussion about this function is made in Chapter 3.

BUSTLE - 1.3

Visit an organisation in your vicinity (or the one in which you are working). Observe the functions of plant engineering department being executed in the organisation. Rate how effectively these functions are carried out or the percentage of time spent on each function per month.

____

____

____

____

Test Your Understanding 1.3

1. List the functions of PEM? Briefly explain each of them.
2. What are the one-time (non-repetitive) functions of plant engineering? Explain their importance. Give illustrative examples.
3. Explain the following functions of maintenance.
   (a) Preventive function (b) Corrective function
   (c) Reconditioning function (d) Installation and commissioning
   (e) Training function (f) Spare parts management
4. Describe the role of maintenance department in improving the machine life.

1.6 PHYSICAL ASSETS MANAGEMENT OR TOTAL PRODUCTIVE MAINTENANCE OR TERO-TECHNOLOGY

The term *physical assets management* (PAM) is synonymous to total *productive maintenance* (TPM) and *tero-technology*. It takes into account all aspects of plant machinery from design to discard, viz., design, manufacture, installation, commissioning, maintenance, replacement, and removal of the plant/ equipment as well as the feedback of performance for the equipment manufacturer. Tero-technology envisages application of the combination of managerial, financial, engineering, and other practices applied to physical assets in pursuit of economic life cycle costs (LCC). It is concerned with the specification and design for reliability and maintainability of plant, machinery, equipment, buildings, and structures. This total LCC concept enables proper equipment evaluation and selection so as to give an overall low LCC. This gives rise to huge potential for savings in terms of cost effectiveness of replacements on considerations of whole life cycle. ‘Design audit’ consists of crucial scrutiny of the designs by the operating and maintenance engineers independent of the design process in order to ensure reliability and maintainability of plant and machinery and to identify weaknesses in designs requiring modifications. The word 'tero-technology' stems from the Greek root 'terein' which means 'to look after', 'to guard over', or ‘to take care of’. The principles of tero-technology as discussed above can be applied, to a greater or lesser extent, to any physical asset in any organisation, no matter what the size of degree or complexity of either asset or organisation is.

1.7 CONCLUSION

Plant Engineering and Management will undoubtedly increase the productivity of an organisation if managed properly. However, any mismanagement or unorganised way of working will result in huge losses. There are quite a few evidences where many organisations blame Plant Engineering and Management in most of the situations for the loss of production. The middle-level production managers feel that their machines are not updated, not made available for longer time due to shutdowns, improper maintenance, waiting for spares, absenteeism of maintenance crew, etc. Therefore, Plant Engineering and Management is usually held responsible for loss of production. This atmosphere frequently leads the managers to look for the excuses to escape at times. In trying to save his/her stake, the managers end up in bitter relations with each other. However, the employees of the organisation should realise that they are working for the growth and prosperity of the organisation. In connection to this, it is everybody's responsibility to upkeep the machinery and equipment and keep them neat and tidy for enabling to preserve, prolong the life, retain and restore the condition of the equipment.

1.8 RECAP

In a nutshell, the Plant Engineering and Management exists in the organisation to preserve assets, prolong the life of the assets, retain their good condition, and restore the acceptable condition of the machinery or equipment or asset if they fail. The plant engineering and maintenance department is often entrusted to execute the functions like corrective (breakdown), preventive, productive maintenance, spare parts management, etc. It also looks after the erection, installation, commissioning of the new equipment, shifting of the machinery, provides utilities like water, compressed air, inter communication services, etc. The maintenance even undertakes the activities like updating the technology and reconditioning of the machinery, etc. It interacts with various internal departments such as production to perform the aforesaid functions, finance in budgetary decisions, accounts in cost analysis, personnel in training and compensation against accidents, PPC in preparing PM schedules. Apart from these, PEM interacts with the outside parties also such as material suppliers, OEMs, vendors, sub-contractors, electricity, telecom, waterworks, fire stations, etc. This planning function is elaborated in Chapter 3 due to its high importance in the present scenario. In present days, the maintenance or plant engineering has taken a new shape and dimension in the form of physical assets management (PAM) or tero-technology (TT) or total productive maintenance (TPM). These concepts are also discussed in Chapter 3.

1.9 KEY CONCEPTS

1. **Plant:** A place where men, material, and machines are brought together to convert the inputs into the desired outputs in a well-defined method.
2. **PEM:** Plant Engineering and Management is that part of organisation that helps to preserve the assets, prolong the life, retain the condition of the equipment/assets, and restore the condition when it fails.
3. **OEM:** Original equipment manufacturer is a person or a group or an organisation that manufactures a machine or equipment or an asset.
4. **Reconditioning:** The activity of regaining the condition of a machine or equipment when its efficiency or performance comes down or deteriorates to nearly its scrap value level.
5. **Erection:** The act of arranging or keeping the machine or equipment in its position as per the layout.
6. **Installation:** The act of arranging the machine in position and keeping it in usable condition.
7. **Commissioning:** The function of keeping the machine in working condition to its full efficiency.
8. **Breakdown:** Failure resulting in non-availability of an item.
9. **Down time:** The period of time during which an item is not in a condition to perform its intended function.
10. **Overhaul:** A comprehensive examination and restoration of an item or major part thereof to an acceptable condition.

REVIEW QUESTIONS

1. Define and explain the term maintenance or plant engineering. What are the activities of this department in a large scale organisation? How do they differ as compared to small scale industry?
2. Explain the objectives of Plant Engineering and Management?
3. What are the possible states a machine can experience in its life? Which state do you think is more disadvantageous or detrimental? Illustrate with examples.
4. What do you understand by the shutdown state of a machine? What are the adverse effects of this status?
5. List the functions of maintenance department. Briefly explain each of them.
6. Explain the role of plant engineering in enhancing the machine life.

CHAPTER 2

Maintenance Planning

PREAMBLE

This chapter discusses the importance of planning in Maintenance and Plant Engineering and explains the functions of planning maintenance and Plant Engineering and Management. The planning functions of maintenance are identified and executed. Various systems such as centralised and decentralised systems for organising maintenance are discussed. This chapter focuses on bringing out and establishing interface between PEM and other departments planning and organising the maintenance activities.

CHAPTER OUTLINE

2.1 INTRODUCTION

Planning function for carrying out the maintenance is a dire necessity. However, its significance and importance in our everyday lives has not been considered seriously over the centuries. This is perhaps due to the following ill notions about maintenance department.

1. For ages, maintenance is considered as 'repairing only' and it has been remembered as and when the equipment seriously breaks down.
2. Further, another ill notion is that there is no special need for planning for the so called 'repairing' maintenance.
3. Generally, when the equipment is in down condition, unless needed for production, it is not offered for maintenance but kept idle. When it is in up condition, until the machine cannot work any longer, it is not offered for maintenance.
4. The plant maintenance is given least and last importance as compared to production and other areas.

However, these notions slowly are wiping out with advent of the concepts of preventive maintenance, condition monitoring, and so forth. Further, a good understanding over these methods and proper training on systematic planning is essential for maintenance personnel for optimal utilisation of equipment. Therefore, implementing these methods has become a challenging job. One can meet and materialise these challenges with better planning activity only.

2.2 PLANNING FUNCTION IN MAINTENANCE

Planning is the determination of a course of action to achieve predetermined goals or formulating the expectations between available resources and required results. In other words, it is deciding in advance what is to be done. Planning involves selection of missions and objectives and the approaches and actions to achieve them. It needs decision making, i.e., selecting from alternatives among future courses of action plans, which can provide a rational approach to accomplish preset objectives. Planning bridges the gap from where we are to where we want to go. In plant engineering department, the scope of planning function is wide spread. The right planning can save time, energy, and material.

2.2.1 Normal Maintenance Activities Requiring Planning

The normal maintenance activities involving the planning actions are enumerated below.

(a) **Recurring or Repetitive Maintenance Activities**: These maintenance activities are to be planned for the entire year. However, preventive maintenance can be planned in long range based on a distinct predetermined time interval or failure frequency or machine health conditions. The other maintenance activities such as repairs and replacements cannot be easily scheduled.

- *Preventive and planned maintenance*: We know that prevention is better than cure. The following are very common and simple activities of maintenance that can be effectively planned so as to save time.
  (i) Lubrication
  (ii) Inspection
  (iii) Overhauls

- *Corrective maintenance*: The corrective maintenance includes repetitive and regular work that can be planned on rolling plans or short-range plans. Such plans may have the following contents.
  - (i) Repairs
  - (ii) Replacements
  - (iii) Attention to breakdowns
  - (iv) Component reconditioning
  - (v) Manufacture of spares

(b) **Non-recurring or Non-repetitive Maintenance Activities**

- *Alterations, updating and modernising*: Under this head, we can include the long-range plans of irregular and occasional maintenance such as—
  - (i) Reconditioning total machines or converting a machine to perform some other functions
  - (ii) Construction and alteration of buildings
  - (iii) Rearranging the layout
  - (iv) Expansions or extensions
  - (v) Designing maintenance activities
  - (vi) Upgrading the configuration of a machine
- *New works*: This is again a one-time plan usually included in long-range plans or perspective plans if an organisation is already established. However, a new organisation in its infancy requires this planning. The activities often found in this category are
  - (i) Installation of new machines
  - (ii) Establishment of new structures
  - (iii) Construction of new building or a floor on already existing building
  - (iv) Creating a new facility
  - (v) Commissioning a new system

2.2.2 Steps in Planning of Maintenance Work

In the planning of maintenance work, the following steps are involved:

- Forecasting or anticipation of maintenance work
- Visualisation of the nature and details of that work
- Determination of the best method to perform the work
- Arranging for the required material
- Securing alternatives in production programme of scheduling maintenance work to conform to production plans
- Allocating work to individuals
- Instructing the individuals about schedules and methods
- Following up and checking of work
- Evaluation of the work and performance

2.2.3 Forecasting and Anticipation of Maintenance Work

The most important function of maintenance planning is forecasting or anticipation of future work. This can be done by information provided by the following sources:

- Instructions and guidance given by manufacturer of the machines
- Technical knowledge of the maintenance and production personnel
- Knowledge of the degree of the utilisation of machine
- Record of the behaviour of the machine and the work done on it
- Complaints and requests from production personnel on the basis of difficulties experienced by them while operating the machine
- Examination of the state of various parts of the machine during their life span

2.2.4 Visualisation of Nature and Details of Work

Planning for materials, manpower, methods, and time required for a job depends on the nature and details of the job. Visualisation of the details of the work can be done on the basis of information provided by the operating personnel and the nature of trouble experienced. Reports of inspectors should indicate the part needing attention and the type of attention required. Drawings of the machines should be used to determine the machines that require repairing, removing, and replacing the defective parts. The discipline of methodical study should be taken in to account while the decision regarding the most effective methods of performing the work is to be taken by maintenance engineers.

2.2.5 Planning and Scheduling Maintenance Activities

It has been said that the tools of good maintenance are planning, scheduling, and cost accounting. The planning and scheduling section of the maintenance department has the basic responsibility for establishing job priorities, for ensuring that necessary tools and materials are available and for preparing and issuing written schedules of jobs to be done.

The size and organisation of the planning and scheduling group will depend on the size and complexity of the maintenance organisation. In a small company, the maintenance supervisor performs all the planning and scheduling. Under such circumstances, the planning and scheduling will probably be handled on quite an informal basis. Large companies utilise a special group to plan and schedule jobs. Although much of the work done in this section is clerical in nature, the supervisor of the group should be a person with wide experience in maintenance methods. Such knowledge frequently is necessary to coordinate the scheduling of the various workmen in the maintenance department.

In order to adopt the necessary day-to-day and long-range planning that is required to make the most effective utilisation of the maintenance personnel, suitable routines are necessary. These will involve:

- The securing of necessary approval from the plant engineer for any projects that are of sufficient magnitude to require his attention;
- The securing of information from equipment records, work orders, work measurement data, craft and shops work force reports, storekeeper's procurement reports, the plant production schedule, and other pertinent sources; and
- Reference to existing master schedules, as is true of all scheduling activities, basic to maintenance scheduling and control is an estimate of the time required for each work order prior to doing the work.

The scheduling of maintenance work involves essentially two steps—a master plan of all jobs that can be predicted in advance and a daily adjustment of this plan necessitated by emergencies. In most plants, there are a number of machines and inspection and overhaul. The frequency of these occurrences will vary with the equipment, but experience figure subjected to suitable mathematical and statistical analysis will indicate the optimum time for the minimum overall costs. After these times have been established, they can be incorporated into the master maintenance schedule.

The master maintenance schedules (MMS) can be broken down into weekly or daily maintenance schedules. If used, the weekly work schedule will provide information to each craft and shop concerning the work to be done on each job for each day in the following week. This schedule is normally issued on every Thursday or Friday and will be the basis for each maintenance foreman to issue daily work assignments to the workers. Some maintenance departments prefer to centralise this function and have daily work assignments issued directly by a dispatcher in the job control section. Whatever system is used, it is sound practice to issue work assignments for the next day to each craftsman prior to the end of the workday.

Accurate information on the number of hours of labour spent on each job or any materials used are essential items for cost-accounting purposes. These charges are made against the pertinent departments and become one of the costs of manufacturing. Further, the costs provide a check on the operating effectiveness of the departments concerned.

BUSTLE - 2.1

1. *Visit and observe the planning activities performed in the plant engineering department of the organisation with which you are most familiar or where you are working. Do you find any lacunae?*
2. *Also observe the planning activity of the production in the same industry. Compare the two.*

____

____

____

____

Test Your Understanding 2.1

1. Enumerate the functions of planning in maintenance. Explain each briefly.
2. Which areas in maintenance and plant engineering require good planning? Why?
3. What do you understand by the term 'maintenance planning'? What are different phases involved in maintenance planning?

2.3 ORGANISING PLANT ENGINEERING AND MAINTENANCE

There is no one-best maintenance organisation that can be used in all cases. The organisation should be tailored to fit the particular technical, geographical, and personnel situations involved. In establishing the maintenance organisations it should be recognised that the basic necessity is to maintain a plant at a level consistent with low cost and high productivity.

Considerable debate takes place among maintenance engineers as to the level to which they should report. Some are of the opinion that they should be under production whereas others think that they should be on equal footing with the production. While others think that they should practice to have the director of the maintenance function report to the level that is responsible for most of the other plant groups that the maintenance engineers serve.

The nature of the industry, that is whether it is primarily electrical, chemical, or mechanical, will have considerable influence on the maintenance organisation. Also, the size of the plant and the scope of activities to be performed are significant factors that will affect the organisation. In a small plant utilising a minimum of mechanical equipment, the maintenance department might consist of an all-round mechanic and a helper or two. In larger companies, the maintenance group will be much more lavish and contain a variety of craftsmen.

The person in-charge of the maintenance engineering activities is usually an engineer and mostly holds the title of plant engineer. Other titles or designations associated include superintendent of maintenance and manager of maintenance. A plant engineer in the maintenance organisation is also expected to perform functions of supervising the power plant operation, the design of special pieces of equipment, the checking of installation specifications, and the planning and scheduling of major maintenance projects.

One of the noteworthy features of most maintenance organisations is the large number of crafts represented in the group. This number will vary considerably depending on the variety of manufacturing processes present and the amount of work involved. Another factor that will influence the number of craftsmen required is the availability of skilled contractors outside and management policy regarding the use of outside maintenance organisations. It is difficult to present any basis for determining exactly how many people are necessary in each craft in the maintenance organisation. The ratio of maintenance to production employees will vary with the level of machinery and capital investment per operating employee and the efficiency of the maintenance organisation.

An important part of the organisation structure is the job and cost control section. This section is responsible for scheduling work for the maintenance personnel and for maintaining cost figures on all jobs. It enables efficient utilisation of the craftsmen anywhere in the plant but at the same time does not prevent localised routine daily assignments.

Machine and repair shops and storerooms provide essential supporting sections within many maintenance organisations. The shops commonly contain machinists and mechanics capable of repairing equipment, making new parts (otherwise not available), and rebuilding equipment. The storerooms maintain spare parts (bearings, shafts, electric motors, pumps, etc.) and general maintenance materials (pipe fittings, standard hardware, light bulbs, switches, lubricants, etc.) needed by the maintenance force.

Other service groups that are commonly under the supervision of plant engineer are the utilities engineering section and the housekeeping and protective service groups. The former is responsible for the operation of the power and heating plant and the other group is responsible for keeping the plant clean and protecting the premises.

2.3.1 Maintenance Organisation

The organisation set up depends upon the system of maintenance adopted and the nature of industry. Maintenance organisation can be centralised, decentralised or a judicious combination of both. Generally, the big industries prefer the combination of centralised and decentralised system. General aspects of this combined system are described here.

Centralised maintenance group (directly controlled by the Chief of Maintenance) is responsible for major overhauls, breakdown maintenance and scheduled maintenance.

Plant maintenance group (working under respective plant manager) is responsible for preventive maintenance (which includes operators' tasks), lubrication schedule, minor repairs, and adjustments.

2.3.2 Systems of Plant Engineering and Management

For performing any function, first establish a systematic approach which not only enhances the efficiency of the work but also provides a good environment. A systematic approach also builds up good relations, demarcates the responsibilities and authorities. An appropriate system also contributes to a good control over the men and their operations. The managers of school of management believe in the principle that "Set up an appropriate system, it will in turn set you up to in correct path".

As mentioned above, the most common systems adopted for PEM are:

- Centralised Plant Engineering System (CPES)
- Decentralised Plant Engineering System (DPES)

Depending on the suitability, feasibility and the requirements, an organisation will adopt one of the systems or in some proportional combination of the two.

(a) **Centralised Plant Engineering System:** Centralised plant engineering is a set up in which the head of PED will have all the maintenance staff under him and will send to attend the breakdown as per the requisitions sent by various departments according to some predetermined rules of priorities. Apart from the repairing activities, the PEM also schedules to take care of failure preventive measures, increasing the machine life, etc., in such a way that it will hardly interrupt the regular production while the available production time is enhanced to a maximum extent. In this system, every matter related to the machines will be executed with the knowledge of the plant engineering.

[Thus, this system is supported by Fayol's Operational School of Management Thought].

(b) **Decentralised Plant Engineering System:** In this system, the plant engineering activities are split into some groups. There are three such ways in general practice. The first way is to split the department into convenient number of wings like mechanical maintenance wing, electrical maintenance wing, electronic maintenance wing, utilities and services maintenance wing, etc. The second way is to divide into groups like breakdown maintenance groups, preventive maintenance group, reconditioning group, etc. The third way is to split the maintenance crew into number of teams equal to that of the number of production shops and to look after all maintenance jobs related to their respective areas. The concept of specialisation and Taylor's scientific management principles are followed in this system.

2.3.3 General Functions of Central Maintenance Group

The following are the general functions of central maintenance group:

- Planning for maintenance and overhaul schedules in consultation with the plant maintenance group and work management.
- Carrying out all heavy repairs and major overhauls.
- Operation and maintenance of the central services.
- Planning for spare parts maintenance material and tools.
- Planning for manpower and their training.

- Preparation for annual maintenance budget and periodical review.
- Recommendation on replacement policy and additional purchase policy.
- Study of defect reports, suggested modifications, suggestions, etc., and making recommendations.
- Surprise inspection to cross check that maintenance at site is carried out as per directions laid down.

2.3.4 General Functions of Decentralised Plant Maintenance Group

Important functions of plant maintenance group are:

- Lubrication as per scheduled programme.
- Carrying out running repairs and preventive maintenance at site.
- Routine inspection of equipment as per checklist.
- Forecasting requirement of spare parts, etc.
- Initiating recommendations about modification, improvisation, etc., for improving equipment efficiency.
- Advising central maintenance group on overhaul schedules.

2.3.5 Decentralisation in Plant Engineering

What is Decentralisation?

Decentralisation is open to a number of interpretations. In administration, it may be used to refer to

- Departmentalising activities
- Location of actual performance

  or
- The dispersal of the centres of authority and decision making at various levels in the organisation.

Centralisation is the systematic and consistent reservation of authority by the top management or the management at the centre (say, in the case of an organisation comprising branches) while Decentralisation consists in the dispersal of authority. It must be remembered that complete centralisation is impossible because a management is not possible without some measure of delegation of authority. Likewise, complete decentralisation involving freedom (in place of autonomy) for the various divisions or departments will amount to virtual disintegration. Between these two extremes where the authority is largely concentrated at the top and mostly only the authority to execute given plans is given to the personnel at lower levels, the organisation is said to be centralised. On the other hand, where much of the authority to plan, direct, execute and control is passed down the organisation and the top management is concerned with broad planning and control, it is described as a condition of a high degree of decentralisation. We are here concerned fundamentally with decentralisation in plant engineering department in the sense of the systematic effort to delegate to the lowest levels of all authority except that which can be exercised at central points. In other words, decentralisation here refers to the dispersal of the centres of decision making throughout the organisation.

2.3.6 Distinction between Delegation and Decentralisation

Decentralisation should be distinguished from delegation. Delegation can take place from one individual to another to complete the process while decentralisation is complete only when the fullest possible delegation is made to all, or most of the people who are delegated a specific kind of responsibility. If the general manager asks one of the plant managers to recruit some executives on his behalf, it is an instance of delegation. On the other hand, if this authority is delegated to all divisional managers for all cases of recruitment, it becomes a case of decentralisation. Up to the authority to their subordinates, decentralisation is carried further down in the organisation. "If delegation is practised systematically in all functions and divisions of the company and for a wide range of authorities and responsibilities, we should say that the company is highly decentralised."

The difference between delegation and decentralisation is highlighted in Table 2.1.

Table 2.1 Difference between delegation and decentralisation

No.	*Delegation*	*Decentralisation*
1	It is a process.	It is the end-result of a deliberate policy of making delegation widespread in the organisation.
2	It takes place between a superior and a subordinate.	It involves delegations between the top management and the departments or divisions of the organisations.
3	It is a must for management as no manager can get things done through people unless he/she passes on the requisite authority for the performance of the work assigned.	It is optional, i.e., the top management may or may not decide to disperse authority as a systematic policy.
4	Control rests entirely with the delegation.	The top management may exercise the minimum of control in a broad way and also delegate the authority for controlling to the divisions or departments concerned.

The extent of decentralisation is judged on three tests:

(a) What kind of authority is delegated? Is it the authority to plan and decide?

(b) How far down in the organisation is it delegated? (Does it go down to the lowest supervisory level?)

(c) How consistently is it delegated?

On the whole, a centralised organisation is one in which detailed and comprehensive planning is done by the chief executive or by a small group of high-level managers. Most decisions are made at the higher managerial level in centralised organisation. On the other hand, in a decentralised organisation operating decisions are pushed down to the lowest level.

2.3.7 Factors Affecting Degree of Decentralisation

The degree and the nature of decentralisation in any organisation will be generally influenced by the following factors:

- *History of the organisation*—whether it has developed as a set of departments and divisions were set up later with the control still resting with the top management.
- *Philosophy of the top management*—whether the top management believes in centralisation or delegation of effective powers of planning and controlling.
- *Availability of competent subordinate managers*—whether competent executives are available to make effective set of the authority passed down by the top management.

- *Dispersion of operations*—whether the operations of the organisation are mostly concentrated at one place or in a region or dispersed over different territories.
- *Costliness and significance of decisions*—whether the decisions generally involve heavy cost or investment making it risky to delegate.
- *Uniformity of policy*—whether the uniformity of policy is of critical importance, if so, decentralisation may be limited.
- *Size*—whether the organisation is such that centralisation will create problems of planning and controlling.

2.3.8 Advantages of Decentralisation

Decentralisation of plant engineering is recommended on account of the following advantages.

(a) **Lesser Burden on Plant Manager:** Centralisation places too heavy a burden on the plant manager who alone is responsible for planning and decision making. In most occasions the decision has to be taken on the field or work spot. In a decentralised set-up, the subordinates share the burden of decision making and leave the plant manager to concentrate on overall planning and control.

(b) **Possibilities of Diversification:** A centralised enterprise with the concentration of authority in single individual at the top will find it difficult to diversify its activities and start new lines of distribution. On the other hand, under a decentralised set-up, diversification is facilitated. This enables the concern to make up for any losses in certain lines by gaining on the other lines and thus reducing the risks of business.

(c) **Executive Development:** Decentralisation gives power and initiative and responsibility to a number of executives. This provides a spur to the development of executives in the organisation.

(d) **Improvement of Morale:** Decentralisation stimulates the formation of small cohesive groups. Since local managers are given a large degree of authority and local autonomy, they weld their people into well-knit groups. With high degree of participation and constant effort, these managers are able to motivate the subordinates and secure a high degree of morale among them which is essential for the employees to work in plant engineering.

2.3.9 Drawbacks of Decentralisation

Decentralisation of organisation, though commendable in many ways, has some drawbacks or limitations, which may be enumerated as under:

1. It creates problems of coordination because authority lies dispersed widely throughout the organisation under such a set-up.
2. It may hamper the development of specialised services like maintenance departments because of diverse thinking.
3. It becomes a handicap of support from top management in case of quick emergency decisions.

BUSTLE - 2.2

1. *Consider the industry where you are working or the institution where you are studying. Observe some practical experiences or problems with reference to centralisation or decentralisation and delegation of power or authority. Conduct a cause and effect analysis of the problems.*
2. *Draw the internal and external organisation charts of plant engineering department in the organisation where you are working or the one with which you are most familiar.*

____

____

____

____

Test Your Understanding 2.2

1. Distinguish between the centralised and decentralised systems of plant engineering?
2. Write a short note on organising maintenance department. How does it differ with reference to overall organisation?
3. Draw a hypothetical organisation chart relevant to maintenance department in an industry and explain it.
4. Distinguish between decentralisation and delegation with suitable examples, with reference to PEM.
5. Do you think that decentralisation is helpful in increasing the efficiency of the plant engineering department? Discuss your answer with supporting evidences or examples.
6. Discuss the advantages and disadvantages of centralised plant engineering department.
7. Discuss the advantages and disadvantages of decentralisation with reference to plant engineering department.

2.4 STAFFING IN PLANT ENGINEERING

Peter Drucker assigns three responsibilities to management viz. managing work, managing workers, and managing managers. The staffing function of management involves the discharge of the second and third of these responsibilities. Staffing function involves in filling the positions in the organisation structure. This is done by identifying requirements of work force and force to managing this force. The staffing encompasses recruiting, selecting, placing, promoting, appraising, career planning, compensating, and training or developing both aspiring candidates and current job holders so that tasks are accomplished effectively and efficiently. This function is very challenging in the maintenance/plant engineering since the staff required for plant engineering must have sound knowledge, be highly skilled, act spontaneously, accept challenges, be willing to risk bearing and be fully aware of all the facets of the job.

Staffing function is continuous and is performed by the maintenance management even in a growing concern too since it covers the processes of developing work skills and executives. The planning of executive or managerial resources, the planning and administration of management development programmes, the appraisal of executive performance, and the administration of executive compensation are all parts of the staffing function of management in an established organisation. In a nutshell, staffing can be defined as "assigning the right job to the right men". Lawrence Appley writes, "Managers would be more skilled and competent if they were carefully selected, specifically trained, continually kept up-to-date in their field of activity, guided in their development for the assumption of greater responsibility and adequately rewarded." And this is to be carefully digested and applied by the top management, particularly to plant engineering department since every manager or engineer and his staff at their own level wilfully takes high risk and greater responsibility.

2.5 DIRECTING IN PLANT ENGINEERING

According to the above statement, the managers have the responsibility not only of staffing (apart from planning and organising the operations) but also of guiding and supervising the subordinates. This is the managerial function called directing.

The function of directing has three essential and elemental activities. These are

- Issuing of orders and instructions,
- Guiding and teaching the subordinates in the proper methods of work, and
- Supervising the subordinates to ensure that their performance conforms to the plans.

The above three activities are performed through the following aspects, which are supposed to be the sub-functions of this important managerial function such as leadership, communication, motivation, supervision, and controlling.

2.6 COORDINATING BY PLANT ENGINEERING AND MANAGEMENT

The accomplishment of group goals becomes impossible if there is no harmony in individual efforts. All the functions of management are just meaningless if there is no coordination. In fact, each of the managerial function is an exercise contributing to coordination. In any organisation, individuals often interpret similar interests in different ways and their efforts towards mutual goals do not automatically match with the views and efforts of one another. It, therefore, becomes the central task of the manager to reconcile differences in approach, timing, effort, and interest and to compromise individual objectives to contribute to organisation goals.

Consider the following case, which can often occur in an industry.

A lathe machine has to be purchased for the production. The general manager conducts a meeting with all the departmental heads working under him to decide the make, type, and whether to buy a new one or second hand and so forth. In the meeting all managers are given a chance to speak.

- The finance manager proposes to procure the machine that works out to be the least in cost.
- The industrial engineer expresses his preference to the most ergonomically designed machine.
- The purchase manager prefers to purchase a machine from a supplier who can offer discounts or free transport or some other advantage or benefit.
- The accounts manager opts to procure the machine from a supplier who can wait longer for payment of the bills owing to the prevailing weak financial position of the company.

- The production manager wants the machine that can produce the highest output.
- The manager of quality assurance department insists upon the process capability of the machine and, hence, suggests the machine that can produce best quality components with high degree of accuracy.
- The human resource development manager stresses on the machine's high degree of safety.
- The maintenance manager asks to go with the machine with trouble-free operations, good condition, and which has longer life.

The general manager has now the real task of coordinating with all these people which is more difficult than planning, organising, staffing, and directing. His decision must be the most convincing to all the managers as the opinions put forward by each manager are equally right and convincing. In such situations, the plant engineer can only help the general manager in achieving good coordination amongst all since maintenance is the only department common to all the other departments. A plant engineer can lay a key role to convince all others to their satisfaction.

Such similar cases often occur in our daily life too. The best coordination results when individuals have a common vision of contributing to the goals of an enterprise. And this is possible if the managers have clarity and understanding of the goals of the organisation as well as of their own individual goals. Otherwise, each will be driven by individual ideas and it will finally result in achieving only the individual goals and missing out on the organisational objectives. To avoid such splintering efforts, the dominant goal of the enterprise should be clearly defined and communicated to all concerned. And obviously, the top management has to define the goals of subordinate departments to contribute to the goals of the enterprise. For all this often the plant manager will be entrusted to coordinate to find the optimum mix of the individual goals to achieve the end-goal of the enterprise.

According to Mary Parker Follett, good coordination can be achieved through the following principles:

- Early beginning (coordination is to be sought at the early stages and later it may become difficult)
- Direct contact (to avoid misleading, misunderstanding, and red-tapism)
- Reciprocity (both the parties are equally responsible and are reciprocal)
- Continuity (coordination is not a one-time job but is a continuous process)

2.7 PLANT ENGINEERING AND MANAGEMENT AS INTEGRATING FUNCTION

To integrate is to join, combine or link together. An important aspect of the function of management of plant engineering department is integration.

To accomplish the objectives of an organisation, the management has to establish two types of integration:

- External
- Internal

External integration means linking the objectives of the organisation to the needs of its external environment consisting of the production shops, customers/consumers, society, labour, suppliers of materials and equipment, investors and financiers, government, and such other bodies on which its survival and growth depends. In order to successfully pilot the aircraft of the organisation, management will involve in maintaining good relations with each of the various sections mentioned above and earning their goodwill and also providing tangible and intangible benefits to them so that they continue to provide support and patronage to the organisation. If, for example, in setting its objectives, the management overlooks payment of fair wages to employees or proper quality of goods and services to consumers or, for that matter, timely payment of taxes to the government, it will lead the organisation into rough weather.

Internal integration refers to proper management of operations by harmonically working of various departments and linking up the allocation of resources (human as well as physical) with the objectives and priorities set as a result of external integration. A very crucial part of internal integration is the integration of employee efforts into company's plans and programmes. This will involve the linking and reconciliation of individual goals with the organisational goals.

In short, management engages in setting and adjusting the organisational goals to meet the needs and demands of various parties in the environment and achieving these goals by creating the right internal environment for people to make their best contributions.

2.8 INTERFACE BETWEEN PLANT ENGINEERING AND MANAGEMENT AND OTHER DEPARTMENTS

Though Plant Engineering is concerned with restoring of machine condition and preventing failures, yet it has relation with each and every department of the organisation. Right from gang man (gm) to

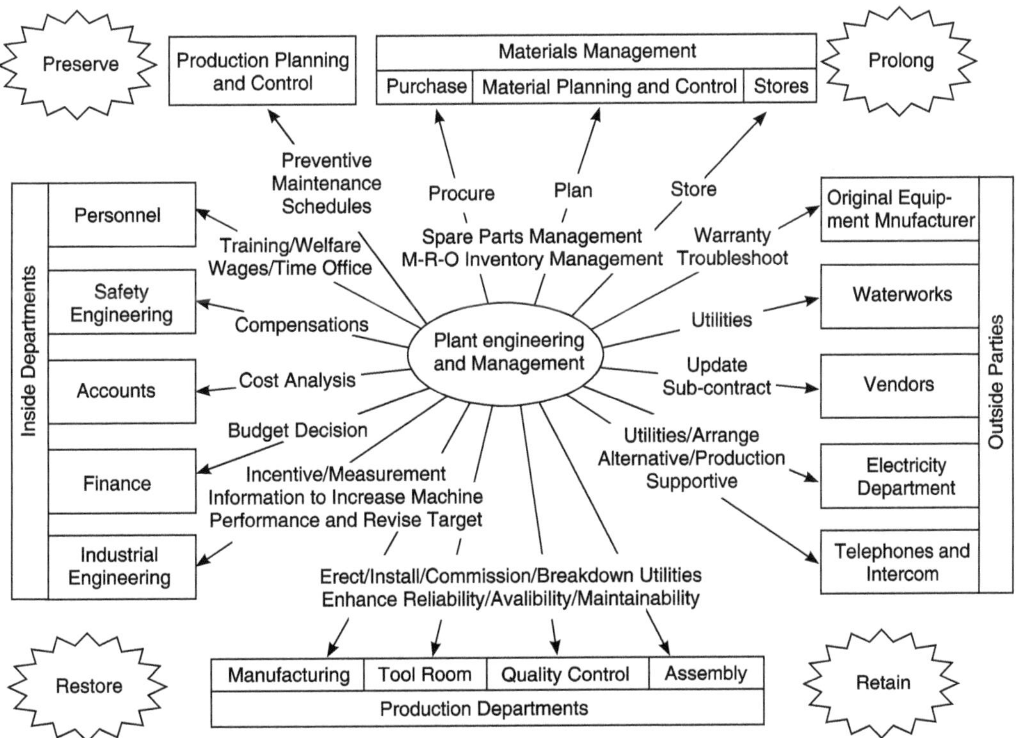

Figure 2.1 Interface between plant engineering and management and other departments

general manager (GM), every employee of the organisation needs the assistance of plant engineering services at one or the other time. Figure 2.1 depicts a few relations and interactions of different departments in various aspects.

Now, we can derive the following relations from Figure 2.1.

1. Plant Engineering Management (PEM) is related to production departments very closely in many aspects. In fact, its very purpose of establishment is for the production departments such as erection, installation, commissioning, overhauls, lubrications, preventive maintenance, shifting of the machinery, updating, and providing utilities like air, water, air conditioning, etc., repair, reconditioning, and upkeeping.
2. PED interacts with PPC to plan the maintenance schedules so that they do not clash with production schedules. It also reports the cost of maintenance, spares, and utilities or services, etc.
3. PED raises its budget requirements in the beginning of the financial year and reports the cost of spares, utilities, day-to-day activities, etc., regularly or periodically to finance and/or accounts department.
4. PED has a relation with personnel department with reference to the wages, office time, welfare, and disciplinary measures apart from the recruitment and training. Often PED looks after the safety engineering activities and it involves the personnel department for compensations and other related benefits.
5. In connection with the work measurement, incentives and target-fixing, etc., the PEM seeks the help of industrial engineering department (IED). Perhaps, it may sound strange to state that IED seeks the help of PEM, but it is the truth that IED requires information about chances of enhancing the production on a machine or revising the targets in relation with the condition of machine and the effect of reconditioning on productivity, etc.
6. Purchase department, materials department and stores are coordinated by PED in case of procurement of spares, tools, consumables, oils, and gases, etc. Usually a store will be managed by PED itself, with the inventory generally referred to as M-R-O (Maintenance-Repair-Operating) inventories. However, the production planning and control (PPC) will generally coordinate with the purchase department and PED in procurements in most of the medium-to large-scale industries.
7. PED will contact and liaise with external parties also to recondition and solve major problems, which cannot be solved by internal personnel due to inadequate facilities or skill/knowledge. It will also coordinate with the OEM to seek its help in troubleshooting especially during warrantee/guarantee period. Sometimes, this coordination is maintained for updating of machinery with the latest technology.
8. PED keeps in touch with material suppliers for spare part management and with the departments connected with the utilities like electricity board, telecom department, water works department, etc.
9. All departments of the organisation seek help from PED failures of their office equipment such as fans, air conditioners, intercommunication telephones, electricity, etc., including the street lights within the organisation premises.

BUSTLE - 2.3

1. *Discuss the reporting system and various records in plant engineering department of the organisation you visited. Suggest some modifications/rectification if required.*
2. *Visit an industry and interview the plant manager about his interactions with various departments and parties. Also interview other managers of various departments about their interactions with the PE department. Write the extra interactions that have been observed.*

____

____

____

____

Test Your Understanding 2.3

1. Explain how plant engineering can integrate all the departments of the organisation.
2. Discuss the significance of coordinating function with reference to the Plant Engineering and Management.
3. Discuss the function of directing with reference to the Plant Engineering and Management.
4. Bring out the interface between PEM and various people outside the organisation.
5. "Mismanaged PEM leads to heavy losses and proves disastrous to the organisation." Comment.

2.9 CONCLUSION

If planned maintenance jobs meticulously, adhered to the plans strictly, and managed properly, Plant Engineering and Management will definitely increase the productivity of an organisation. But any mismanagement or unorganised method of working will result in huge losses. Several organisations claim that the major contribution to any loss of production is attributed to the Plant Engineering and Management. The shop floor level production managers generally claim that their machines are not updated, not made available for more time due to shutdowns, improper maintenance, waiting for spares, absenteeism of maintenance crew, etc. Obviously, the Plant Engineering and Management will usually be held responsible for loss of production. This atmosphere frequently leads the managers to search for the reasons to escape at times. Plant Engineers! Beware of these Managers who are said to have the ailment called "Throw it Out of the Wall Syndrome (TOW Syndrome)".

So, *plan the work and work the plan.*

2.10 RECAP

In Chapter 1, we have learnt that the existence of Plant Engineering and Management in the organisation is to preserve assets, prolong the life of the assets, retain their good condition, and restore the acceptable condition of the machinery or equipment or asset if they fail. To execute and achieve these objectives, PEM organises either centralised or decentralised system or a mix of both (based on the advantages and disadvantages the organisation may derive out of the system) to execute the functions like corrective (breakdown), preventive, productive maintenance, spare parts management, etc. It also looks after the erection, installation, commissioning of the new equipment, shifting of the machinery, providing utilities like water, compressed air, inter-communication services, etc. The maintenance even undertakes the activities like updating the technology and reconditioning of the machinery. In this connection it interacts with various internal departments such as production in performing the aforesaid functions, finance in budgetary decisions, accounts in cost analysis, personnel in training and compensation against accidents, PPC in preparing PM schedules. Apart from these, PEM interacts with the external parties also such as material suppliers, OEMs, vendors, sub-contractors, electricity, telecom, water works, fire stations, etc. The planning function is elaborated due to its high importance in the present scenario. The staffing, directing, and coordinating are also discussed with reference to the functions of maintenance or Plant Engineering and Management. These concepts are discussed in this chapter. To execute these functions effectively, we need to have certain strategies. Various strategies and their philosophies will be discussed in Chapter 3.

2.11 KEY CONCEPTS

1. **Plant:** A place where men, material, and machines are brought together to convert the inputs into the desired outputs in a well-defined method.
2. **PEM:** Plant Engineering and Management is that part of organisation that helps to preserve the assets, prolong the life, retain the condition of the equipment/assets, and restore the condition when it fails.
3. **OEM:** Original equipment manufacturer is a person or a group or an organisation that manufactures a machine or equipment or an asset.
4. **Breakdown:** Failure resulting in the non-availability of an item.
5. **Down time:** The period of time during which an item is not in a condition to perform its intended function.
6. **Overhaul:** A comprehensive examination and restoration of an item or major part there of to an acceptable condition.

REVIEW QUESTIONS

1. How is maintenance planning different from general planning and production planning? Give examples.
2. What is maintenance planning? Explain different stages involved in maintenance planning.
3. Explain the merits and demerits of centralised and decentralised plant engineering department.
4. Explain the factors affecting the degree of decentralisation. Give suitable examples with plant engineering department.
5. Explain the organisation of maintenance activities in a large scale industry.
6. Explain the relationship between plant engineering and other departments within and outside the organisation.
7. Plant engineering concerns and serves everyone of the organisation. Provide points in support of the statement.

CHAPTER 3

Maintenance Strategies

PREAMBLE

This chapter lays down emphasis on various methods or policies of maintenance from which an organisation may select one or more to meet its requirements. The policies are designed depending on prevailing factors in a given organisation. The details on various maintenance methods, policies, and strategies are clearly stated in this chapter. The merits and demerits of various methods or policies are discussed in detail. The systematic procedures for carrying out maintenance and setting a maintenance strategy are also discussed in this chapter.

CHAPTER OUTLINE

3.1 INTRODUCTION

One of the critical areas of any organisation is maintenance. In the present-day context of the advancement and rapid growth in technology, the adoption of this technology is also increasing at a very fast rate in industries. Consequently, the maintenance of such production machinery and equipment is also getting more significant. Yet, in most of the industries in India, owing to various reasons, the capacity utilisation is less than 60%. The reason may be mostly attributed to the act of running the maintenance department without any systematic policy/strategy. Perhaps, many small to medium scale industries, least important to this department as they feel, it is non-productive. But, it is understood from many practitioners that a correct policy of maintenance and systematic implementation could keep the machines in a state of maximum efficiency with economy.

3.2 MAINTENANCE STRATEGIES

Different experts have classified the maintenance strategies and methods in numerous ways of which the following are the most comprehensive.

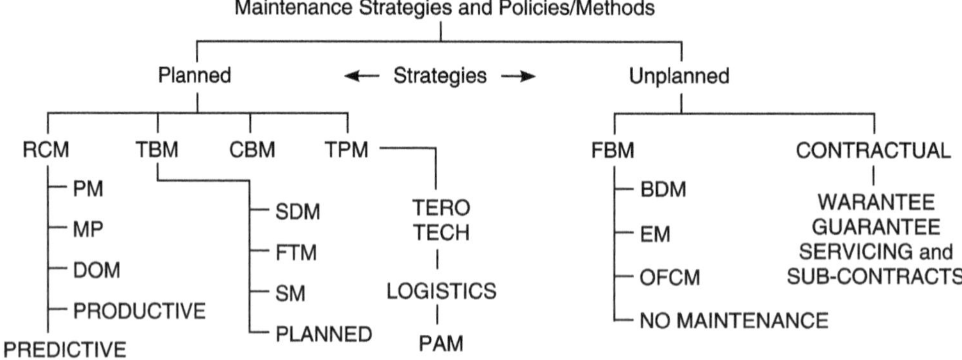

Figure 3.1 Classification of maintenance strategies/policies

To understand various maintenance methods, they can be discussed under two heads, unplanned and planned strategies. The planned strategies are designed on the basis of certain maintenance characteristics such as number of failures or frequency of failures or durations of failures, time between failures, time to repair, machine condition, machine life, reliability, maintainability and availability and so on. Summing up all these, planned strategies can be classified into four methods/ techniques: reliability-centered maintenance (RCM), time-based maintenance (TBM), condition-based maintenance (CBM), and total productive maintenance (TPM). Under unplanned strategies, there are two techniques, failure-based maintenance (FBM) and contractual maintenance (CM). These are further classified into various methods depending on their dominant features.

(a) **Failure-based Maintenance (FBM):** This includes breakdown maintenance (BDM), operate to failure and corrective maintenance (OFCM), emergency maintenance (EM).

(b) **Contractual Maintenance:** Service contracts, sub-contracts, maintenance provided by the OEM, maintenance in the warrantee/guarantee period, etc.

(c) **Reliability-centred Maintenance (RCM):** Maintenance prevention (MP), design out maintenance (DOM), productive maintenance, and predictive maintenance are included in this.

(d) **Time-based Maintenance (TBM):** This includes preventive maintenance (PM), scheduled maintenance (SM), shutdown maintenance (SDM), and planned maintenance.

(e) **Condition-based Maintenance (CBM):** Dynamic predictive maintenance (DPM) or condition-based maintenance (CBM).

(f) **Total Productive Maintenance (TPM):** TPM, tero-technology, and physical assets management (PAM).

3.3 FAILURE-BASED MAINTENANCE (FBM)

According to Murphy

- *If a failure can occur, it will.*
- *The failures occur at the most inconvenient times.*

Since failures are considered to occur at the most inconvenient times, a sufficient thought is to be given to overcome if not prevent such inconveniences. The failure-based maintenance techniques (such as breakdown maintenance, emergency maintenance, operate failure, corrective maintenance, etc.) are difficult to plan since the breakdowns and emergencies occur at unknown or unpredictable times (if they can be predicted, they cannot be called emergencies). Thus, the failure-based maintenance is assumed as an unplanned maintenance.

Failure-based maintenance philosophy can be applicable on the basis of failures. Therefore, an understanding of the different types of failure before using these strategies is necessary. The failures are classified in different ways. A brief outline of classification of failures is given below for ready reference.

(a) **Based on Machine Life Cycle**

- *Early failures*: Immediately or in the preliminary days of purchase of equipment, in the early days of its life, the machine has some design failures or failures due to unsuited environment or due to user's carelessness, etc. The early failures are usually taken care by the OEM as they are covered under the warrantee/guarantee. (Refer Figure 3.2.)
- *Random failures*: These failures (also known as chance failures) occur randomly. The emergency maintenance and breakdown maintenance are suitable depending on the significance (critical/normal/unimportant) of the operation carried out on the machine. (Refer Figure 3.2.)
- *Wear out failures*: These failures occur on age old machines due to worn out parts. In such failures, the OFCM, reconditioning, etc., are used. (Refer Figure 3.2.)

(b) **Based Cost of the Failure**

- *Minor failures*: These are the failures that cost very less. The maintenance mangers do not bother about such failures. However, a small failure if neglected may lead to catastrophe at times. Therefore, it is better to attend to such issue in its nascent stage before it turns into real problem.

- *Medium failures*: These failures may not cost high but considerable loss may result in the form of lost production. Therefore, a timely address to the problem will save from heavy loss.
- *Major failures*: These failure cause considerable production loss besides being costly to reinstate the condition.
- *Catastrophes*: These result in huge loss and may even claim lives, causing loss to production and equipment.

(c) **Based on Frequency of Failures**

- *Recurring failures*: Same or similar failure occurring repeatedly comes under this category.
- *Rare event or chance failures*: These are the failures that seldom occur.

(d) **Based on Length (or Duration) of Failures**

- *Short failures*: These are the failures that take little time to correct and restore the condition of the machine. An uncorrected or pending short failure results in long failure.
- *Medium failures*: These are the failures that do not take much time but considerable duration to correct and restore the condition of the machine.
- *Long failures*: These are the failures that take longer time to correct and restore the condition of the machine.

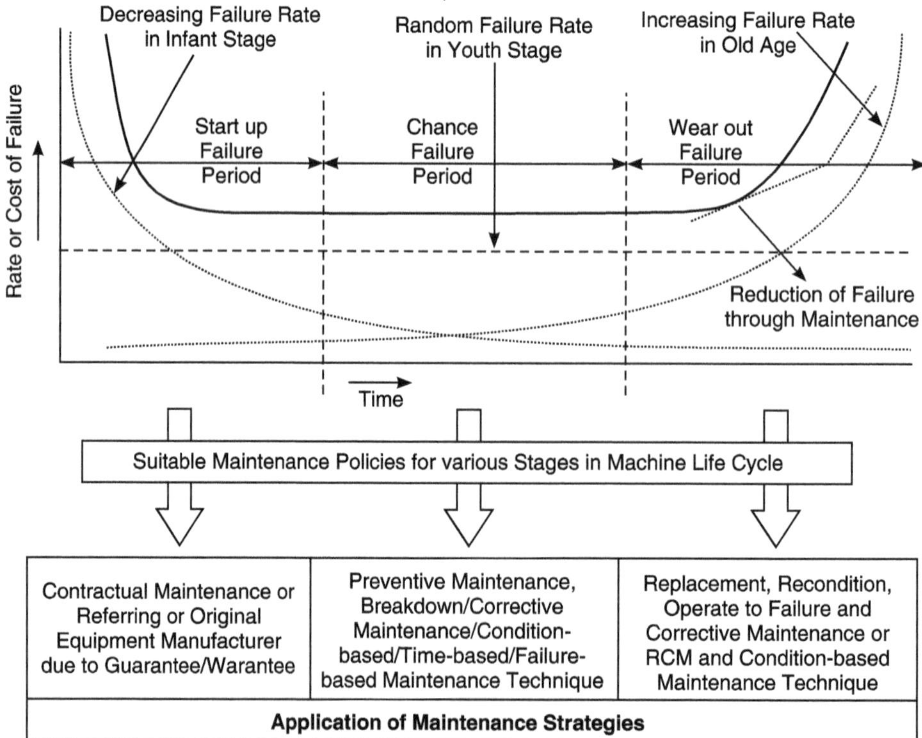

Figure 3.2 Bathtub curve (machine life cycle) showing early failures, random failures, and wear out failures

(e) Based on Intensity of Failures

- *Progressive failures*: The failure intensity gradually increases and causes deterioration in the machine part or equipment. (Refere Figure 3.3a.)
- *Constant failures*: The failure intensity neither increases nor decreases. (Refere Figure 3.3b.)
- *Retrogressive failures*: The failure intensity gradually increases and the condition of machine improves with time. Usually these are called teething problems and vanishes as the machine gets adjusted to the environment. (Refere Figure 3.3c.)

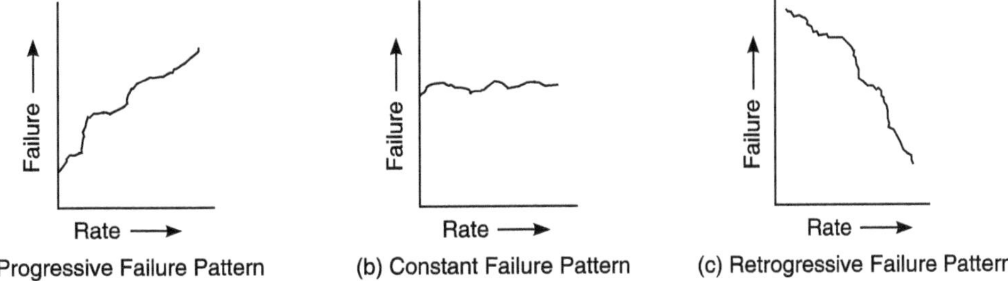

Figure 3.3 (a) Progressive, (b) constant, and (c) retrogressive failures

3.3.1 Breakdown Maintenance (DM)

As stated by Murphy, if a machine can fail, it will. Hence, no machine is completely immune to failures. Every machine experiences breakdown some day or other, no matter how carefully it is run, adhering to its specifications and ambient operating conditions. These breakdowns are believed to occur due to gradual wear and tear of the machine parts or any unpredictable failures. In the event of such failures, breakdown maintenance policy is usually adopted. This policy is considered to be suitable in two cases:

(i) when the continuous running of machines is not compulsory, and

(ii) on those machines which do not affect any production flow due to the shutdown.

However, in general practice this policy is adopted in combination with preventive maintenance.

The breakdown maintenance starts on the machine from the moment it fails and ends when it is repaired and handed over to the production personnel (i.e., machine is put under use). Except the normal lubrication and cleaning, this system is no way concerned to prevent the occurrence of the breakdown but gets alert when the breakdown has occurred. Hence, in most of the cases, this system reduces the machine utility directly and indirectly and, thus, results in low machine efficiency because this type of maintenance usually takes a long time to repair.

The moment the machine fails it is said to be in down condition. When the machine is repaired, i.e., when the machine is put into operation, it is said to be in up condition. The time taken between any two consecutive failures i.e., from the moment a machine is kept in use (up) to the moment it next fails, is called the time between failures or TBF. The time taken to repair and put it into use (up condition) is called time to repair or TTR. These are shown pictorially in Figure 3.4.

The terms TBF and TTR seem to carry no meaning, in the case of non-repairable items, i.e., the items which will not be repaired or cannot be repaired but replaced. Since there will not be any repair work on the item, the term time to repair has no meaning as far as the item is considered. But, usually the item will be a small part of the main system and the system gets rectified when the item is replaced by the new one. Therefore, the term TTR is suitable to describe the system status. However, TBF (time between failures) is not appropriate even for the system also (*why?*). Therefore, we often use the term time to fail (TTF) in stead of time between failures (TBF) in the case of non-repairable items.

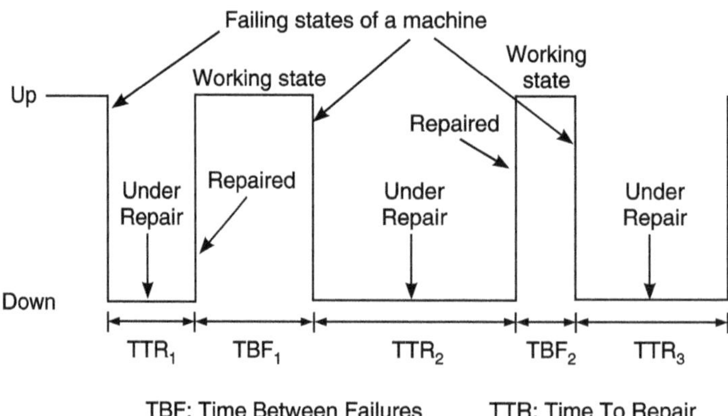

Figure 3.4 Up and down conditions of machine with TBFs and TTRs

The average of such times is referred to as mean time between failures (MTBF), mean time to fail (MTTF), and mean time to repair (MTTR). The basic level formulae are given below.

Mean Time Between Failures (MTBF)

= Total time the machine is in up condition/Number of failures

or = Worked time/Number of failures

Mean Time To Repair (MTTR)

= Total time the machine is in down condition/Number of failures

or = Total repair time/Number of failures

3.3.2 Emergency Maintenance (EM)

Although this policy of maintenance is used with the same sense as that of breakdown maintenance, a little difference exists. The action in emergency maintenance is taken up immediately when the complaint is received while in the case of the breakdown maintenance (BDM) we have a breathing time and is carried out after deciding the order of priority by some sequencing rules. This policy is applied in the emergency wards of a hospital, fire stations, police stations, underground mines, accident planning of an industry, etc. When the nature of work in an organisation is hazardous or the probability of occurrence of accidents is high, such a policy is adopted. A group of people is kept for the purpose when this policy is adopted and will not be disturbed even if they are idle when there are no complaints.

The above discussed methods, EM and BDM seem to be similar but they differ in some aspects which are depicted in Table 3.1.

Table 3.1 Distinction between breakdown maintenance and emergency maintenance

No	*Breakdown Maintenance*	*Emergency Maintenance*
1.	Slow response, i.e., action may not begin immediately, but as per the convenience of the maintenance personnel or situation.	Quick response, i.e., action begins immediately after the failure occurs.
2.	No efforts are made to prevent failures before or during the occurrence of failure.	Efforts are made to prevent further damages or catastrophic failures.
3.	Suitable for non-hazardous, calm, or serene area of operation where the equipment does not cause any damage or loss.	Suitable for hazardous or dangerous areas.
4.	No separate team is maintained; even if maintained they may be utilised for other functions of maintenance when they do not have work.	A separate crew is maintained who will not be used for any other purpose though they are idle.
5.	Non-emergency areas such as open cast mining, in hospital wards where immediate attention is not necessary such as out patient clinical equipment.	Underground mining, fire station, aircraft maintenance, hospital (equipment at acute medical ward, casualty, emergency, etc.).

3.3.3 Operatre to Faliure and Corrective Maintenance (OFCM)

In this maintenance policy, no predetermined action is taken to prevent failure and the emphasis is given on efficient corrective maintenance. It is carried out on the machine whose guaranteed life is already utilised. Any amount of production achieved on these machines (whose resale value has reached or nearing the scrap value) is treated as bonus to the organisation. The equipment is run till it fails without any attempts being made to prevent the failures. When failure occurs, it is rectified if it is minor and economical, otherwise scrapped or sold by making necessary replacement analysis. For instance, a bike or scooter is not used after 15 years (its life is stated as 15 years).

3.3.4 No Maintenance Technique (NMT)

The theme of this maintenance policy is to avoid maintenance of the equipment. In other words, under this policy, the maintenance of equipment is not required. In general terms, this is often called *use and throw system,* where the equipment is used to its full length of life and then scrapped or replaced with a new one, e.g., some furnaces or kilns, etc., will be dismantled after its utilisation. Now-a-days, many items are available in this form in the market. The syringes, razor blades, etc., were reused by sterilising or cleaning earlier but now they are disposable and are insisted upon this policy to use and throw in order to prevent the spreading of contagious diseases. This system is adopted when

- reusing/recycling is harmful or dangerous or not advisable or is risky (e.g., syringes, razor blades, chemical, etc.).
- recycling is not possible (e.g., non-rechargeable batteries, items made up of some chemicals, products made up of plaster of Paris, and cement, etc.).
- repairing is not economical as compared to buying a new item (some types of ballpoint pens, electronic goods, and the equipment whose spares are very costly).

The two techniques, OFCM and NMT may look alike but they differ in several aspects as shown in Table 3.2.

Table 3.2: Distinction between NMT and OFCM

No	Operate Failure and Corrective Maintenance	No Maintenance Technique
1.	This policy encourages to use as much as possible.	This technique encourages to use and throw.
2.	After stated life is completed, maximum advantage is obtained before scrapping.	After stated life is completed, it is discarded or scrapped.
3.	Efforts are made to reuse or correct the failure if it is not much expensive.	No efforts are made to correct or reuse after completing life span.
4.	It may cause inventory costs, but contributes additional production after its complete useful life which is considered as bonus in the production.	No advantage is sought after using it, therefore, thrown out and hence does not cause any inventory cost.
5.	This policy is meant for capital intensive equipment and they are reusable and resalable.	This policy is meant for the types of products which are so designed as not reusable or recycled.
6.	Suitable for repairable equipment.	Suitable for non-repairable (replaceable) equipment.
7.	Normally, this is applied on costly equipment where the new one is costlier as compared to repairing of the existing or older one.	Normally, this policy is adopted on less expensive equipment or parts (or the cost of new one is less than or equal to repair cost).
8.	An old lathe machine, a boiler in an industry.	Syringes, electronic gadgets, electric bulbs.

BUSTLE - 3.1

Sometimes it becomes inevitable to work in an unplanned way. Can you imagine such situations? Support the statement with reference to PEM with some practical/hypothetical evidences.

____

____

____

____

Test Your Understanding 3.1

1. Classify and discuss different types of failures with suitable examples.
2. Distinguish between breakdown maintenance and emergency maintenance?
3. Explain the concept behind the following maintenance methods:
   (a) Breakdown Maintenance (b) Emergency Maintenance
   (c) No Maintenance Technique (d) Operate Failure and Corrective Maintenance
4. Distinguish between no maintenance technique (NMT) and operate to failure and corrective maintenance (OFCM).
5. Explain on what bases the maintenance policies are planned? Explain a few.

3.4. CONTRACTUAL MAINTENANCE

Under this strategy, an outside personnel or agency is made responsible for the maintenance of the equipment. The agency will come and repair the equipment whenever called or visit time to time as required or contracted. This outsider may be the original equipment manufacturer (OEM) or a consultancy or experienced personnel or an agency. Generally, a Memorandum of Understanding (MoU) or a contract is executed between the two parties for a certain period usually one year. This policy is in practice in most of the industries and organisations under the title, Annual Maintenance Contract (AMC).

Usually this system is adopted

- on machines which do not interrupt the production line at all, e.g., type machines, time card punching machines, computers, etc.
- if it is cheaper to get repaired by an outsider than in-house repair.
- if the repairing needs high skill, knowledge and technology or special accessories or measuring instruments.
- when maintenance of spare parts is not economical or perishable or having high degree of obsolescence.
- when the machines on which breakdown rarely occur.
- on the machines which are under guarantee/warrantee period.
- as a temporary arrangement when PEM becomes handicapped due to some problems like insufficient manpower or unavailability of required spares in the open market, etc.
- by sub-contracting or making service contracts due to heavy work load in PEM. Wasting the experience of the maintenance personnel is not at all advisable for repair work of smaller jobs which could be done by sub-contractor. For instance the winding of motors, repair of valves, etc., will be usually given to sub-contractors for this reason.

Merits of Contractual Maintenance:

- No risk
- Time saving
- No cost of inventory

Demerits of Contractual Maintenance:

- High dependability
- Logistic problems
- Lack of timely action

BUSTLE - 3.2

1. *Consider any process or product industry of your knowledge that is adopting the contractual maintenance. Investigate the reasons why they are opting so.*
2. *Most of the software industries are opting contractual maintenance or annual maintenance contract (AMC). Explore the reasons for this.*

____

____

____

____

Test Your Understanding 3.2

1. Why and when does an organisation choose contractual maintenance? Explain with examples. What are its merits and demerits?
2. What are the situations warranting the contractual maintenance? On what type of machines do you suggest this policy? Explain with suitable examples.

3.5 RELIABILITY-CENTRED MAINTENANCE (RCM)

In a sample survey of maintenance management effectiveness for complex and large capacity equipment, it is found that more than one third of all maintenance costs are wasted as a result of unnecessary or improperly carried maintenance over maintenance. It is seen that task selection in maintenance need not be optimal but still very subjective in approach. The reliability-centred maintenance (RCM), suitably blended with predictive maintenance and maintenance prevention can help in reducing maintenance costs.

The RCM is defined as a process or method used to determine maintenance of equipment in its environment after proper evaluation of failure consequences. The RCM is a systematic approach to quantitatively assess the need to perform or revise preventive maintenance tasks and plans. The RCM focuses on the system function, functional failures, dominant failures and their effects. The RCM uses Decision tree to classify the criticality based on the consequences of analysis to identify applicable and significant tasks. Once the significant maintenance tasks are identified, their maintenance intervals can be estimated either by using condition monitoring devices or by reliability analysis of the failure statistics. Close analysis of RCM approach indicates that it concerns with system effects (safety, economic, social) on consequences without any reference to human aspects are advocated and treated as key to success in business and maintenance. The philosophy of RCM focuses on enhancing the probability of the machine not failing in the given time span and assuring or building up the confidence of using the equipment in the specified time. The concepts and the philosophy of RCM are exhaustively discussed in Chapter 12 and 13.

3.5.1 Predictive Maintenance

Predictive maintenance aims at the earliest forecasting or prediction of failures by judicious application of condition monitoring or trend checking techniques and instrumentation. The objective of such a philosophy is to predict failures in advance, so that maintenance actions can be planned to take place at a time convenient to man and machine and, hence, to reduce the unplanned interruption of production system.

3.5.2 Preventive Maintenance (PM)

This maintenance philosophy is based on the proverbial sayings “A stitch in time saves nine” and “Prevention is better than cure”. Preventive maintenance is performed to reduce the possibility of failures. The functions of preventive maintenance will be such as oiling, greasing, cleaning, voltage, and current checkup, etc. On the other hand, it may be dismantling and assembling a part or entire machine in order to overhaul completely based on certain time scale. Thus, it covers the entire range of involvement of careful analysis of maintenance requirements like cost relationships, maintenance planning and control, corrective actions, spare part management, etc. Thus, the objectives of the PM can be summarised as follows.

Objectives of Preventive Maintenance:

- To enhance the availability of the equipment and to keep it ready to use at any moment, thus, to enhance the reliability of the equipment.
- To minimise the probability of unanticipated interruption of production or major failure by uncovering or locating any condition that may lead to it.
- To reduce the work content of maintenance job as it tries to prevent the major failures.
- To minimise the maintenance cost (by achieving the minimum repair cost).
- To assure safety of life of the operators.

As a matter of fact, there is no readymade or on the shelf preventive maintenance programme for any organisation. It is tailored to fit the requirements of the plant. Perhaps, the reason is that the plants differ in size, age, location, manpower, resource, layout and machinery. However, the preventive maintenance is carried out in the following steps, which will appear more or less same in any industry.

Procedural Steps in Preventive Maintenance:

- The PM is carried out on the equipment on which we require more available time of operation. Therefore, such equipment should be first identified. Most common practice is that the identification of such machines is done by a committee consisting of members from both production and maintenance departments.
- To carry out PM, the parameters have to be identified by maintenance department (or plant engineering department). Based on these parameters, a checklist is prepared for each and every machine. These may vary from one another.
- The next step of PM is to determine the interval of the checking or inspection of the machine condition. This interval is fixed on the basis of past records of the frequency of failures or experience.
- The parameters in the checklist of the machine are inspected and overhauled or made minor repairs or dismantled partially or totally based on the machine condition.

Although the basic concept is derived from preventive maintenance only, due to their difference in the structure or dominant factors of the system the PM has taken some improvised shapes/ advancement such as the scheduled maintenance, shutdown maintenance, planned maintenance, productive maintenance, maintenance prevention, etc.

3.5.3 Maintenance Prevention (MP)

This maintenance policy concentrates on the prevention of maintenance of the equipment rather than the prevention of the failures. The objective is to eliminate or to reduce the frequency of maintenance work to the extent that the cost of improvement can be justified. Skill and knowledge of operators, usage of equipment, and their cost, all are considerable factors in providing efficient or reliable maintenance standards at economic cost. The focus of this policy is on improvements or upgrading or updating rather than repairs and reuse. It also advises the machine upkeeping by the user, minor changes in design, substitution of more suitable components, or improved materials. The importance will be given to replacement with the new parts than repairing them. The older parts will be disposed of as "second-hand sales". This may be due to the reason that the new parts are believed to give trouble-free performance with high reliability and the life of the equipment is, thus, enhanced.

The failed part of the equipment when replaced with a new one is considered "as good as new" while if the failed part of the system is repaired and put back is termed "as bad as old". In short, this policy believes in keeping the equipment in "as good as new" condition than keeping it in "as bad as old".

Further, it has another advantage that the updated and technically improved parts can be incorporated into the machine to improve the productivity. However, this is possible if the equipment is so flexible, compatible and can match/adjust with the new parts. Obvious is that the cost of maintenance rises due to the spare parts.

Merits:

- Strives to keep the equipment in "as good as new" condition.
- Failure costs and effects are less.
- Majorfailures/losses, catastrophes are prevented.
- Though not intentionally, the equipment gets updated and upgraded frequently and, hence, the productivity is enhanced.
- The equipment cost/resale value will not come down and remains at reasonably good level due to updating.

Demerits:

- Cost of spare parts increases because even if the part is repairable, it is preferred to replace.
- Unnecessary inventory of spares without knowing the compatibility.
- Reselling the equipment is easier than reselling parts. Thus, second hand sale of parts is not easy and, hence, the replacement proves costly. Further, the cost of maintaining such removed parts increases.

3.5.4 Design Out Maintenance (DOM)

In contrast to the preventive maintenance and maintenance prevention policies, which aim to minimise the effect of failure, this maintenance strategy focuses on the updating function of maintenance. Under this policy, the maintenance department conducts a study on the equipment working less efficiently or causing high maintenance cost. It checks whether the machine performance and life can be improved by replacing or upgrading some parts. Clearly, this requires support from Engineering department, Planning department, Production department and Purchase department, and the top management. The maintenance department has to convince and seek the approval by justifying the need for upgradation or modernisation of the equipment of interest. This is usually a policy for areas of high maintenance cost due to poor design or the equipment is out of its design specifications. An appropriate choice of tribological materials might eliminate the need for subsequent lubrication frequencies. Thus, this can be regarded as the research and development activity of PEM to increase the productivity or ease of operation or reducing the time of operation by changing or modifying the designs. In few industries, this activity is performed by the R&D department but it is always advisable to assign this to PED or at least to a committee formed by the members from PED, R&D and production.

3.5.5 Productive Maintenance

The stress of this policy is on productivity and aims at the reduction of volume of maintenance work by better designs and better use of the materials. The equipment is so designed that the units or assemblies can be easily replaced as and when necessary before a breakdown occurs. Such replaceable items or units or sub-units are usually referred to as least replaceable assemblies (LRA). The replaceability should be with great ease so that the maintenance policy is to aim at abolishing or minimising repairs and upkeep by replacements or modifications. Even if it is to be repaired, it is done with minimum disturbance to the production. This can be achieved by a large number of standard types of machine tools or interchangeable units, e.g., the sub-units in metal cutting machines such as head stock, fraction mechanisms, gearbox, turret, spindle units, etc.

Another good example can be quoted here with personal computer systems (Figure 3.5). A computer system is composed of various alternatives of data storing/retrieving sub-systems like RAM, Internal HDD, External HDD, 3½" FDD, 5¼" FDD, CDROM, CDRW, CDRAM, USB Drive (Pen Drive), etc., which enables the replacement of these sub-systems in the event of their failures without affecting others working and overall performance.

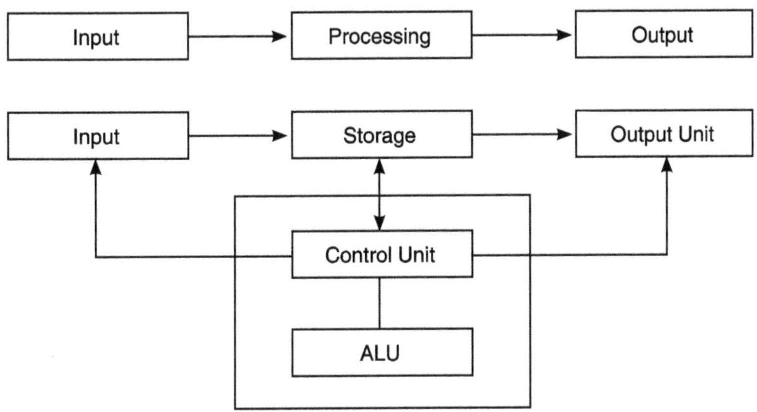

Figure 3.5 A block diagram of computer system

Although the cost of spare parts increases, here, the duration of repair time and loss of productive hours are reduced considerably and machine utilisation is increased. Also the production shop can work at full capacity and the quality of product is improved.

BUSTLE - 3.3

In the organisation where you are working or the institution where you are studying, collect the data of configurations of all the computers and check if you can suggest any improvements or upgradation for better performance.

____

____

____

____

Test Your Understanding 3.3

1. Describe various types of preventive maintenance techniques?
2. "A stitch in time saves nine". How would you support with reference to preventive maintenance?
3. "Prevention is better than cure". How would you support with reference to plant engineering functions?
4. Write notes on the following maintenance techniques:
   (a) Design Out Maintenance
   (b) Preventive Maintenance
   (c) Planned Maintenance
   (d) Productive Maintenance
   (e) Predictive Maintenance
5. Critically bring out the differences between maintenance prevention and design out maintenance.
6. Distinguish between preventive maintenance and maintenance prevention.
7. What are the objectives of preventive maintenance?
8. Discuss the merits and demerits of preventive maintenance.
9. Distinguish between the productive and predictive maintenance techniques.
10. Describe the common procedural steps in designing a preventive maintenance programme.
11. What do you understand by the term RCM? Explain how predictive maintenance can help in implementing RCM?

3.6 TIME-BASED MAINTENANCE (TBM)

Before selecting the right policy of maintenance, it is essential to fix up the basis such as the failures or the time or the reliability or the condition of the machine. Perhaps, choosing the machine failure as the basis for adopting the maintenance is the oldest and most general. However, on analysing the failures it is found that mostly the failures can be fitted in some regular pattern such as Normal or Poisson distribution, etc. Based on the frequency of occurrence of these failures, it is observed that most probable failures are expected to occur at regular intervals. These calculations give rise to a thought of preventing the failure since they are expected to occur regularly at specific intervals. Once these intervals are identified, a plan or schedule of maintenance can be made. As the emphasis is on the time intervals, these preventive maintenance techniques are classified under time-based maintenance (TBM) models.

3.6.1 Shutdown Maintenance (SDM)

This type of maintenance is carried out when the equipment is out of service. Whenever the machine is not in use or remains idle for reasons other than breakdown, like shortage of materials, huge decoupling inventory piled up, non-availability of manpower, absenteeism, holiday, etc., it is taken up for preventive maintenance/overhauling. Thus, the interest of this policy is not to disturb the production for want of maintenance and is done in leisure period of the machine. Perhaps, this method is suitable for the production system where the machines are not run continuously. If the machines have to be run continuously, the planning of this method will be difficult and will disturb the regular schedules of the maintenance. In such cases, it is better to attach this activity to the group that looks after breakdown maintenance.

3.6.2 Fixed Time Maintenance (FTM)

This policy recommends replacing the item when its promised or guaranteed life is completed or it is taken up for repair after a fixed interval repair irrespective of its condition. This programme is effective if the failure mechanism of the item is clearly time dependent, i.e., the item is expected to wear-out within the life of unit and the total cost of such replacement is less than those of failure replacement repair. Thus, the item is simple replaceable. There are two main reasons for not adopting this policy for complex replaceable and adopted for simple replaceable items.

- Firstly, the more complex item is assumed less likely to exhibit failure and is time dependent. Due to simple item failure, it may become complex.
- Secondly, the complex items are expensive to replace or repair and subsequently exhibit "foreign maintenance" problems.

3.6.3 Scheduled Maintenance (SM)

The focus of this maintenance is to keep the equipment clean, presentable, and in proper functional condition. The maintenance schedules are prepared in the beginning of the year in such a way that the schedules will least affect the production. Once these schedules are approved and agreed by the production department and PEM, irrespective of the condition of the machine, it is taken up for maintenance as per the schedules. A few functions of this maintenance system are enumerated hereunder.

- Using protective coatings to retard wear and tear on the machine.
- Cleaning of parts that are subjected to intensive wear due to dirt/dust arising out of operating process.
- Adjustment or accuracy checking so that the jobs produced is within tolerance limits.
- Testing and calibrating electric/electronic/chemical equipment.
- Lubrication or replacement of oil, etc.
- Whitewashing, cleaning of tanks, etc.
- Cleaning of boilers.

This maintenance is adopted on the equipment where it affects the production adversely, e.g., generators, painting, air compressor, A/C units, dust extractors, duct lines blowers, etc. These cannot be planned in the regular working hours as the standby is not available and not affordable. Hence, this should be scheduled in a time such that it will not affect the production.

3.6.4 Planned Interval Maintenance (PIM)

This maintenance strategy is in practice in most organisations either in ditto form or with little variations with the names systematic maintenance or regular plant maintenance. The duration of the maintenance and the checking parameters are predetermined in the following four different intervals:

(a) **Inspection (I)**: It is the external inspection carried out at very short intervals (usually on daily basis) which takes a very little time (about 5 to 10 minutes approximately).

(b) **Small or Minor Repairs/Overhaul (S)**: Small or minor repairs/overhauls (S) are carried out on the spot or on weekly basis (short intervals) without affecting production time. Simple adjustments, lubrication, etc., are performed and these activities take very less time, ranging from 15 minutes to 2 hours.

(c) **Medium Repair/Overhaul (M)**: Medium repairs (M) are carried out at long intervals such as monthly or quarterly. Partial dismantling, replacing worn-out parts, readjustments, accuracy checkup, etc., are taken care. This type of work generally takes 4 to 8 hours.

(d) **Complete or Major Repair Overhauls (C)**: Complete or major repair/overhauls (C) are largest planned/preventive maintenance, which involves complete dismantling of the equipment and replacement/reconditioning, etc., are carried out. They are assembled again and tested for accuracy. This maintenance is taken up relatively at very long intervals, e.g., on half yearly or yearly basis and usually takes 1 to 3 days.

Apart from the above regular activities, some occasional activities such as the alterations and installations of new machines are also included in this maintenance.

Regular plant maintenance is suitable where the standby machines are available or when there is no work or insufficient work on the machine. This type of maintenance gives many advantages by reducing the cost of breakdown maintenance, downtime, and spare parts control, rejections and catastrophic failures. This maintenance is performed on time scale irrespective of the condition of machine and uses the checklists prepared for the purpose. When such maintenance is carried out, it may so happen that some machines, which are in good conditions, may also have to be handed over to the maintenance and suffer production. And if the alignment or assembly is not adjusted properly, the machine breaks down within short duration.

3.7 CONDITION-BASED MONITORING (CBM) OR DYNAMIC PREDICTIVE MAINTENANCE (DPM)

A better alternative concept is that the correct time of taking up the equipment for maintenance is determined by monitoring *condition* and/or *performance*.

The desirability of this policy, the monitoring technique and its periodicity depends on the determinability of characteristics of the equipment. This is due to the fact that determinability of machine characteristics is found to be in two extremes. At one extreme, the items (simple replaceable) can be checked visually and by other simple techniques. At the other extreme, the items (complex replaceable) require the techniques like liquid penetration, vibration monitoring, shock pulse monitoring, oil analysis, defect analysis, debris analysis, thermography, etc. The use of instrumentation may be justified by high repair and unavailability costs.

The detailed procedures of these techniques are discussed in Chapters 9 and 10.

3.8 TOTAL PRODUCTIVE MAINTENANCE (TPM), TERO-TECHNOLOGY, LOGISTICS AND PHYSICAL ASSETS MANAGEMENT (PAM)

Total productive maintenance (TPM) developed in Japan, tero-technology in U.K., logistics in U.S.A. have the same goals.

TPM aims to maximise equipment effectiveness. In fact, its goal is same as tero-technology's goal of attaining an economic life cycle cost.

Tero-technology, according to BSI, is a combination of management, financial, engineering, and other practices applied to physical assets in pursuit of economic life cycle costs (LCC). Its practice is concerned with the specification and design for reliability and maintainability of plant and machinery, equipment, buildings, and structures with their installation, commissioning, maintenance modification, and replacement with feedback of information on design performance and costs.

Logistics is an old military term referring to support frontline through procurement, storage, transportation, and maintenance of manufactured goods and systems. In current methods, logistics consists of LCC, reliability engineering and maintenance engineering. Thus, the goal of TPM, tero-technology, and logistics is same as goal of LCC and they differ in terms of preuse target location and responsibility.

The TPM concepts are explained in Chapter 14 in detail.

All the above explained concepts are summarised and tabulated in Table 3.3.

Table 3.3 Maintenance strategies and their central themes

Sl. No.	*Maintenance Strategy/Policy*	*Central Theme*
1.	Failure-based Maintenance	To correct or rectify the failure and restoring the usable condition of the equipment.
2.	Time-based Maintenance	To maintain the machine after certain predetermined period of time.
3.	Contractual Maintenance	To assign maintenance to outside party to save time in-house, or save maintenance cost or due to warrantee/guarantee period.
4.	Reliability-centred Maintenance	To enhance the reliability of the equipment and build the confidence of the user.
5.	Condition-based Maintenance or Dynamic Predictive Maintenance	To improve the condition and performance of the equipment.
6.	Breakdown Maintenance or Corrective Maintenance	To repair the machine after failure occurs.
7.	Emergency Maintenance	To attend the breakdown or failure without delaying and to prevent further loss or catastrophe or control the situation.
8.	Operate to Failure and Corrective Maintenance	To use unto death deriving maximum possible outcome without (much) expense (in other words, to get maximum out of available one even after the life span is completed).
9.	No Maintenance Technique	To use and throw after completing its specified or intended life period.
10.	Preventive Maintenance	To prevent the occurrence of failures and, hence, save the maintenance work, cost, to improve safety.
11.	Maintenance Prevention	To prevent the maintenance by replacing with new parts when old parts fail.
12.	Productive Maintenance	To improve the productivity by carrying out the maintenance on machine parts by providing alternative without shutting down the machine.
13.	Predictive Maintenance	To predict the occurrence of failures in advance and attempting to forestall or stop the failures or reduce the effect.

Sl. No.	*Maintenance Strategy/Policy*	*Central Theme*
14.	Planned Maintenance	To maintain systematically at regular intervals with a checklist and fixed maintenance durations such as minor/ major overhauls.
15.	Scheduled Maintenance	To maintain machine condition by schedules predetermined with the consent of user department irrespective of the condition of the machine.
16.	Shutdown Maintenance	To maintain the equipment when it is not in use so as not to interrupt the production.
17.	Fixed Time Maintenance	To maintain the machines at fixed time intervals presuming the machines as strictly time dependent.
18.	Design Out Maintenance	To find the root cause of failures and to update the equipment with changing technology and eliminating poor design features and aiming to eradicating the failures by design changes.
19.	Tero-technology or Total Productive Maintenance or Logistics and Physical Assets Management	To maintain in all respects over entire life cycle i.e. 'design to discard' of the equipment by preventing the failures, preserving the condition, prolonging the life, and restoring the condition through out their life for best productivity.

BUSTLE - 3.4

1. *Prepare a comparative master chart (as shown below) of various methods of maintenance including the principle, areas of application, examples, merits/demerits, etc.*

S.No.	*Maintenance Method*	*Principle/Cnetral Theme*	*Suitability/Area of Application*	*Merits*	*Demrits*	*Example*

2. *List out the policies/methods of maintenance those are followed for various machines in the organisation where you are working or the one, which you are familiar with. How are these policies executed in the organisation?*

3. *What are the bottlenecks/roadblocks you noticed in implementing the various methods of maintenance policies in the (above referred) organisation/institution?*

____

____

____

____

Test Your Understanding 3.4

1. Briefly explain any ten kinds of maintenance policies of your choice.
2. Explain on what bases the maintenance policies are planned? Explain any five.
3. What do you understand by planned interval maintenance? How do you plan the intervals for maintenance? Discuss the types of maintenance durations under this policy applied to checking the parameters.
4. Distinguish between scheduled maintenance and shutdown maintenance.
5. Write short notes on
   (a) Fixed Time Maintenance (b) Shutdown Maintenance
   (c) Scheduled Maintenance (d) Planned Maintenance
6. What is no maintenance and maintenance revention? How do they differ?
7. Explain the terms TPM, tero-technology, logistics, and physical assets management.

3.9 FACTORS INFLUENCING THE SELECTION OF MAINTENANCE POLICY/STRATEGY

We are now aware of the following maintenance policies and various methods under them.

- Contractual Maintenance
- Failure-based Maintenance
- Time-based Maintenance
- Condition-based Maintenance
- Reliability-centred Maintenance
- Total Productive Maintenance

Depending upon the requirements, we can blend these policies in suitable proportions. We can even find an appropriate combination of various methods (as discussed earlier) under each of these policies. One policy can be considered for a few machines and another on some other. For example, if contractual maintenance is chosen on computers, failure-based maintenance can be selected for production machinery of the same industry. Similarly, the electrical part can be considered for time-based maintenance while mechanical parts can be monitored by condition-based maintenance for a diesel generator. However, when such policies are laid down, a great exercise is required to be done on every machine to study all the characteristics that influence the decision. The policy, thus, selected for majority of the machines or on long-term plan can be then stated as the maintenance strategy.

The most common factors that influence the decision in laying down a maintenance policy or strategy are listed below.

- Maintenanceobjective
- Cost of machinery and maintenance
- Plant condition
- Age of the machine (both calendar age and running age)
- Quality policy of the company
- Guarantee/warrantyby original equipment manufacturer (OEM)

- Internal resource levels such as tools, spares, material, time, etc.
- Contract resources
- Statutory regulations (licenses from Pollution Board/Government/permits)
- Safety
- Production requirements (both short-term and long-term)
- Mode of operation, method used
- Material and their quality levels
- Cycletime or duration the machine is run uninterrupted
- Machine characteristics and specification
- Operating conditions
- Environmental conditions
- Availability of spares
- Technological competence
- Availability of skilled/trained/suitable manpower

BUSTLE - 3.5

What maintenance policy do you choose for your household equipment like a car or bike, TV/DVD player, refrigerator, a personal computer, etc.? Why?

____

____

____

____

Test Your Understanding 3.5

1. List the factors influencing the maintenance policies.

3.10 MAINTENANCE STRATEGY—FORMULATION

The maintenance strategy is to be formulated to achieve the maintenance objective and directing the resources under the given constraints in the form of availability and requirement. Normally, the maintenance strategy is formulated for long run. However, a due consideration is given to short-term production requirements also. More or less, all the above listed factors have their own influences on formulating the strategy.

The main decision is usually concerned with establishment of the level of preventive maintenance needed to bring the output characters and maintenance resource levels under control. Thus, various possible strategies will be results of combinations in different proportions of preventive and corrective maintenance, ranging from 100% preventive and 0% corrective to 0% preventive and 100% corrective. The problem of identifying and establishing the right maintenance mix to achieve the maintenance objectives poses the following points. These are the bottlenecks for formulating the maintenance strategy, which can be better derived from Figure 3.6.

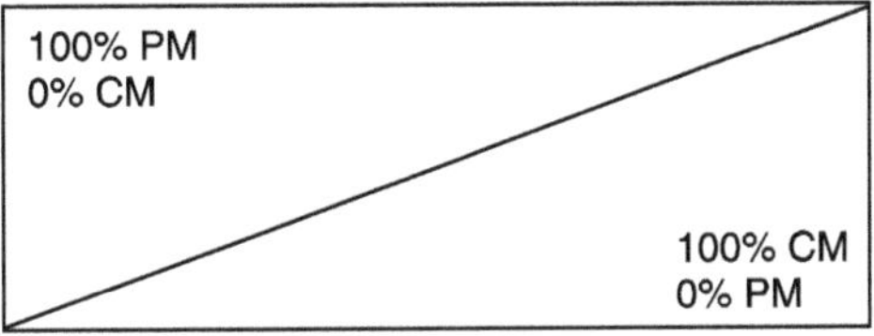

Figure: 3.6 Maintenance mix of PM and CM

3.11 HURDLES IN FORMULATING MAINTENANCE STRATEGY

- The maintenance strategy is a function of relationship between production and maintenance and, hence, in most of the cases it will be subjected to variations.
- The maintenance strategy/policy varies from machine to machine, place to place, and situation to situation. It becomes very difficult at times to decide what policy will be more suitable to a given machine. Often, a common formula does not work out for all the machines simultaneously.
- If preventive maintenance work is neglected for a period of time, it results in increased subsequent corrective maintenance work. Consequently, more pressure is built up on limited maintenance resources to further neglect preventive maintenance. This in turn may lead to 100% corrective maintenance, i.e., operate to failure and corrective maintenance (OFCM) policy.
- When a machine is operated to failure for a period of time, it has to be brought back to an acceptable condition through corrective maintenance before any strategic decision on preventive maintenance is taken or implemented. Otherwise, it can weaken the PM or the PM may become ineffective.
- There is a level of preventive and corrective maintenance that minimises the sum of resource costs and lost output. It is very difficult to locate this point of optimum level otherwise may become over-maintained (unnecessary costs) or under-maintained (lost output).
- The optimum level of preventive maintenance is very sensitive to cost of lost output. Therefore, as far as possible the maintenance work should be carried out during the hours in which the machine is not under operation or not run or under outage.
- The maintenance strategy is to be determined by suitable parameters and policy such as time based, failure based, age based, condition based, etc. A preventive maintenance can also fit in any frame of the parameters. Thus, a suitable design is to be made for the conduct.
- Apart from all these, the factors that are mentioned earlier also have their own impact while selecting and designing the maintenance strategy/policy. A due consideration is to be given to each of these factors while establishing strategy/policy.

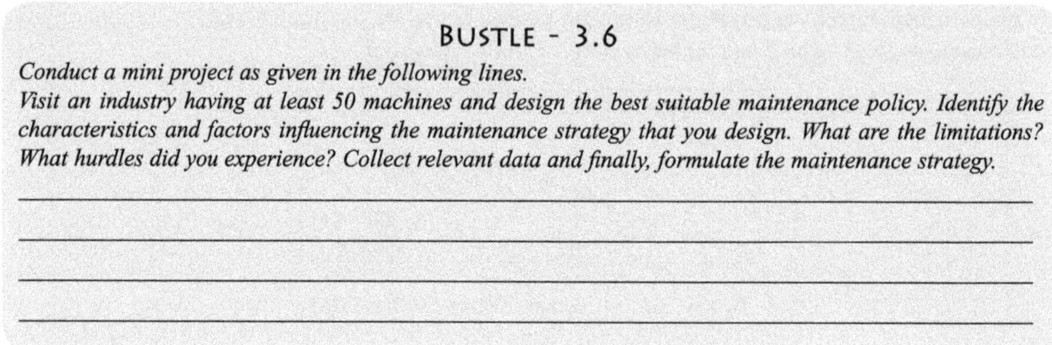

BUSTLE - 3.6

Conduct a mini project as given in the following lines.

Visit an industry having at least 50 machines and design the best suitable maintenance policy. Identify the characteristics and factors influencing the maintenance strategy that you design. What are the limitations? What hurdles did you experience? Collect relevant data and finally, formulate the maintenance strategy.

Test Your Understanding 3.6

1. How do you formulate the maintenance strategy? Explain.
2. What are the bottlenecks you find while formulating the maintenance strategy?

3.12 MAINTENANCE PROCEDURE AND THEIR SELECTION

After setting up an appropriate policy with a suitable maintenance mix, the maintenance operators can be trained according to the norms set. Usually, the procedure followed by plant engineers in almost all the industries irrespective of their policy/strategy is as follows.

Step 1: Adjust

Usually, a machine performs out of its specifications due to continuous running. The machine is to be re-adjusted properly in its settings. In few cases this may be termed as calibration also. With simple adjustments, the machine will function properly. To carry out this operation, the plant engineer or the maintenance operator should have thorough knowledge on the design of the machine as well as its performance characteristics. It will be an added advantage if plant engineer has the history and past data of the machine if the machine is old. This job is entrusted to the production department itself in some industries.

Step 2: Simple Repair

Often the machine can be repaired with a minor repair unless otherwise the machine experiences a fatal accident or subjected to severe wear or misoperation. In any of the latter cases, the fault can be found very easily since it is seen with the naked eye. But in the case of small repairs only, it is very difficult to locate and identify the problem. Once the problem is identified, it will be very simple to repair. And simple may become complex if not identified in time. The simple repair also includes periodic check up or checking of some parameters at regular or irregular intervals based on hour, miles, output of production or some combination of these and making necessary repairs.

Step 3: In Situ Repair/Replacement

A machine may fail due to failure of a component or part, which is often to be replaced immediately by an alternative or a spare one. This is often referred to as *in situ replacement or in situ repair* in plant engineering terminology. In certain circumstances, it is considered with least replaceable assembly (LRA) in which the total assembly of the part is taken out of the machine and replaced with a standby assembly. The removed assembly is taken to the plant engineering department and repaired for next use or standby. This repairing requires sufficient stock of components or spare parts or standby units, which is likely to fail that is statistically predictable.

Step 4: Design Out Maintenance

As indicated by cause investigations after major or recurring failure, a thought may be given towards updating the part or component with the latest technology and replace the item. A suitable planning and follow up is necessary to do this. A regular watch and statistical analysis with the concept of probability and reliability analyses are the helpful tools for design out maintenance.

However, besides the above procedure, a preventive maintenance programme may be scheduled and adhered to. The procedure for scheduling the preventive maintenance is discussed in Chapter 4.

3.13 CHARACTERISTICS OF MAINTENANCE STRATEGY

While establishing the best strategy, the maintenance manager is required to study the following characteristics.

(a) **Failure Characteristics**

- Mean time to fail (MTTF)
- Mean time between failures (MTBF)
- Mean life
- Uncertainty of item life (average, range, standard deviation, Weibull a, b, etc.)
- Detectability of failure (feasibility and possibility of inspection, lead time, etc.)
- Replacement and reparability (MTTR – mean time to repair)

(b) **Cost Characteristics**

- Cost of preventive actions (cost of resources, cost of lost output)
- Consequence of failure (damages, lost output, resource cost, quality, etc.)
- Cost of labour on repair
- Cost of spares
- Indirect costs
- Other costs

(c) **Safety Characteristics**

- Damages to the machine
- Damages to the humans
- Damages to the environment
- Statutory safety regulations

(d) **Operating and Production Characteristics**

- Process
- Product
- Type of production
- Output

- Resources and their properties
- Quality level

Out of the above characteristics, the last one is not considered in certain organisations while in some other organisations (such as service organisations) it may not physically exist. These are discussed in detail in the subsequent chapters.

3.14 THE TOP DOWN–BOTTOM UP APPROACH

A systematic approach is required to establish an efficient and effective maintenance strategy. Further, this system should also consider the following points:

- The operational and structural complexity of the plant.
- The relationship between production and plant engineering and its dynamic nature.
- The relationship between maintenance strategy, maintenance objective, workload, and resource availability.
- Requirements, bottleneck, and influencing factors.

The top down–bottom up approach accommodates all the above points for consideration. This system is briefed out here under.

Step 1: Understand the characteristics of plant operation.

(i) Construct a process flowchart.
(ii) Understand the type of production and production policy.
(iii) Rank the units according to their criticality.
(iv) Identify the maintenance scheduling characteristics.
(v) Identify the prevailing maintenance strategy and maintenance objective.

Step 2: Establish a maintenance plan for each unit.

(i) Analyse the plant units into their 'maintenance causing items'.
(ii) Determine the best maintenance method for each item and divide into 'on-line' and 'off-line' methods.
(iii) Establish an off-line plan for each unit and frequency.
(iv) Identify the corrective maintenance guidelines.

Step 3: Prepare a maintenance schedule for the plant.

(i) Prepare the list of all machines or units.
(ii) Prepare the on-line schedules and frequency.
(iii) Prepare the off-line schedule and frequency with the help of Step 2 (iii).
(iv) Estimate the planned workload for on-line and off-line schedules based on the information in Step 3 (ii) and (iii). Also forecast the expected and unexpected unplanned workload by experience.
(v) Review the schedules once again in accordance with the actual resources available.

Step 4: Establish the initial policy for spare parts and for reconditioning.

(i) Identify the parts that are to be maintained as spares with the help of Step 2 (ii) and Step 3 (i), (ii), and (iii).

(ii) Consider the overall plant demand for spare parts and repairable items to determine the spares inventory policy.

(iii) Set a reconditioning/replacement policy by calculating the replacement periods for each unit.

(iv) Consider all the safety and quality standards to be maintained.

BUSTLE - 3.7

Suggest or conduct a systematic maintenance by top down–bottom up approach in the industry where you are working or the one which you are familiar with. Write a detailed report of the same.

____

____

____

____

Test Your Understanding 3.7

1. Describe the systematic procedure of maintenance.
2. List the characteristics of a maintenance strategy.
3. Explain the systematic maintenance procedure by top down–bottom up approach. What are the considerations it accommodates in it?

3.15 CONCLUSION

The fundamental job of maintenance in most of the above methods seems to be more or less similar. Though the job is similar, the intention and the central theme behind carrying out the job are different. It is difficult to distinguish the methods unless we do not understand the concept of various methods of maintenance by distinct factors. Perhaps, the confusion is due to the reason that one is inter-linked with another. The failures, the intensity of the problem/failure, the intervals of maintenance, the duration of the maintenance, the application area, etc., are some of the factors those differentiate the maintenance philosophies. In this chapter, an effort is made to give a clear picture on different methods on the basis of the above factors. Truly speaking, there is no hard and fast rule for applying these methods in an organisation in the ditto form as mentioned above. These can be changed according to one's own requirements, resources and availability. In fact, no single strategy or method can work in an industry all the times. It is observed that in most industries a suitable mix of two or more of the above methods will be employed.

3.16 RECAP

To perform various functions of PEM, one needs to know the method. Various methods and the philosophies or principles behind them are discussed at length. The unpredictable machine failures are rectified by a team of breakdown maintenance if they are of random in nature. If the failure requires urgent attention it is included in emergency maintenance. If the failures are due to gradual wear and tear or old age failures, this is operated to failure and then corrective maintenance is taken up depending on the cost consideration. The early failures, as they are covered by the guarantee/warrantee, will be maintained by contracts apart from other sub-contracts. In a few cases, use and throw system will work better than rectifying the failure.

Much better policies for prevention of failures by lubricating, overhauling, etc., and reducing the workload of the PEM are the themes of PM. A systematic way by short-range to long-range maintenance is done in planned maintenance. *Plan the work and work the plan*—is the caption that exactly suits to scheduled maintenance in which schedules are prepared and followed. Off-time is best utilised in shutdown maintenance; on-time is used in running maintenance. Using LRA system to facilitate uninterrupted production in productive maintenance and expecting the probability of failure much before and acting accordingly in predictive maintenance are central ideas. MP and DOM are improvement and updating oriented policies while RCM, CBM, TPM, tero-technology, logistics, and PAM are the advanced versions of maintenance models, which take care of the equipment from design to discard.

The details on various maintenance methods, policies, and strategies are clearly discussed in this chapter. The systematic procedures for carrying out maintenance and setting a maintenance strategy are also discussed in this chapter.

3.17 KEY CONCEPTS

1. **Failure-based Maintenance:** An unplanned maintenance strategy carried out by repairing the equipment whenever failures occur.
2. **Time-based Maintenance:** The planned maintenance strategy based on the time intervals of maintenance or the period of operation.
3. **Reliability-centred Maintenance:** The maintenance carried out to enhance the reliability of the equipment.
4. **Condition-based Maintenance:** The maintenance techniques based on the condition and performance of the machine.
5. **Dynamic Predictive Maintenance:** Same as the above. This is another name of condition-based maintenance.
6. **Breakdown Maintenance:** The maintenance by attempting to correct or rectify the failure when occurred on a machine.
7. **Emergency Maintenance:** The immediate maintenance action when failure occurs.
8. **Operate to Failure and Corrective Maintenance:** The technique applied on the machine whose resale value has reached the scrap value. The machine is operated till it fails and the corrective actions are taken if they are not costly.
9. **No Maintenance Technique:** Use and throw system.
10. **Preventive Maintenance:** The actions taken up to prevent occurrence of failures by proper lubrication, oiling, greasing, and overhauling, etc., periodically.
11. **Maintenance Prevention:** Machine updating and reliability enhancement by replacing with new parts when failure occurs on old parts.

12. **Productive Maintenance:** Maintenance by dividing the machine into LRA units to see that machine is not shutdown for want of repairing.
13. **Predictive Maintenance:** Maintenance by expecting (predicting) the probable time of failing a machine and maintaining to prevent failure or reduce the effect.
14. **Planned Maintenance:** A systematic and regular plant maintenance activity at regular intervals with a clear-cut plan apart from the occasional activities like alterations and installations.
15. **Scheduled Maintenance:** Maintenance based on the pre-determined schedules irrespective of the condition of the machine.
16. **Shutdown Maintenance:** The maintenance carried out on a machine only when it is out of service.
17. **Fixed Time Maintenance:** When the failures are clear time dependent, the machine is overhauled or replaced, based on the belief that more failures occur in simple replaceable items and less in complex replaceable items.
18. **Design Out Maintenance:** The activity of maintenance to update or improve the machine performance by finding the root causes at the design stage itself and eliminating these poor design features.
19. **Tero-technology:** The concept of maintenance developed by U.K. to take care of the equipment from design to discard.
20. **Total Productive Maintenance:** A similar concept as that of tero-technology, developed by Japanese to look after the machine from design to discard.
21. **Logistics and Physical Assets Management:** The management of all the physical assets of the plant to prevent the failures, preserve the condition, prolong the life, and restore the condition throughout their life for best productivity. This is also a similar concept as that of tero-technology and TPM.

REVIEW QUESTIONS

1. Define the term 'failure'. Discuss various types of failures based on cost, time, frequency, and intensity (or severity). Give some examples
2. Distinguish between
   (a) Breakdown Maintenance (BDM) and Emergency Maintenance (EM)
   (b) No Maintenance Technique (NMT) and Operate to Failure and Corrective Maintenance (OFCM)
   (c) Productive Maintenance and Predictive Maintenance
   (d) Planned Maintenance and Scheduled Maintenance
3. What are the situations demanding adoption of contractual maintenance? Discuss its advantages and disadvantages.
4. Discuss various types of maintenance strategies that are aiming to prevent the occurrence of failures.
5. Differentiate between the following:
   (a) Planned Maintenance and Maintenance Planning

(b) Scheduled Maintenance and Maintenance Scheduling
(c) Preventive Maintenance and Maintenance Prevention
(d) Design Out Maintenance (DOM) and Maintenance Prevention

6. Explain the concept of
   (a) Breakdown Maintenance (BDM)
   (b) Preventive Maintenance (PM)
   (c) Condition-based Maintenance (CBM)
   (d) Reliability-centred Maintenance (RCM)
7. Give step by step procedure of designing and implementing of preventive maintenance.
8. Describe various maintenance strategies and their central themes.
9. Classify various strategies on the basis of failures, time machine condition, and reliability. Briefly describe each of them.
10. Discuss the systematic procedure of conducting maintenance (or repair).

CHAPTER 4

Maintenance Scheduling

PREAMBLE

This chapter focuses on the importance of scheduling and various methods of scheduling. Various scheduling procedures for breakdown maintenance and preventive maintenance are discussed here. The sequence of the jobs can be determined depending on the type of the organisations and their requirements and availability of resources. The basic procedure for scheduling starts with the sequencing of the jobs and then assigning them to the available manpower and materials. A few numerical problems are also solved on sequencing and assignment of maintenance jobs. The scheduling formats of a product industry are also discussed for preparing maintenance master schedule (MMS).

CHAPTER OUTLINE

4.1 INTRODUCTION

As and when equipment is procured, the plant engineering department has to take up the responsibility to see that the machine is kept in production state and then to upkeep, update and/or maintain it. This activity involves various maintenance functions probably in the sequence of erection, installation, commissioning, preventive maintenance scheduling, corrective maintenance, overhauling, lubricating, etc. Often, the maintenance managers or plant engineers schedule these jobs by their experience and commonsense. But it is not always successful and economic. It is not so easy to perform all these jobs at times so that the production is not affected. The knowledge of scientific methods in scheduling of these jobs will definitely add a feather to the cap of the plant engineer for his effective utilisation of the resources.

4.2 MAINTENANCE SCHEDULING AND ITS IMPORTANCE

Scheduling means assigning time to the planned job. In the context of maintenance work, scheduling usually means to set the expected starting times (and ending time also in many cases) for the maintenance functions such as repair or overhauling work. Note that scheduling is not determining how long a job will take. Suppose you estimated that a repair activity of a milling machine will take eleven hours, this is a measure of the workload. But when you say that the repair work of milling machine will start say, at 11 a.m. on August 11, you have scheduled it. Scheduling also implies deciding ahead of time and not deciding to a repair force on a machine only a few minutes before the job starts.

Maintenance scheduling deals with the complaints (for repair) with reference to machine status, the requirements (for improvement) and the availability constraints against achieving planned production on that machine. When there are several complaints, then scheduling them judiciously according to certain rule such as first in first out (FIFO), the job significance in production chain, etc. becomes necessary. These scheduling aspects are discussed in detail in this chapter.

The sequencing, scheduling, and loading of maintenance jobs often pose a challenge to the plant engineer for the following reasons:

- The maintenance jobs require the pre-requisite knowledge of the technological and managerial aspects to arrange them in the correct order.
- Sometimes the jobs will have the fixed order while at other times they will have flexibility. Hence, the plant engineer should know the characteristics of the jobs.
- The plant engineer should know thoroughly about the probable time of performing each job to find optimal schedule of the jobs so as to minimise the total elapsed time (i.e., the time from the start of the first job to the completion of last job as well as the idle times).
- The plant engineer should have sound knowledge to judge the parameters for fixing the sequence and schedules of the jobs. Sometimes the sequence is to be found for the jobs on machines or men and in some other times it is men or machine on jobs.

The maintenance schedules are prepared in the following stages:

(a) **Sequencing:** All the jobs those are to be done in a certain period (day/week/month, etc.) are written in an order according to predetermined priority rules, i.e., how and when and in which order to do. Analytical methods using Johnson's and Jackson's algorithms and graphical methods such as Gantt chart can be used to solve such problems.

(b) **Assignment:** The jobs are allocated to the crew so as to complete in lowest possible time, i.e., who to do what. Hungarian technique can be used to solve such problems.

(c) **Work Scheduling:** The allocation is assigned with the (starting/ending) time, i.e., when (what time) to do.

Depending on the requirements and needs of the situations, the maintenance scheduling can be of one or two or three stages as stated above viz. sequencing, assignment, and scheduling.

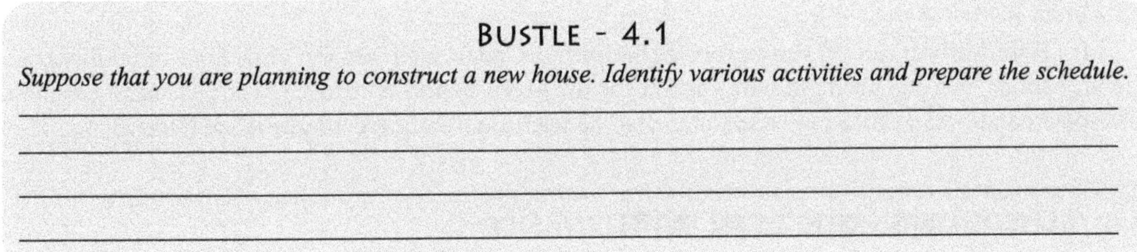

BUSTLE - 4.1

Suppose that you are planning to construct a new house. Identify various activities and prepare the schedule.

Test Your Understanding 4.1

1. Distinguish between scheduled maintenance and maintenance scheduling.
2. Explain the importance of scheduling in Plant Engineering and Management.

4.3 SCHEDULING STRATEGIES IN MAINTENANCE

Maintenance scheduling depends on the maintenance strategy that the top management establishes for its equipment. How to choose i.e., on what basis the equipment is to be taken up for maintenance really matters while scheduling. The frequency and the interval of maintenance are first formulated by some basis as discussed in Chapter 3.

The scheduling strategies generally in practice are given here.

(a) **Time-based Maintenance Scheduling:** Under this strategy, the equipment is taken up for maintenance in cyclic order with a fixed interval, irrespective of the condition, or cost of the equipment. For instance, a machine is scheduled on 10th day of every month say, it will be taken up on 10th of the month irrespective of its condition. Usually the routine checkups will be scheduled using this basis.

Example: In a bus depot, every bus will be taken up for water servicing (washing) on daily or weekly basis.

(b) **Cost-based Maintenance Scheduling:** Under this strategy, all the machines are divided into three groups viz. high cost machines, medium cost machines, and low cost machines. The high cost machines are maintained utmost care; condition monitoring techniques are employed on them to determine the frequency of the maintenance while preventive maintenance techniques are used for medium cost machines. The low cost machines are not much cared and even operate failure and corrective maintenance techniques can be employed.

(c) **Failure-based Maintenance Scheduling:** This strategy uses the failure data. The failure data can be used in several ways such as failure frequency, failure cost, failure duration (MTBF or MTTF), etc. Once the criterion is fixed, the maintenance frequency is determined and scheduled accordingly.

(d) **Condition-based Maintenance Scheduling:** The condition of the machine is the criterion for deciding the maintenance interval. Usually this strategy is considered for costly equipment, hazardous areas and high failure costs. The trend monitoring and condition checking methods are used to fix the interval.

(e) **Reliability-based Maintenance Scheduling:** Under such strategy, machine reliability, i.e., the probability that the machine will not fail for the given period is focused. Various statistical tools and some condition monitoring techniques may be useful in fixing the maintenance interval.

4.4 SCHEDULING POLICIES IN MAINTENANCE

If the maintenance crew is idle because of no breakdowns, it is an indication that the plant is in good condition with full availability and good efficiency. But this is a rare event. However, many a times the crew may be found idle though the equipment is under breakdown or undergoing maintenance. Such situation arise because of non-availability of spares, absence of required technology, right tools not being available, improper planning and scheduling, etc. In such situations, maintenance manager will have to answer the following questions.

- What work is to be allocated to the maintenance crew in slack periods, such as non-availability of spares, tools or technology, etc.?
- What are the forward planning requirements when heavy jobs are to be performed?
- What should be the action plan when demand exceeds the capacity?
- What should be the strategy in the case of the emergency?
- What should be the policy decision on overtime or under time?
- Should the work be allocated to an individual or group?
- Is the management of spare parts going on smoothly?

The above forces have their own impacts and exertion of high influence on the scheduling of maintenance jobs. To answer the above questions satisfactorily, it is essential to scan various scheduling policies in practice. A few scheduling policies more suitable to the maintenance department are discussed here, which are in practice in many of the industries in India.

4.4.1 Relaxed Scheduling Policy

Under this policy, the bottom line maintenance personnel like supervisors or foremen are given free hand in scheduling the maintenance interval or frequency. The foremen can choose the equipment for maintenance with their knowledge and experience and awareness on the machine failure behaviour. The maintenance scheduling responsibility is shouldered completely by the foreman.

Generally, this is a very short-term scheduling policy. In many small companies, most of the scheduling jobs are left to foremen, owing to their practical experience. They are likely to do their scheduling on their own backs or old envelops. Some of them hardly have a fixed schedule as far ahead as one week and the 'real schedule' can be found only in the scheduler's head.

This type of scheduling amounts to no more than responding to a foreman's hurry-up call "½ inch screws are out of stock or not available in the stores" or 'simply waiting for complaints to arrive and then searching or investigating the stock on hand'.

Merits:

- Scheduling is simple and easy.
- This policy puts no burden on top management.
- Effective planning is possible since the scheduler is more practical than theoretical.
- Since the shop floor men are involved, they know the day-to-day problems and, hence, they can plan efficiently.

Demerits:

- Scheduling has no scientific basis.
- The personal prejudices and opinions may affect the scheduled jobs.
- If the foreman is not available due to any reason, work gets disturbed or stopped.
- Spare part management is difficult because forward planning is missing.
- Behavioural problems affect the schedules.

4.4.2 "No-work" Pay Guarantee Policy

This policy is related to the labour contracts. Today's labour contracts usually provide that men must be told ahead of time if there is work shortage. If there is no work for men who report in the morning, they can be sent back home but they have to be paid for four hours in lieu of notice. Work plan well ahead for the individuals must be ensured. Therefore, before scheduling it is required to estimate the work content exactly and thence the required manpower. Wrong estimation leads not only to losses but also to legal complications.

Merits:

- This policy ensures better forward planning.
- Workers are informed well ahead of either work or their minimum pay.
- Legally valid agreement, so good relation exists between the employee and employer.
- The scheduler or employer will carefully plan the work ahead, since he is worried of penalty for no work otherwise.

Demerits:

- Reputation is lost, if 'no-work' happens several times.
- When workers come to work and if they are not assigned work for the day, they will lose interest in working.
- Workers when sent for no-work, and paid for four hours of that day, they may not get any work for another four hours. Good workmen, therefore, will not show interest to work under such scheduling practices because of uncertainty of income inflow.

4.4.3 Overtime Work or Compensatory Off Policy

If no-work pay guarantee is one side of the coin; overtime is the other side that poses problem to the maintenance manager. Men must be warned or informed about overtime in advance or they will be free to refuse to stay. So, if overtime is anticipated the maintenance manager or the scheduler has to inform the foreman so that he can tell his men ahead in time. In maintenance department, such situations are common that demand overtime due to unexpected breakdowns and unsolved puzzles while repairing. The maintenance manager should be very cautious while adopting this overtime policy, because there is every chance of regular work getting slowed down intentionally by workmen with a view to earning double by working overtime. Industries have realised this problem and thence have implemented a new policy called compensatory off in lieu of overtime. In this policy, the extra time that workmen worked after his regular duty is credited as a leave which can be utilised on any other day. However, this encourages absenteeism and creates a problem if this absence occurs on a day when the resource is really needed. Therefore, the maintenance manager is advised to set the labour contracts in such a way that these overtime and undertime problems will not disturb the regular work.

Merits:

- Workers can get job satisfaction.
- Workers are motivated to work more.
- Availability and maintainability of the machines can be improved.
- A job (particularly repair job) if kept pending or transferred to another person (or group of persons), it gets delayed because the new person needs time to understand the problem. Sometimes it may not be solved also. Therefore, it is better to allow the same person who started to continue and complete the work.

Demerits:

- The workmen may take a chance to slow down the job intentionally so that they can get a compensatory off if done in extra time.
- It encourages absenteeism and may affect the normal day work as the workmen may take leave when it is really needed.
- Sometimes a little extra time over the normal time may be sufficient to complete the job, but the worker may demand full day compensation in lieu of this.
- It creates conflicts among the workmen because a particular workman may get more compensatory offs while other may get less due to the nature of repair/maintenance work carried on the machines depending on the age of the machine, nature of the failure, criticality of the failure, etc.
- It may induce prejudices in the scheduler or even if the scheduler plans work efficiently for workmen in the interest of giving the best to the organisation, it may give scope to others to say that the scheduler is partial. It affects the important principle of management, i.e., principle of equity.

4.4.4 Short Work Schedules

What would you do, when you don't have sufficient work to keep everyone busy for a full day? This poses certain problem on decision making about what best can be done in such situations. Do you cut work hours for all? Or lay off some men and keep others on regular hours? Or do you pay all equally, even if they don't work, as they don't have work to do? Of course, the answers may be written into the labour contracts but it is quite a brain teasing and difficult problem too to schedule the jobs in line with the decisions of top management.

Usually, in job work, it is customary to let differing workloads work themselves out. But in maintenance jobs it is not so easy to take such decision. If there is enough work which can be finished in five or six hours a day, everybody will opt for such work only and nobody will choose the work that is difficult or takes longer duration. This inequality and unevenness of jobs, but inevitable in maintenance, throws the foremen in hell of the human relation problems in his day-to-day work.

Therefore, in scheduling such jobs a lot of attention is to be paid towards the load for the week as well as for the day. If there is likely to have 32 hours of work for a week, it is probably better to work for five short days (five or six hours) instead of four full days to keep some flexibility for emergency jobs. Sometimes, it is necessary to pay attention to the need of full day working crews. Perhaps, this is commonly observed in preventive maintenance scheduling and is less common in breakdown maintenance.

Merits:

- This policy helps develop good human relations.
- It creates flexibility in work schedules.
- Work will be done at a faster pace.
- The policy motivates workmen to complete the job early. Thus, the jobs get finished faster.
- It provides scope for job rotation and hence job satisfaction and motivation to the workmen.

Demerits:

- Some workers may do hotch-potch work to complete the work early.
- Workmen once accustomed to more flexibility cannot easily return to their normal way when needed. This induces additional human relation problems.
- The adoption of this policy creates conflicts between departments, particularly, between production and maintenance departments, due to difference in their timings of work.
- Urgent work gets affected.

4.4.5 Period-End Pile Ups

Most commonly, the maintenance schedules list the machines to be overhauled monthly or fortnightly or weekly except in breakdown maintenance where the planning will be on daily basis. If the required spare parts are not available in time, these will be piled up at the end of the period, and cannot be completed in time. Even it may occur with the breakdown maintenance also. This may affect the availability of the machines adversely. The production manager will obviously put the blame on the maintenance manager and the maintenance manager will find a way to escape by putting the blame on materials manager or planning manager for want of spares. But what is the outcome of such boomerangs? Sometimes this pile up may result out of the non-availability in the local market or due to poor planning of maintenance manager. Apart from these technical and managerial problems, the pile up may result from PM schedules, laidback attitude in the beginning of the period due to less pressure and rushing up at the end of the period under pressure to meet the target date. Perhaps, this brings down the quality of the work and affects the reliability of the maintenance.

What should be done to avoid period-end pile ups? The simplest thing is to schedule for shorter periods. A periodic review and regular follow up may also help in reducing the pile up. The skill, dedication, team spirit, coordination, and motivation are other significant factors that can control the pile up.

Merits:

- It necessitates the periodic review to control the period-end pile up.
- Workmen will have full work and 'no-work' days can be avoided.
- It builds up team spirit and coordination while controlling the pile up.

- As there is less idle time, conflicts will come down and behavioural problems can be avoided.

Demerits:

- Pressure on workmen develops certain human relation problems.
- Leisure attitude in the beginning of the period creates pressure to complete on the target date during the end of the period. This may pile up the work or bring down the quality.
- The pressure and rush up work will affect the reliability, availability, and maintainability of the machines adversely.
- Pressure on workmen creates an unsafe work environment and even unsafe to the health of individuals also.

BUSTLE - 4.2

Conduct a mini project under the following guidelines.
Visit (a) process industry and (b) product industry having plant engineering department and observe the policies of scheduling followed by them. Compare them. Point out some merits and demerits in implementing these policies with some practical examples.

____

____

____

____

Test Your Understanding 4.2

1. What is relaxed scheduling policy? What are its merits and demerits?
2. What do you understand by "no-work pay guarantee policy"? Is it suggested for small industries or large industries? Discuss.
3. Write a brief note on short work schedules.
4. Explicate overtime work or compensatory off policy with reference to the maintenance departmental jobs.
5. List the merits and demerits of period-end pile up scheduling policy.

4.5 SCHEDULING MAINTENANCE JOBS BY SEQUENCING TECHNIQUES

It will be most convenient to study the sequencing of jobs of maintenance in the following models.

- n jobs × 1 machine (or man or department)
- n jobs × 2 machines
- n jobs × 3 machines
- n jobs × m machines
- 2jobs × n machines

These are explained as follows.

4.5.1 Scheduling n Jobs × 1 Machine

Suppose there are five production departments from which five different requisitions have come to plant engineer for repairing of their machines. It is now the turn of the engineer to decide the order of these jobs and allocate them to his men. If there is only one operator to perform these five jobs, it becomes further critical and crucial. He can adopt one of the following methods in such situations.

Criteria for *n* jobs × 1 machine sequencing

(a) **First In First Out (FIFO) or First Come First Serve:** When the jobs do not require any preferential treatment this rule is considered. Repair complaints at telecom department, electricity departments usually follow such principle.

(b) **Last In First Out (LIFO):** When a machine is being opened for repair or overhaul, the parts are put back in LIFO system. However, pipeline laying in water works, electrical wiring maintenance system will follow this model.

(c) **Earliest Due Date (EDD):** Most of the times, the production departments ask for the probable completion time of the repair job and based on the commitment of the plant engineer they plan their production. To facilitate this, the plant engineer has to arrange all the jobs according to their due dates of delivery/handing over.

(d) **Shortest Process Time (SPT):** This is another policy adopted for plant engineering jobs with the concept that the jobs that take less time to repair can be handed over and, thus, facilitate maximum availability to the machinery.

(e) **Longest Process Time (LPT):** In contrast to the above, the job that takes longer time will be taken up first and the jobs which take smaller time will be taken up last.

(f) **Pre-emptive Priority Rule:** A job that is very urgent will be taken up on priority basis by attending immediately and stopping all other jobs. Under this rule, the highest priority job is allowed to enter into the service immediately even if another job with lower priority is already in service.

(g) **Non-pre-emptive Priority Rule:** In this case, the highest priority goes ahead in sequence but service is started immediately after completion of the current service.

(h) **Priority Service:** Priority is given to certain jobs by virtue of their urgency or production loss due to their shutdown losses, etc.

(i) **Service In Random Order (SIRO):** Under this rule, jobs are selected for service at random irrespective of their arrival, urgency, due date, etc.

The first five criteria are static in nature and the rest are dynamic. A numerical example is given below to understand the above concepts.

ILLUSTRATION 4.1:

In a repair shop, there are five orders P, Q, R, S, T to repair and the due dates are as follows.

Jobs	*Repair time (Days)*	*Due date (Days from now)*
P	10	18
Q	8	22
R	5	25
S	11	15
T	6	30

Prepare a sequence of repairing the above jobs under the following rules:

(a) SPT (b) FCFS (c) LPT (d) EDD (e) treating job Q is to be done on top priority and others in LIFO.

From the above set of jobs, calculate total (1) completion time, (2) average-completion time, (3) average number of jobs in the system and (4) average-job lateness.

Solution.

(a) **By SPT Rule**: As per this rule, the job with the shortest time is scheduled first, immediately followed by the next lowest repair time and so on. In each calculation we compute the flow time for the jobs, i.e., the amount of time the job spends in the system to get completed. Flow time for a job is the addition of waiting time and the repair time for that job. Job lateness is the positive difference, if any, between its flow time and its due date.

Job sequence	*Repair time (Days)*	*Flow time (Days)*	*Due time (Days from now)*	*Job lateness (Days late)*
R	5	5	25	0
T	6	11	30	0
Q	8	19	22	0
P	10	29	18	11
S	11	40	15	25
Total	40		104	36

(i) Total completion time = 40 days.

(ii) Average completion time = Total flow times/number of jobs = 104/5 = 20.8. Days

(iii) Average number of jobs in the system = Total flow times/total repair time = 104/40 = 2.6 jobs

This result can be calculated in another way as follows:

Repair time	*Jobs in sequence*
First 5 days	5
Next 6 days	4
Next 8 days	3
Next 10 days	2
Next 11 days	1
Total repair time = 40	

Average number of jobs in the system = [5(5) + 4(6) + 3(8) + 2(10) + 1(11)]/40

= 104/40 = 2.6 jobs.

(iv) Average job lateness = Total of job lateness/number of jobs = 36/5 = 7.2 days.

(b) **By FCFS Rule:**

Job sequencing	*Repair time*	*Flow time*	*Due time*	*Job lateness*
P	10	10	18	0
Q	8	18	22	0
R	5	23	25	0
S	11	34	15	19
T	6	40	30	10
TOTAL	40	105		29

(c) **By LPT Rule:**

Sequence	*Repair time*	*Flow time*	*Due time*	*Job lateness*
S	11	11	15	0
P	10	21	18	3
Q	8	29	22	7
T	6	35	25	10
R	5	40	30	10
Total Time	40	136		30

(d) **By EDD Rule:**

Sequence	*Repair time*	*Flow time*	*Due time*	*Job lateness*
S	11	11	15	0
P	10	21	18	3
Q	8	29	22	7
R	5	34	25	9
T	6	40	30	10
Total Time	40	135		29

(e) **By Non-pre-emptive and LIFO:**

Sequence	*Repair time*	*Flow time*	*Due time*	*Job lateness*
Q	8	8	22	0
T	6	14	30	0
S	11	25	15	10
R	5	30	25	5
P	10	40	18	22
Total Time	40	117		37

Summary

Priority rule	*Total completion time*	*Agerage completion time*	*Average No. of jobs*	*Average job lateness*
SPT	40	20.8	2.6	7.2
FCFS	40	21	2.625	5.8
LPT	40	27.2	3.4	6
EDD	40	27	3.375	5.8
NPE-LIFO	40	23.4	2.925	7.4

Generalised comparison statement cannot be made on the basis of only one example. However, it appears that SPT is efficient in average job completion time and in reducing the average number of jobs waiting, however, it increases the average job lateness as some long process time jobs may be pushed behind in queue. FCFS rule seems to function fairly well on all counts.

BUSTLE - 4.3

List the jobs those you have to do today in your institute/office. Write the probable time required in completing these jobs and sequencing them according to the various rules discussed above.

Test Your Understanding 4.3

1. Based on the condition of five milling machines the estimated time of maintenance (maintainability), their reported time in maintenance logbook as on 31 December 2009 and the due time as per maintenance master schedules (MMS) are given below. Prepare the correct sequence by examining the following rule (a) SPT (b) reported time (c) LPT, and (d) EDT with reference to (i) total completion time (ii) average completion time (iii) average number of machines in the system, and (iv) average delay in repair.

Machine	*Reported Time from log book on 31.12.09*	*Estimated Maintenance Time (in Hours)*	*Due Time (Hours from now as per MMS)*
Milling-1	9.00 am	19	55
Milling-2	9.35 am	17	56
Milling-3	10.00 am	15	35
Milling-4	10.15 am	21	65
Milling-5	10.45 am	16	50

2. Chakki Car Care Centre, the authorized service centre of Benz Cars, takes up the car servicing based on the number of cars arrived till the morning of every Monday and promises the delivery accordingly. The estimated repair time, the arrival and the promised delivery dates are given in the following table.

Car Number	*Date Arrived*	*Estimated Servicing Time*	*Promised Delivery Date*
AP10K-4848	10-06-09, 18 hrs.	5 days	04-07-09, 18 hrs.
SN20J-1995	14-06-09, 17 hrs.	6 days	23-07-09, 17 hrs.
DL11M-1965	13-06-09, 14 hrs.	11 days	28-06-09, 14 hrs.
KS29D-1201	09-06-09, 13 hrs.	9 days	25-06-09, 13 hrs.
HP20M-2000	10-06-09, 12 hrs.	7 days	30-06-09, 12 hrs.

What will be the better priority rule if Benz Company insists the service centre on minimum delay to the customer? Compare the results in (a) SPT (b) FCFS (c) LPT (d) EDD. Also compare the total completion time, average completion time, average number of cars in the system and average delay in service.

HINTS

1. SPT and EDT are better policies as shown in the following comparison table.

Criterion Rule	*Sequence*	*Total Completion Time*	*Average Completion Time*	*Average Number of Machines in the System*	*Average Delay in Repair*
SPT	3-5-2-1-4	88	49.8	2.83	7
RT	1-2-3-4-5	88	53.2	3.02	12.2
LPT	4-1-2-5-3	88	58.8	3.34	9.4
EDT	3-5-1-2-4	88	2.85	2.85	6.8

2. FCFS will be better option.

	Priority Rule Total Completion Time	*Average Completion Time*	*Average Number of Cars*	*Average Delay*
SPT	38	19.8	2.61	6.8
FCFS	38	23.2	3.05	3.4
LPT	38	25.8	3.39	4.8
EDD	38	25.6	3.37	3.6

4.5.2 Scheduling n Jobs × 2 Machines Using Johnson's Rule

S. M. Johnson gave the following algorithm for finding sequence of a situation where there is a group of 'n' jobs to be processed through two successive work centres. The rule ensures minimum completion time for the group of 'n' jobs by minimising the total idle times of the work centres.

Assumptions: The following assumptions are made to use Johnson' sequencing algorithm.

(a) **No Passing Rile:** This is the most important and fundamental to apply 2 × n Johnsons's algorithm. It means that "*any job cannot be loaded on the second machine unless it is processed by first machine*".

(b) **The Process Time is Estimated or Known:** The process time for each job is known or estimated in the beginning.

(c) **Negligible or No Tool/Job Seting Time:** The tool or job setting time or job transfer time from one machine to another is assumed to be negligible or the given process inclusive of these.

(d) **Equal Efficiency:** All the machines work with equal efficiency without breakdown. Further, all the workmen on these are also expected to work with same efficiency.

(e) **Job cannot be Unloaded until it is Completed:** A job once loaded on a machine should not be removed unless the operation on the machine is completed.

(f) **All Jobs at the Begining:** All the jobs that are to be sequenced should be available at the beginning of the day or a given period. No job is inserted later after the sequence is determined.

The rule (sequencing algorithm) is,

1. Identify the job with lowest processing time among all the jobs ($2n$) on both the machines.
2. If this shortest processing time belongs to the first machine (or work centre), then this job is placed first in sequence, if the shortest processing time appears under the second work centre, this job is put last in the sequence.
3. Having placed the job, now delete it from any further consideration.
4. For the remaining jobs, repeat step 2, until all the jobs are placed in sequence.

Following example illustrates Johnson's Rule.

ILLUSTRATION 4.2:

Sai Automobile has six cars for repair. The repair consists of two step procedure viz. dent removing and painting. The time estimates are as follows:

Car Number	1	2	3	4	5	6
Time Estimate (dent removing)	16	10	11	13	8	18
Time Estimate (painting)	15	9	15	11	12	14

Find the correct sequence of operations and prepare Gantt chart and schedule chart so as to minimise the idle time.

Solution.

Sequencing:

Step 1: The most minimum time among all the 12 jobs (6 denting jobs and 6 painting jobs), is taken by car number 5 (say C_5) for denting, i.e., 8 hours.

Step 2: As the least time 8 for C_5 falls under the first operation, this is to be sequenced at the extreme left.

C_5					

Now, this C_5 is deleted for further sequencing.

Step 3: Step 2 is to be repeated till the sequencing is completed.

As the next shortest processing time, i.e., 9 for C_2 belongs to the second work centre (i.e., painting) this is to be sequenced to the extreme right.

C_5					C_2

The next minimum is 10 for again C_2, which is already sequenced and deleted from further consideration. Then next minimum is 11, which appears in C_3 for work centre 1 as well as in C_4 for work centre 2. Therefore, C_3 goes to left and C_4 goes to right.

C_5	C_3			C_4	C_2

Similarly, the next shortest processing times are 12 for C_5 for work centre 2 and 13 for C_4 for work centre 1, but these are already allocated (deleted for the consideration). Thus, C_6 occupies the next position towards the right side of the left out positions as this has the next least process time, i.e., 14 hours at work centre 2.

C_5	C_3		C_6	C_4	C_2

Obviously, C_1 comes in the left out position. Thus, the sequence is…

C_5	C_3	C_1	C_6	C_4	C_2

Remarks:

- If same minimum processing time is assigned for two jobs, one at work centre 1 and the other at work centre 2, then the one with the shortest time at work centre 1 is kept at the most left and the other at the most right.
- If there is a tie in the shortest processing time for any two jobs in the first work centre, then, the sequence is to be determined by checking their minimum time at the second work centre. The job having minimum time is to be allocated to most right and the other is kept just before it.
- If the tie is for two jobs at the second work centre, check their minimum time at the first work centre and the job having minimum time is to be given the left position and the other just after it.
- If there is a tie in processing times of two jobs with both the machines and work centers then the sequencing is done by selecting arbitrarily. In this case, the problem will lead to multiple optimal solutions (more than one optimal solution).

Schedule Chart:

Car No.	*First Work Centre (Dent Removing)*			*Second Work Centre (Painting)*			*Idle time*	*Remarks*
	Job time	*Time in*	*Time out*	*Job time*	*Time in*	*Time out*		
C_5	8	0	8	12	8	20	8*	Painter can start at 8th hr only.
C_3	11	8	19	15	20	35	0	C_3 waits for 1 hr.

C_1	16	19	35	15	35	50	0	No waiting for job and painter.
C_6	18	35	53	14	53	67	3	Painter waits for 3 hrs.
C_4	13	53	66	11	67	78	0	C_4 waits for 1 hr.
C_2	10	66	76	9	78	87	0	C_2 waits for 2 hrs.
Total elapsed time			76			87	Total idle time = 11	Painter may be asked to come 8 hrs late to the duty.

*Some treat this as slack time but not the idle time, as painter may be asked to come late to the duty 8 hours hence to the dent remover. If this duration is not considered as the idle time or waiting time for painter, the idle time is only 3 hours.

Thus, the delivery schedule may be given as C_1 on 50th day, C_2 on 87th day, C_3 on 35th day, C_4 on 78th day, C_5 on 20th day and C_6 on 67th day so as to minimise the idle times for the machines or the repair men.

TEST YOUR UNDERSTANDING 4.4

1. There are 7 jobs each of which has to through m/c m1 and m2 in order M1, M2 processing times (in hrs) are given below. Determine the sequence that will minimise the elapsed time and prepare the time schedule chart and Gantt chart.

Job	1	2	3	4	5	6	7
M1	3	12	15	6	10	11	9
M2	8	10	10	6	12	1	3

2. In a pipe fabrication shop, the time required for cutting and bending operation for 6 different sizes of pipes are given. Determine the order in which those pipes should be processed in order to minimise the total time for processing all the pipes.

Pie	1	2	3	4	5	6
Cutting Time	3	12	5	2	9	11
Welding Time	8	10	9	6	3	1

3. Find the sequence that minimises the total elapsed time (in hours) required to complete on the following two machines.

Task	A	B	C	D	E	F	G
Machine I	2	5	4	9	6	2	7
Machine II	6	8	7	4	3	9	3

4. Find the sequence that minimises the total elapsed time (in days) required to repair and test the following seven jobs.

Activity	1	2	3	4	5	6	7
Rapairing time in days	3	5	9	9	6	3	3
Testing time in days	4	5	1	6	3	3	4

HINTS

1. Optimal sequence is 1-4-5-3-2-7-6

Job	*M1*		*M2*	
	time in	*time out*	*time in*	*time out*
1	0	3	3	11
4	3	9	11	17
5	9	19	19	31
3	19	34	34	44
2	34	46	46	56
7	46	55	55	59
6	55	66	66	67

Total time elapsed time = 67 hrs.
Idle time on M1 = 1 hr.
Idle time on M2 = 17 hrs.

2. Optimal sequence is 4-1-3-2-5-6

Job	*M1*		*M2*	
	time in	*time out*	*time in*	*time out*
4	0	2	2	8
1	2	5	8	16
3	5	10	16	25
2	10	22	25	35
5	22	31	35	38
6	31	42	42	43

Total elapsed time = 43 hrs.
Idle time on M1 = 1 hr.
Idle time on M2 = 6 hrs.

3. Optimal sequence is FACDEG
4. Optimal sequences are
   3-1-7-6-2-4-5 or
   3-1-7-2-4-6-5 or
   3-7-1-6-2-4-5 or
   3-7-1-2-4-6-5

4.5.3 Scheduling n Jobs × 3 Machines Using Jackson's Rule

For sequencing and scheduling n jobs on three machines, Jackson proposed a simple rule by which the $n \times 3$ is converted to $n \times 2$ model for preparing the sequence. And then schedule may be prepared as usually for the three machines individually either by time chart or Gantt chart.

To convert the $n \times 3$ to $n \times 2$, the given problem has to satisfy one or both of the following conditions:

Condition 1: The minimum process time among all the jobs to be performed on the first machine is greater than or equal to the maximum process time among the jobs to be performed on the second machine.

Condition 2: The minimum process time among all the jobs to be performed on the third machine is greater than or equal to the maximum process time among the jobs to be performed on the second machine.

In other words, the maximum of second machine is greater than or equal to the minimum of the first and/or third machine timings.

If one or both of the above conditions is/are satisfied, the 3 × n sequencing problem is converted to 2 × n sequencing problem by adding the job timings of the first machine with the corresponding job timings of the second machine to make it as the first work centre and similarly adding job timings of second machine with their respective job timings of third machine to make it as the second work centre. Now, the sequencing is made for the two work centres with these new timings according to Johnson's rule.

After finding the sequence, the Gantt chart and time chart calculations are done by considering three machines with the timings given in the original problem.

A numerical example is shown here:

ILLUSTRATION 4.3:

A company divides its maintenance crew into three teams for its preventive maintenance of heavy vehicles. The first team looks after the replacement of worn out parts, the second oiling and resetting, and the third checking and tests running. The estimated time for maintenance of each of these vehicles is given in hours in the following table and passing on is not allowed. Find the sequence and schedule them so as to minimise the total elapsed time and idle time.

Team/Vehicle No.	1	2	3	4	5	6	7
Replacement team	3	8	7	4	9	8	7
Resetting team	4	3	2	5	1	4	3
Inspection team	6	7	5	11	5	6	12

Solution.

Seven vehicles are given to be processed through three repair teams, say, R_1, R_2, R_3, in the order R_1, R_2, R_3. Now, the minimum time in R_1 series is 3, and that in R_3 is 5 while the maximum in R_2 is 5. As minimum $R_3 \geq$ maximum R2, Condition 2 is satisfied, Hence, the problem can be converted to that of 7 jobs × 2 work centres.

Thus, WC1 = $R_1 + R_2$ and WC2 = $R_2 + R_3$

The problem is rewritten as

WC\V.No.	1	2	3	4	5	6	7
WC1	3 + 4 = 7	8 + 3 = 11	7 + 2 = 9	4 + 5 = 9	9 + 1 = 10	8 + 4 = 12	7 + 3 = 10
WC2	4 + 6 = 10	3 + 7 = 10	2 + 5 = 7	5 + 11 = 16	1 + 5 = 6	4 + 6 = 10	3 + 12 = 15

Using Johnson's optimal sequence algorithm, the correct sequence is

V_1	V_4	V_7	V_2	V_6	V_3	V_5

For calculating the total time elapsed and the idle times, we draw the time chart.

Team Vehicle No.	*Replacement Team Timings*			*Resetting Team Timings*			*Inspection Team Timings*			*Idle Time/Waiting Time*		
	In	*Repair*	*Out*	*In*	*Repair*	*Out*	*In*	*Repair*	*Out*	R_1	R_2	R_3
V_1	0	3	3	3	4	7	7	6	13	-	3	7
V_4	3	4	7	7	5	12	13	11	24	-	-	-
V_7	7	7	14	14	3	17	24	12	36	-	2	-
V_2	14	8	22	22	3	25	36	7	43	-	5	-
V_6	22	8	30	30	4	34	43	6	49	-	5	-
V_3	30	7	37	37	2	39	49	5	54	-	3	-
V_5	37	9	46	46	1	47	54	5	59	13	7 + 12	-
TOTAL		46	46		22	47		52	59	13 + 37 + 7 = 57		

The table indicates that the total elapsed time is 59 hours. Idle time of replacement team is 13 hours at the end of project. Resetting team gets 37 hours of idle time of which 3 hours at the starting and 12 hours at the ending and 22 hours in middle of the project. The inspection team will have to wait 7 hours at the beginning of the project.

4.5.4 Scheduling n Jobs × m Machines Using Jackson's Rule

Sequencing the jobs may not be always in the form of n × 2 or n × 3. In fact in real practice, we often come across with the n jobs to sequence on m machines or men or work centres. However, it is not very difficult to sequence in such case. The method we apply here is again based on Jackson's conditions as explained below.

Let there be n jobs, of which each is to be processed through m work centres/machines.

Say, $W_1, W_2, W_3, \ldots W_m$.

Now, the iterative procedure is as follows:

Step 1: Find minimum process time of W_1 and W_m series, and maximum process time of all the middle series.

Step 2: Check whether

Min. of W_1 series $\geq$ Max. of middle series.

Min. of W_m series $\geq$ Max. of middle series.

Step 3: If the inequations of step 2 are not satisfied, the method fails. Otherwise, move to the next step.

Step 4: Convert m work centre problem to two-work centre problem by two fictitious machines say A and B.

A = The algebraic sum of process time of W_1 and all corresponding timings of the mediocre series.

B = The algebraic sum of process time of W_m and all corresponding timings of the mediocre series.

(However, in general practice we can find the sequence by taking the timings of W_1 and W_m if the sums of the mediocre series are fixed positive constants.)

ILLUSTRATION 4.4:

Five machine parts are to be reassembled after yearly maintenance in the order P, Q, R, S, and T on four machines A, B, C and D. Find the optimal sequence for the machines when passing is not allowed of which the repair time (in hours) is given below. Also find the total elapsed time, and individual idle times for the repairmen on each machine and machine part. Consider that the machine will be handed over to the production department immediately after reassembling all the parts.

Machine/Part	*P*	*Q*	*R*	*S*	*T*
A	7	6	2	3	9
B	6	6	4	5	10
C	5	4	5	6	8
D	8	3	3	2	6

Solution.

Checking sequencing algorithm rules:

Minimum repair time of P series [Min. $(t_{P,j})$] = $(t_{P,C})$ = 5

Minimum repair time of T series [Min.$(t_{T,j})$] = $(t_{T,D})$ = 6

And Maximum repair time of Q, R, S series [Max($t_{Q,j}$ $t_{R,j}$ $t_{S,j}$)] = { 6, 5, 6} respectively.

Since the condition Min.$(t_{T,j})$ >= Max($t_{Q,j}$ $t_{R,j}$ $t_{S,j}$) is satisfied, the given problem can be converted into a four jobs and two machines (say X and Y) problem as

$$t_{X,j} = (t_{P,j} + t_{Q,j} + t_{R,j} + t_{S,j})$$
$$t_{Y,j} = (t_{Q,j} + t_{R,j} + t_{S,j} + t_{T,j})$$

Machine/Job	*A*	*B*	*C*	*D*
Machine X	17	21	20	16
Machine Y	19	25	23	14

Now, using the optimal sequence algorithm, the following optimal sequence can be obtained.

A	B	C	D

The total elapsed time corresponding to the above sequence can be computed as shown in the following table, using the individual processing times as given in the original problem:

Machine/Part	*P*	*Q*	*R*	*S*	*T*
A	0-7	7-12	12-14	14-17	17-26
B	7-12	12-16	16-21	21-27	27-35
C	12-18	18-24	24-28	28-33	35-45
D	18-26	26-29	29-32	33-35	45-51

From the above table, it can be derived that the minimum total elapsed time is 51 hours.

The idle times for the man who reassembles the machine parts are as follows:

For P = 51 –– 26 = 25 hrs.

For Q = 7 + (18 – 16) + (26 – 24) + (51 – 29) = 33 hrs.

For R = 12 + (16 – 14) + (24 – 21) + (29 – 28) + (51 – 32) = 37 hrs.

For S = 14 + (21 – 17) + (28 – 27) + (51 – 35) = 34 hrs.

For T = 7 + (27 – 26) = 18 hrs.

The idle times for the repairmen working on the machines are as follows:

For A = No idle time and will be handed over after 26 hours.

For B = 7 idle hours and will be handed over after 35 hours.

For C = 12 + (35 – 33) = 14 idle hours and will be handed over after 45 hours.

For D = 18 + (33 – 32) + (45 – 35) = 29 idle hours and will be handed over after 51 hours.

BUSTLE - 4.4

Identify a few situations at your office/organisation/institution where you can apply the sequencing methods. Observe what methods are in practice. Are they scientific and optimal? If not, modify them and prepare a report.

___

___

___

___

TEST YOUR UNDERSTANDING 4.5

1. A mechanic has to assemble 5 machines, in which each assembling takes three stages, for which he employed three persons. The sequence of the stages is fixed as stage I, stage II, and stage III. Determine the optimal sequence for the machines to be assembled to minimise the time elapsed from the start of first machine to the end of last machine. The time required, in days for each machine at each stage is given in the following matrix. Also find the time that the mechanic could promise to deliver each machine. If the mechanic can de-link a person from the present process if his stage work is completed or not started, what will be the net idle time that the process demands?

Machine code	*Stage I*	*Stage II*	*Stage III*
P11081975	6	8	14
R11051965	16	10	18
S20061995	14	2	10
H20032000	10	4	12
K29122001	8	6	20

2. Five spare parts are made, each of which must go through machines M1, M2, M3 in order M1, M3, M2. Processing times are given below. (i) Determine the optimal sequence, total elapsed time, and idle time of each machine. (ii) If it can be processed by the sub-contract services by three outside parties, P1, P2, and P3 to process on M1, M2, and M3 respectively, schedule the parties optimally, under the conditions that the parties may be called on any day but the contract once started should be continued till the last job of

the respective party is completed and the payment should be made for the process delays also. (iii) What will be the amounts paid to each party if it costs ₹ 10/- for each working hour and ₹ 5/- for each waiting hour?

Spare part	1	2	3	4	5
M1	8	5	4	6	5
M2	10	13	11	10	12
M3	6	2	9	7	4

3. Find the sequence that minimises the total elapsed time in hours to complete the following jobs on 3 machines. Prepare Gantt chart.

	A	B	C	D	E
M1	3	8	7	5	2
M2	3	4	2	1	5
M3	5	8	10	7	6

4. Find the sequence that minimises total machining time. The data is:

Tasks	A	B	C	D	E	F
Time on M/C 1	4	9	8	5	10	9
Time on M/C 2	5	4	3	6	2	5
Time on/C 3	7	8	6	12	6	7

5. Find the sequence that minimises total elapsed time.

Tasks	A	B	C	D	E	F	G
Time on M1	3	8	7	4	9	8	7
Time on M2	4	3	2	5	1	4	3
Time on M3	6	7	5	11	5	6	12

HINTS

1. Optimal sequence is: P-H-K-R-S

Machine	*Stage I*		*Stage II*		*Stage II*		*Delivery day from now*
	time in	*time out*	*time in*	*time out*	*time in*	*time out*	
P	0	6	6	14	14	28	28
H	6	16	16	20	28	40	40
K	16	24	24	30	40	60	60
R	24	40	40	50	60	78	78
S	40	54	54	56	78	88	88

Total elapsed time is 88 hrs and idle time for Stage I – 34 hrs (whom the mechanic can de-link in the last 34 hrs.), for Stage II - 58hrs (in which 6 + 32 = 38 hrs can be de-linked) and for Stage III – 14 hrs (in which all the 14 hrs can be de-linked).

Net idle time due to process is 20 (Stage II operator).

2. Optimal sequence is: 2-5-3-4-1

Job	*M 1*		*M 2*		*M 3*	
	time in	*time out*	*time in*	*time out*	*time in*	*time out*
2	0	5	5	7	7	20
5	5	10	10	14	20	32
3	10	14	14	23	32	43
4	14	20	23	30	43	53
1	20	28	30	36	53	63

Total elapsed time = 63 hrs.

Idle time on M1 = 35 hrs, on M2 = 35 hrs and on M3 = 7 hrs.

P1 starts at 0th hour and ends at 28th hour (28 hrs),

P2 starts at 5th hour and ends at 36th hour (31 hrs) and

P3 starts at 7th hour and ends at 63rd hour (56 hrs)

Payments for P1 – ₹ 280/- (all are working hrs);

for P2 – ₹ 295/- (28 working hrs and 3 waiting hrs) and

for P3 – ₹ 560/- (All working hours).

3. Optimal sequence is A-D-E-C-B

Job	*M 1*		*M 2*		*M 3*	
	time in	*time out*	*time in*	*time out*	*time in*	*time out*
A	0	3	3	6	6	11
D	3	8	8	9	11	18
E	8	10	10	15	18	24
C	10	17	17	19	24	34
B	17	25	25	29	34	42

Total elapsed time = 42.

Idle time on M1 = 17 hrs.

Idle time on M2 = 27 hrs.

Idle time on M3 = 6 hrs.

4. Optimal sequence is: A-D-B-F-C-E

Job	*M 1*		*M 2*		*M 3*	
	time in	*time out*	*time in*	*time out*	*time in*	*time out*
A	0	4	4	9	9	16
D	4	9	9	15	16	28
B	9	18	18	22	28	36
F	18	27	27	32	36	43
C	27	35	35	38	43	49
E	35	45	45	47	49	55

Total elapsed time = 55 hrs.
Idle time on m/c 1 = 10 hrs.
Idle time on m/c 2 = 30 hrs.
Idle time on m/c 3 = 9 hrs.

5. Optimal sequence is A-D-G-B-F-C-E

Job	*M 1*		*M 2*		*M 3*	
	time in	*time out*	*time in*	*time out*	*time in*	*time out*
A	0	4	4	9	9	16
D	4	9	9	15	16	28
G	9	18	18	22	28	36
B	18	27	27	32	36	43
F	27	35	35	38	43	49
C	35	45	45	47	49	55
E	37	46	46	47	54	59

Total elapsed time = 59 hrs.
Idle time on M1 = 13 hrs.
Idle time on M2 = 37 hrs.
Idle time on M3 = 7 hrs.

4.6 SCHEDULING PROCEDURE FOR BREAKDOWN MAINTENANCE

Breakdown maintenance resembles with the job production. Therefore, the scheduling of breakdown maintenance jobs is similar to those of job production with little changes. Basically, the breakdown maintenance is the repair function of Plant Engineering. Thus, when various jobs for repair are received by a plant engineering department, they are sequenced, assigned, and scheduled according to some predetermined rules. Again, the procedural steps differ slightly according to the system of maintenance (centralised or decentralised) followed in the organisation. The common procedural steps are given below.

(a) All the repair jobs are collected at the beginning of the shift/day. If any reports of breakdowns are received during the shift they will be kept pending for the next shift/day or taken up after the present jobs are completed. However, if the breakdown is critical or the production is going to be stopped in the absence of the availability of the machine or equipment, the preferential treatment is given to those machines using pre-emptive rule of sequencing.

(b) The repair jobs are sorted out based on the importance of the machine in the production chain. If the non-availability of the machine leads to significant losses or stoppage of the production chain, they are kept on top order.

(c) A probable time to repair and restore its condition will be estimated.

(d) The repair jobs will then be sequenced according to the priority rules. Most commonly used rule is SPT particularly when the machines are of nearly same importance in the production chain. Perhaps the reason for using this rule is that the machine that takes lesser time to restore its condition can be handed over for production at the earliest, and thus its availability is increased. This rule varies according to the production process, production system, available resources with plant engineering department like manpower, spares and materials, tools and accessories, material handling equipment (such as cranes, chains, pallet trucks, lifters, forklifts, etc.). For instance, in intermittent production system, SPT, FIFO rules are preferred while in continuous production system, pre-emptive and non-pre-emptive rules may be preferred.

(e) Then the availability of the manpower for doing these jobs will be noted. Based on the efficiency, the probable time that each man will be able to complete the job is written against each job.

(f) The assignment of the jobs will be made so as to minimise total time.

(g) The schedule chart will be then filled up mentioning the start timings or/and end timings and process timings.

(h) A report will be prepared at the end of the shift/day on the actual performance (achieved) vs scheduled (planned) work along with the reasons for incompletion or remarks. The left out jobs will be carried forward to the next shift/day.

BUSTLE - 4.5

Narrate the method/procedure of scheduling the breakdown maintenance followed by the organisation where you are working/studying. Is it scientifically befitting or suitable to the situation? Justify if so. Rectify if not.

____

____

____

____

4.7 SCHEDULING PROCEDURE FOR PREVENTIVE MAINTENANCE

As we know that preventive maintenance (PM) is adopted in numerous ways in the industries of which the most common in use are planned maintenance, scheduled maintenance, maintenance prevention, etc., the scheduling depends on the PM policy adopted by the organisation.

The importance goes to the uninterrupted production in scheduled maintenance while the priority rule is based on the frequency of failures or rate of failure, mean time between failures (MTBF) in planned maintenance, and the age/utilised life of the equipment in maintenance prevention. The general guidelines for scheduling procedure of preventive maintenance are as follows:

- A suitable preventive maintenance policy is laid down depending on the requirements and the availability of the organisation.
- Understand the priority rule that predominates in the PM policy adopted.
- List all the machines and identify the machines those require regular preventive maintenance.
- Determine frequency of preventive maintenance on each machine.

- Design the duration of PM in terms of inspection (I), small repair (S), major overhaul (M), and complete overhaul (C) or weekly (W), monthly (M), quarterly (Q), half yearly (H), and Yearly (Y), etc.
- Allocate these I, S, M, C or W, M, Q, H, Y or any other suitable set to various days in the maintenance horizon (the period for which the maintenance policy is observed) such that jobs are distributed evenly and uniformly.
- Correlate with the production schedules and remove the overlaps, if any.
- Reschedule by using step 6.

BUSTLE - 4.6

Narrate the method/procedure of scheduling the breakdown maintenance followed by the organisation where you are working/studying. Is it scientifically befitting or suitable to the situation? Justify if so. Rectify if not.

____

____

____

____

4.8 CHECK SHEETS

Every machine has got its checkpoints where the maintenance personnel have to concentrate. These points are usually given by the OEE or may be suitably selected. A thorough study is to be made before laying down the norms for checking these points. The required check sheet for such checking and inspection or overhaul is to be written in a suitable format for each machine.

For example: A lathe machine has got various points like lubrication points, cleaning points, performance check parameters, etc. These are to be listed. A report is to be written after the PM is done and is to be filed.

4.9 SCHEDULING FORMATS

A suitable format may be designed so as to accommodate all the machines and their PM schedules. These are to be prepared at the beginning of the year for a certain period usually one year (called preventive maintenance horizon) and these schedules are to be strictly adhered. The schedule formats are to be approved by the top management and the copies of approved schedule formats are to be sent to each production department and to all the persons concerned so as to keep those machines available and to make necessary arrangements for plant engineering department in those schedules. The results are to be recorded after every PM activity. The scheduled format duly approved by the top management is reframed in a larger size and is treated as maintenance master schedule (MMS). For the horizon (usually the year), the MMS is regarded as regulatory and reference document. Any deviation or discrepancy reported can be solved by using this MMS only. After completion of each PM activity, the respective cell in MMS is marked or filled with a green colour. If any major problem is noticed during PM, it will be filled up with a red colour. One may choose different colours for indications of such information about various problems, which occur during PM activity. The designed time and the actual time may also be indicated, for further analysis and as guideline for the preparation of the MMS of the next year.

Test Your Understanding 4.6

1. Compare and contrast scheduling of breakdown maintenance and preventive maintenance.
2. Give the procedural steps of scheduling breakdown maintenance.
3. Give the procedural steps of scheduling preventive maintenance.
4. What do you mean by master maintenance scheduling (MMS)? What is its significance in an industry?
5. Describe various scheduling policies adopted for maintenance and examine their merits and demerits with some practical examples.
6. What are the assumptions made while solving sequencing problems by Johnson's algorithm? Describe the procedure of optimal solution for processing n jobs through two machines.
7. Discuss Jackson's conditions and describe the procedure of optimal solution for processing n jobs through three machines and hence for n jobs through m machines.

4.10 CONCLUSION

Any work, if planned strategically in advance and worked out systematically, it surely yields better results as compared to that done without proper plan. The scheduling strategies and techniques will help the maintenance engineer in two ways. Firstly, it will enhance the productivity and the maintenance efficiency (maintainability). Secondly, these schedules or plans prepared will become the path showers or guidelines for future work and act as regulators (or controllers) against the actual performance.

Hence, it is always better to *plan the work and work the plan.*

4.11 RECAP

The need for scheduling the maintenance is focused in this chapter. The basic procedure for scheduling starts with the sequencing of the jobs and then assigning them to the available manpower and materials. A few numerical problems are also solved on sequencing and assignment of maintenance jobs. The scheduling procedures for breakdown maintenance and preventive maintenance are given. These procedures, however, may change slightly depending on the type of the organisation and their requirements and availability of resources. At the end, the scheduling formats of a product industry are also included so that the reader has the guidelines for preparing the master schedules to suit his/her organisation.

4.12 KEY CONCEPTS

1. **Sequencing:** The order of writing various jobs to be performed.
2. **FIFO:** First In First Out; it is also called First Come First Serve (FCFS).
3. **LIFO:** Last In First Out.

4. **SIRO:** Select In Random Order.
5. **SPT:** Shortest Process Time.
6. **LPT:** Longest Process Time.
7. **EDD:** Earliest Due Date.
8. **Gantt chart:** A graphical representation of the jobs to be performed in bar graph showing the start time, end time, and idle time (if any).
9. **Assignment:** The optimal matching of the jobs to men or machines.
10. **Scheduling:** Assigning time to sequenced jobs.
11. **Allocation:** The allotment of work or jobs or materials.
12. **PM Horizon:** The period for which the preventive maintenance jobs are scheduled.
13. **MMS:** Maintenance Master Schedule.

REVIEW QUESTIONS

1. Discuss the significance of scheduling in plant maintenance.
2. List out and discuss various policies of scheduling applicable to maintenance.
3. "Master maintenance scheduling (MMS) can be used as the yardstick for measuring the effectiveness of the preventive maintenance (PM) in an organisation". Critically appreciate the statement.
4. Explain the concept of the following scheduling strategies in the backdrop of Plant Engineering.
   (a) Relaxed Scheduling
   (b) No-Work Pay Guarantee
   (c) Short Work Schedules
   (d) Overtime Work Schedules
   (e) Period-End Pile Ups
5. Discuss various criteria used in sequencing under 1 machine × n jobs.
6. List the assumptions used for sequencing under 2 machine × n jobs.
7. Describe Johnson's algorithm of the sequencing under 2 machine × n jobs.
8. Describe Jackson's conditions and the algorithm of the sequencing under 3 machine × n jobs.
9. Explain Jackson's conditions and the algorithm of the sequencing under m machine × n jobs.
10. Describe the algorithm of the sequencing under m machine × 2 jobs.

CHAPTER 5

Spare Parts Management

PREAMBLE

This chapter mainly focuses on the classification and controlling of the spare parts and solving the problems related to spare parts inventory. This chapter discusses three inter-related topics of spare parts management, viz. selective inventory control techniques, codification systems and standardisation of the spare parts. Various selective inventory control systems such as ABC, VED, SDE, HML, SOs, GOLF, FSND, VEIN, CIN are exhaustively dealt. Discussions are extended to the maintenance system optimisation and other relevant applications. The codification and classification systems are then explained including parts classification, Opitz classification, multi-class and general classifications. The standardisation, its significance, methods, levels, advantages and barriers are described with a special reference to the maintenance department. A list of some applications of computers in maintenance department is provided at the end of the chapter.

CHAPTER OUTLINE

5.1 INTRODUCTION

A spare part is the part (or component) identical to the part (or component) of the equipment that needs replacement, usually due to wear and tear. Generally, the life of the part (particularly, moving parts or parts which are subjected to stress) is less than the operating life of the equipment. Therefore, such a part experiences more wear and tear which in turn, needs replacement. Large-scale industries maintain the inventory of such spare parts, for which there is higher probability of failures or lower life span.

Spare parts, which are treated as stock items of Plant Engineering are commonly classified under MRO (maintenance, repair, operating) inventories. Some industries simply call them 'maintenance inventories'. The spares are usually common to more than one type of machine. These are normally of standard specification e.g., bearings, belts, contactors, connectors, cables, fuses, pipes, springs, handles, knobs, clutch plates, brake liners, cylinder, piston rings, filters, regulators, lubricants, oil seals. A component/sub-assembly/assembly for substitution is a spare part. Therefore, items like plates, sheets, rods, strips, etc., which are required to be fabricated for manufacturing complete units/ sub assemblies, are also considered as spare parts.

5.2 FEATURES/CHARACTERISTICS OF SPARE PARTS

The spare parts inventory management is different from the general inventory management. Perhaps, this difference is due to the fact that the spare parts unlike general inventory items have specific distinguishable characteristics/features. In general, the spare parts will possess the following characteristics/features:

- Uncertain demand (or requirement).
- Very less demand.
- Uneconomical to manufacture because of low demand.
- High tendency to obsolescence.
- Non-uniformity in consumption.
- Excessive stock in distribution channels.
- Requirement of very large variety but lesser quantity.
- Small range of items also can meet a large percentage of requirements.
- Cost due to non-availability is more than that of the spare part.
- Poor availability of exact specification.
- Increase in demand with the age of the equipment.

5.3 FUNCTIONS OF SPARE PARTS MANAGEMENT

Although spare parts inventory management is a part of materials management, the maintenance department is often entrusted to manage this inventory autonomously. The is because spare parts have distinctly different features as listed in Section 5.2, and they need special attention and the management also has to deal in different manner, unlike the regular materials management.

The functions of spare parts management are boundless in a large-scale industry. However, three major parts, viz. selective inventory control, codification, and standardisation are discussed at length in this chapter. Others are self-explanatory and, hence, left to the readers' understanding (moreover, they are inter related with materials management and production management and, hence, the redundancy is avoided).

The general functions of spare parts management are listed below:

- Identify the spares to be procured or maintained in stock.
- Make decision on the quantity and time of procurement of spares and other maintenance inventory such as consumables, repair tools, safety equipment, utilities, fuels and lubricants, etc.
- Procure spares and other inventory in association with materials department or purchase department.
- Receive the spare parts and inspect to confirm the correct specification.
- Conduct a suitable selective inventory control analysis and follow.
- Arrange the stocks of spares, consumables and other relevant inventory in appropriate places in the storeroom according to the adopted rule of selective inventory control. This helps in quick traceability and assessment for the next period.
- Codification of the spare parts and maintenance inventory.
- Standardisation of spare parts, particularly if they can be used for more than one machine.
- Issue the right spares in right time for right machine.
- Maintain records for issues, receipts, and specifications.
- Carry out stock checking and auditing at predetermined intervals.
- Control obsolescence and expiries.

BUSTLE - 5.1

1. *Conduct a mini project using the following guidelines:*
   (i) *Select an organisation having a plant engineering department maintaining considerable number of productive machines.*
   (ii) *Enumerate the spare parts used in the organisation.*
   (iii) *Compare the features listed in Section 5.2.*
   (iv) *Also compare the functions of the spare parts management and write down the differences, if any.*
   (v) *Analyze the performance of the spare parts management in respect of their availability, demand, utilisation, expiry dates, obsolescence, etc., and give your report.*

____

____

____

____

Test Your Understanding 5.1

1. List the functions of spare parts management in plant engineering department.
2. Define the term 'spare part'. Enumerate the features of spare parts inventory.
3. Do you think that the management of spare parts inventory is different from the general inventory management? If so, why? If not, give your reasons.

5.4 CLASSIFICATION OF SPARE PARTS

Spare parts are usually classified as per the convenience of the department or the industry. There are no rigid rules for classification of spare parts. However, the general nomenclature in classification of spare parts is given below.

(a) **Complete equipment**

- *Major assemblies*: Usually these are costly and have more life than the main equipment also.
- *Minor assemblies*: These are comparatively less expensive and have lesser life.

(b) **Capital repair spares or insured spares**

These are generally critical items. Non-availability of such items leads to very heavy downtime cost or may prove catastrophic or may lead to permanent damage with huge stock out cost. Lead time of procurement is very high.

(c) **Regular maintenance/breakdown spares**

These are regularly needed and, hence, kept in stock in considerable numbers. Both reliability and unit cost of these items are not so high, e.g., belts, bearings, oil seals.

(d) **Rotary/floating spares**

The old parts when replaced by the new parts are not always scrapped. But can be repaired and reused. Such spares are called floating or rotary spares. Costly spares like pumps, engines, motors propellers, etc. are repaired and stored because the stock out cost is high. This is observed with good practice in road transport departments, railways, shipping, and airline sectors.

(e) **Overhauling spares**

The spares that need overhauling at regular intervals, e.g., valves, coupling, gasket, etc., are classified under overhauling spares.

(f) **Contingency spares**

These are interchangeable or alternate methods/technologies that are available to take care of the situation, e.g., rivets, bolts, valves, gears.

Often, the spare parts are classified for maintaining and controlling them in stores. They can be classified based on their consumption, cost, significance of the item in the machine, importance of the machine in which it is used, availability in the market, and so forth. The more scientific and proven methods of analyses and classifications, known as selective inventory control techniques, are discussed in this chapter in the following sections. These are viewed from the angle of maintenance inventories and the explanation is given with reference to the maintenance stores only. Therefore, for further details, if required the readers may refer other relevant materials/subjects on Materials Management/Inventory Control/Production and Operations Management/Operations Research, etc.

5.4.1 ABC Analysis

This is the most common management principle employed for almost every inventory management problem. In this approach, the inventory, i.e., the spare parts are categorised according to their annual consumption cost in terms of money (rupees). As the name suggests, the inventories are classified into three categories, **A**, **B**, and **C**.

- **A Class** inventories are the most valuable but are few in number and require more attention because the consumption cost is about 70% of the total.
- **B Class** inventories are consumed moderately and cost intermediate (not too high as that of A class nor too low as that of C class). They constitute about 20% of the total cost and require moderate care.
- **C Class** inventories are the least valuable and so do not need much attention. The unit costs less and consumption is more. However, the consumption cost of these items is about 5% to 10% of the total.

ABC analysis takes care of two aspects—(i) the annual consumption and (ii) the cost. According to Pareto's law, "Vital Few—Trivial Many". Most generally, the valuable (costly) items are very less in number while the cheaper items will be many. So also in maintenance spares, the costly items are rarely consumed (replaced), while trivial ones will be frequently consumed.

Consider an example of motor bike spare parts management in a service centre. At a service centre, the bikes come for repairs and servicing. The service centre examines the bike arrived, and makes necessary repairs and adjustments. If required, it replaces the parts that are malfunctioning or not working or worn out. In view of this, the service centre has to maintain an inventory store with various spare parts. Now, let us observe the complaints that the service centre usually receives, and the type of spares they replace.

We can observe that usually they get the complaints to change gear wire, brake wire, accelerator wire, tyre failure problem, colour coating, replacement of head lamp or indicator lamps, seat cover, spark plug, etc. Think! How much do such problems cost? They cost very less. But these are more in number. Let us classify such items under 'C' class.

We also can find that the spare parts such as gears, seat, rim, etc. may also be changed, now and then. The cost of such items is neither too high nor too low. These spares can be categorized as 'B' class items.

The engine, piston-cylinder, etc., are rarely changed. So such spares can be considered as 'A' class items. How many such items do we find? Well, the answer is very few. May be, we can count on fingers, say three, four, or five in a year. But think of their cost, the cost of each such part is high.

Summarily, if an industry consumes 100 types of spare part items each year and the consumption of these items in terms of rupees is, say, ₹ 1,00,000, it can be noticed that usually 10% of the items only (i.e., 10 items in this case) value up to 70 – 80% of the total consumption, i.e., ₹ 70,000 – 80,000. In this case here, let us take 70%, i.e., ₹ 70,000. Next, about 20% of the items (therefore, say 20 items) would value 20% of the total consumption, i.e., ₹ 20,000. And the rest 70% of the items value only 10% of the total consumption in terms of rupees, i.e., ₹ 10,000 only in this case.

Thus, most generally we can assume Figure 5.1 for ABC analysis of any inventory.

Category	*Cost of the item*	*Cumulative cost*	*Consumption of items*	*Comulative consumption*
A Class	10%	70%	10%	10%
B Class	20%	90%	20%	30%
C Class	10%	100%	70%	100%

On Cost Scale	*A Class items (70%)*	*B Class (20%)*	*C Class (10%)*
0	70	90	100

On Consumption Scale	*A Class (10%)*	*B Class (20%)*	*C Class (70%)*
0	10	30	100

Figure 5.1 ABC analysis of any inventory

The steps to conduct ABC analysis are as follows:

1. Calculate the annual consumption of each type of elements (quantity in terms of numbers only).
2. Get the annual consumption in monetary terms, i.e., rupees. This is obtained by multiplying the number of items used per year (annual consumption) with their respective unit costs.
3. Rank the items from highest to lowest annual consumption cost in terms of rupees.
4. Calculate the cumulative annual consumption cost.
5. Calculate the percent cumulative annual consumption cost.
6. Plot the graph of number of items versus cumulative annual consumption cost.

Table 5.1 shows the annual usage in rupees of 10 types of items.

Table 5.1 ABC analysis: annual usage in rupees

Item Code	*Annual Consumption*	*Cost of item in ₹*	*Annual Consumption cost in ₹*	*Rank*
A20	50	20	1,000	VII
A21	300	10	3,000	V
A22	3,000	15	45,000	I
A23	20	200	400	X
A24	600	10	6,000	IV
A25	50	3	15,000	III
A26	3,000	10	30,000	II
A27	100	5	500	IX
A28	70	10	700	VIII
A29	200	10	2,000	VI

Table 5.2 shows the annual usage in terms of rupees, arranged according to their rank and categories.

Table 5.2 ABC Category

Item Code	*Annual Consumption (₹)*	*% of annual Consumption*	*Cumulative % Consumption Cost*	*Category*
A22	45,000	43.4	43.4	A
A26	30,000	29.0	72.4	A
A25	15,000	14.5	86.9	B
A24	6,000	5.8	92.7	B
A21	3,000	2.9	95.6	B
A29	2,000	1.9	97.5	C
A20	1,000	0.9	98.4	C
A28	700	0.7	99.1	C
A27	500	0.5	99.6	C
A23	400	0.4	100.0	C
Total	1,03,600	100.0		

Let us assume that 20% of items will be kept in **A Class**. So the first two items (i.e., item A22 and A26) which are most valuable and represent 72.4% (43.4 + 29) of total value of the inventories are categorized as **A Class** items.

Next 30% of items are kept in category **B Class**. So items A25, A24, and A21 are items of **B Class** and represent 23.2% (14.5 + 5.8 + 2.9) of the total annual usage and the rest 5 items belong to category **C Class** representing only 4.2% of the total annual usage in rupees.

Inventory means blockage of money. No industry can afford to keep valuable and costly items in store for very long. Therefore, once the items are categorised, more attention can be paid towards **A Class** items for their quick use and safety.

5.4.2 SDE Analysis

This analysis is generally used to ascertain lead time for procurement and helps in purchase decisions (i.e., materials procurement) strategies.

S-D-E analysis is based on the availability of the items in the market. S-D- E means:

- **S** stands for 'scarce items'. These items are in very short supply and generally imported. Procurement of such items generally requires a lot of exercise and investment.
- **D** refers to 'difficult items'. Such items may be locally available but cannot be procured that easily, as there may be very few suppliers, or shortage of materials or they may be available at far-off places only.
- **E** means 'easily available items'. These items are easily available even in local market. There will not be any problem in procuring such materials because they are 'just go and bring' types.

5.4.3 FSND (or FSN) Analysis

This analysis is done with a view to controlling obsolescence of consumable spare parts and takes into account the consumption pattern of the items. FSND means as follows:

- **F** means 'fast moving' items.
- **S** refers to 'slow moving' items.
- **N** stands for 'non-moving' items.
- **D** denotes 'dead' items.

In this exercise, quantity and consumption rates of all items are analysed and accordingly classified. Such classifications help in arranging (storing) the items in the stores. For example, the fast-moving (i.e., frequent or fast consumable) items should be kept near the exit of the stores so that these are easily available (located) and can be handled out without disturbing other items. Likewise slow-moving and non-moving items should take back seats in the stores. The non-moving (i.e., rare-moving) items should take positions at the top racks of the stores.

To carry out this analysis, the date of receipt and date of issue of all items are noted down and their movement rates (i.e., consumption rate) are ascertained.

5.4.4 VED Analysis

This analysis or classification is mainly used for maintenance (spare parts inventory management) and is based on the significance or criticality of the item or spare part. The abbreviations V, E, and D stand for the following terms:

- **V** stands for 'vital' items. These items are highly significant, no matter whether they are costly or cheap. Non-availability of such items will halt the production and running of the machine. For instance, the piston and cylinder are vital items for a bike. Of course, these are very costly items. However, a valve for the tyre of the bike is also vital item though it is very cheap.
- **E** stands for 'essential' items. Non-availability of such items renders temporary dislocation or loss of production. In the case of stock-out of such items, expensive procurement may have to be made, if no stand-by facilities are available. For example, the brake for bike is an essential item.
- **D** stands for 'desirable' items. These items are necessary but do not cause any immediate loss in production. The stock-out of desirable items entail very nominal expenditure in procurement and cause minor disruption for a short duration. Examle, the seat covers, rear view mirror for a bike may be categorized under desirable items.

Note that ABC analysis categorises items to their consumption value (cost oriented), while VED analysis determines the effects of items on production and other services (i.e., function oriented).

5.4.5 CIN Analysis

This classification is also based on the criticality. This is similar to VED analysis. The only difference is that the VED is used for spare parts or components while CIN classification is often used for machinery and equipment. The abbreviations C,I, and N stand for the folowing terms:

- **C** stands for 'critical' machines.
- **I** stands for 'important' machines.
- **N** stands for 'normal' machines.

Criticality of the equipments/spare parts is often decided based on their usage, failure cost, reliability, or availability, etc., or a combination of two or more of such criteria. Table 5.3 gives the generally used guidelines for such decision making in plant engineering departments of large industries.

Table 5.3 Factors of spares to decide criticality of machine

S. No.	*Factor*	*Degree of Variation*	*Weightage (on 10 point scale)*
1	Rate of consumption or rate of usage or demand	Sporadic	2–4
		Regular	5–7
		Fast moving	7–9
		Continuous	9–10
2	Cost of failure or loss of production	No loss	1–2
		Marginal loss	3–5
		Major loss	5–7
		Heavy loss	7–8
		Catastrophic	9–10
3	Reliability	Highly reliable	2–5
		Reliable	5–8
		Unreliable	8–10
4	Availability	In local market (easily)	1–3
		In local market (difficultly)	3–5
		In open markets (easily)	5–7
		In open market (difficultly)	7–9
		In foreign markets (scarce)	9–10
		Only in government market (scarce)	9–10

Generally, if a spare part, unit, component, or machine scores 7–10 points it is categorised under *critical*; if the score is between 4–6, then it is *important,* and if the score is below 4 points, then it is *normal.*

5.4.6 VEIN Analysis

Judging certain components or spare parts becomes very difficult when they are classified based on their criticality. For example, in an automobile, the engine is categorised as vital or critical item, while a seat cover is categorised under desirable or normal item. But the brake liners and a head lamp have their difference of criticality. Brake liners are more critical—we can run the automobile at least during daytime without head lights. Therefore, it is better to classify these into two categories such as essential and important. Thus, this classification is a blend of **VED** and **CIN** analyses.

- **V** stands for 'vital' items. Vital equipment has high criticality.
- **E** stands for 'essential' items. Essential equipment has less than high criticality.
- **I** stands for 'important' items. Important equipment has more than (normal) low criticality.
- **N** stands for 'normal' items. Normal equipment has low criticality.

Further, VED, CIN, or VEIN analysis can be extended to maintenance system optimisation. This is discussed in the following sections.

5.4.7 SOs Analysis

This system is based on the seasonality of the items.

- **S** stands for 'seasonal' items.
- **Os** stands for 'off seasonal' items.

Again here different cases can be observed for a company with reference to the seasonality of their production and product as given below. The spare part management obviously will be different in such companies.

Case – 1: The production is seasonal and is also used seasonally.

Suppose a company purchases certain agricultural products (such as mangoes) in raw form, processes or ripens artificially, and makes in ready-to-eat form. Such a product is seasonal and the production also will be seasonal.

Case – 2: The production is seasonal and is used continuously throughout the year.

Products like sugar are used throughout the year but can be produced in the season of cane production only. The maintenance department in such companies will have full work of preventive maintenance in the off season while breakdown maintenance during the season and, hence, should plan accordingly.

Case – 3: Both the production and the product are continuous throughout the year.

This case is usual with many companies such as a manufacturer of say automobile. Therefore, no special planning is required.

Case – 4: The production is continuous and the product is used seasonally throughout the year.

Crackers are manufactured throughout the year. Woolen cloths (winter wear) are manufactured throughout the year but used (sold) only seasonally.

5.4.8 GOLF Analysis

This analysis is based on the market in which the item or spare part is available.

- **G** stands for 'government' market.
- **O** stands for 'open' market.
- **L** stands for 'local' market.
- **F** stands for 'foreign' market.

Thus, the market in which the spare part is available is the criterion for decision making on its stock quantity and care in storage. It is obvious that more care is required on G and F items.

5.4.9 HML Analysis

This analysis is based on the cost of the item.

- **H** indicates 'high' cost items.
- **M** means 'medium' cost items.
- **L** stands for 'low' cost items.

5.4.10 XYZ Analysis

This is based on the value of the item. One often gets confused between value and cost. Cost is the expenditure incurred to manufacture or price paid to acquire while value is the worth to pay for the product. For example, the documents of appointment orders of employees of the company may cost less but are valued more.

In this system, the abbrevations X, Y, and Z stands for the following:

- **X** indicates 'low' value items.
- **Y** indicates 'medium' value items.
- **Z** indicates 'high' value items.

Example, Z-type security is given for high valued individuals.

5.4.11 Two-Dimensional Classification

In all the above analyses, we can use only one system at a time. If the analysis requires classifying in two directions, say ABC and VED, then the items can be arranged in two-dimensional way in stores as shown in the following tabular representation.

	Vital	*Essential*	*Desirable*
A Class	A-V items	A-E items	A-D items
B Class	B-V items	B-E items	B-D items
C Class	C-V items	C-E items	C-D items

5.4.12 Music 3-D model

Further, the two-dimensional model can be extended to conduct three analyses and arranged in the store room in three dimensions. Consider the two-dimensional model as shown above. In each cell of this model, imagine a drawer in which there will be three or four slots. In the first, i.e., the nearest slot, we place fast moving items, in the middle slot we keep slow moving items, while the third slot contains non-moving items. Thus, suppose we pull the drawer of B-V cell; we find B-class, vital fast moving items in the first slot; B-class, vital slow moving items in the second slot; and so on. This is pictorially shown in Figure 5.2.

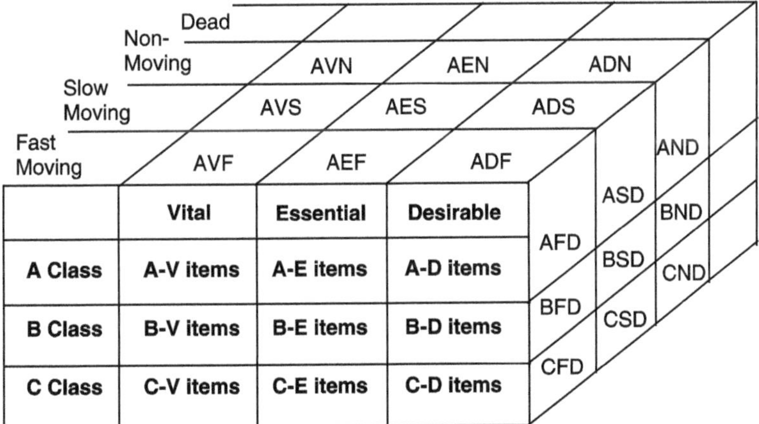

Figure 5.2 Music 3-D model for spare part inventory control

5.5 MAINTENANCE SYSTEM OPTIMISATION

We know already that the maintenance system can be broadly divided into the following two categories:

(a) Breakdown maintenance (BM)

(b) Preventive maintenance (PM)

It is very clear that no machine is completely immune to failures. According to Murphy's law, if a failure can occur, it will. In view of this, certain amount of breakdown maintenance is necessary for any machine, even though the best preventive maintenance practices are being observed. On the other hand, preventive maintenance is not required (or much close monitoring may not be necessary) on some machines which are not critical in the production chain (for example, low-cost machines, stand-by equipment). The cost of preventive maintenance will be more on such machines and it is not worthwhile. Therefore, it is important to establish a maintenance system with a blend of these two maintenance practices (BM and PM) in a right proportion. Thus, optimisation of maintenance system can be done by classifying the machinery into certain groups based on their criticality or any other characteristics. Usually, VEIN, CIN, or VED is used to classify the machinery. Once the machines are classified, the amount of breakdown maintenance and the amount of preventive maintenance can be decided and planned according to their categories.

For equipment or machinery which has low criticality, such as a lathe or a milling machine, in a process layout can be considered as 'not very important' (i.e., normal category). This is because the breakdown on such machines will not affect the total production. Of course, it will have very minor effect as there are other stand-by machines that can be used or the same operation can be done by some other machine by little adjustments or design changes of tools. In the worst-case scenario, the work can be sub-contracted with any local parties. So for such machines, one can even go for breakdown maintenance.

Now, consider a C.N.C. machine on the same shop floor. The items finished on it are of high precision, and breakdown of the C.N.C. machine will stop the production totally. Hence, this

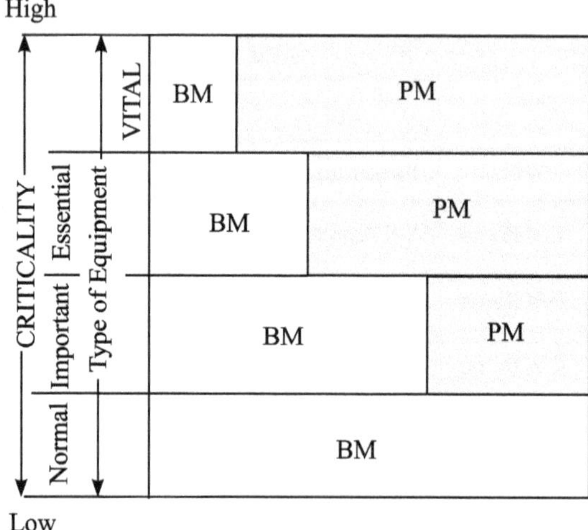

Figure 5.3 VEIN analysis for maintenance system optimisation

machine comes under high critical zone, and, therefore, for such machinery reventive maintenance is worthwhile and must be planned with good amount of time. Similarly, power generation system, on which running of all the machinery depends, must be under preventive maintenance system. Thus, the percentage of breakdown and preventive maintenance is determined for various grades of machinery depending on their importance or criticality.

In Figure 5.3, the VEIN analysis for maintenance system optimisation is interpreted. For vital equipment, some percent, say 5–10%, of breakdown maintenance may be planned and the rest of the time may be allocated to preventive, or condition-based maintenance (5–10% of BM is due to the fact that in spite of best preventive maintenance, a system might fail).

For normal equipment, no preventive maintenance has been shown. There is only breakdown maintenance. However, if required up to 10% PM can be allocated.

For important and essential equipment, the PM and BM should be planned in right combination ranging between 30–70 percentage. The right maintenance mix can contribute to the organisation in the following way.

5.5.1 Advantages of Maintenance System Optimisation by VEIN Analysis

- Reduced maintenance cost
- Reduced troubles
- Reduced maintenance hours and maintenance crew become free
- Increased production
- No production interruption during peak season
- Improved reliability and availability

5.5.2 Requirements for Maintenance System Optimisation

The following data for each machine are necessary for maintenance system optimisation through VEIN (or VED or CIN) analysis:

- Frequency of breakdown—daily/weekly/monthly.
- TimeBetween Failures (TBF).
- TimeTo Repair (TTR), i.e., repair/replacement time for each of the breakdown.
- Type of breakdown or failure.
- Size of maintenance crew.
- Production loss.
- Spares consumed and their costs.

The above data will change as the machines are ageing. Hence, while planning the percentage of breakdown maintenance and preventive maintenance, the age of the equipment also must be taken into consideration and planned properly for each type of equipment. Both breakdown and preventive maintenance incur cost to the organisation, but in different proportion. Thus, the organisation has to design a proper mix of the two BM and PM for minimum loss or maximum gain.

5.5.3 Implications of Improper Maintenance Mix

The criticality of the equipment has to be determined by proper maintenance engineering analysis, taking all the factors into consideration. In fact, this is a challenging job for any organisation. Improper maintenance mix may lead to the following disadvantages:

- Extra cost will be incurred in maintenance.
- More maintenance personnel (i.e., larger maintenance crew) will need to be employed.
- Extra spare parts will need to be stored.
- The equipment will need to be stopped to carry out (unnecessary) maintenance.
- There will be loss in production.

BUSTLE - 5.2

1. *Which type of selective inventory controlling technique is used in the organisation you have chosen for Bustle 5.1? What are the practical difficulties or drawbacks you find? What measures do you suggest to help overcome the difficulties?*
2. *Classify the spare parts in the most suitable way and justify your method of classification.*
3. *Conduct ABC analysis.*

Test Your Understanding 5.2

1. What is ABC analysis? Narrate the procedural steps to conduct ABC analysis.
2. Discuss FSN analysis in detail with reference to the application to maintenance of spare parts.
3. Explain SOs, SDE, and GOLF analyses. Explain with suitable example related to spare parts management.
4. What do understand by VED, CIN, and VEIN analyses? Discuss their relevance to the plant engineering department.
5. Distinguish between HML and XYZ analyses. Explain them with suitable examples taken from the maintenance department.
6. Describe Music 3-D model applied to inventory control of spare parts.
7. List the merits of maintenance system optimisation by VEIN analysis.

5.6 CODIFICATION

Maintenance spares are usually available in plenty. They are repetitive in nature. The basic difficulty with the spare part management is that the time of the requirement of a particular item is not known. And the non-availability of the item can cause severe or unpredictable loss also. Therefore, whether used or not, such items are kept in stock.

In course of time, the items become obsolete or expire, or the shelf life can be completed. Sometimes, the place where the spares are stored cannot be located. In some other occasions, the spare parts controller or stores clerk may be absent or transferred. In all such cases, if the spare parts are not maintained systematically and recorded properly, it can result in huge loss that will affect the organisation adversely. This problem can be sorted out by coding all the spare parts of the company and maintaining by adhering to the codes. Further, the stocks can be reviewed easily if the data base is maintained in the computer and updated regularly as and when used.

5.6.1 Advantages of Coding System

The major advantages of well-designed coding and classification system for group technology are as follows:

- It reduces design duplication.
- It reduces the inventory and storage cost.
- It gives reliable spare parts utilization.
- It helps in rationalization and improvement in tool design.
- It helps better utilization of machine tool, cutting tool fixtures, and manpower.
- It facilitates NC part programming.
- It facilitates the formation of machine cells and part families.
- It facilitates accurate estimation of spare part requirement and logical decision on procurement.
- It aids maintenance planning and scheduling procedure.
- It facilitates quick retrieval of spare parts and their designs, drawings, and process plans.

5.7 TYPES OF CODING AND CLASSIFICATION SYSTEMS FOR SPARE PARTS

There are several systems of classifying and coding spare parts. But none has been adopted universally. One system may be suitable for one organisation while another system suitable for another organisation. In the same organisation, the existence of many systems in various departments is also observed.

However, codification helps in simplifying the work content of the maintenance. In many occasions, it is essential to reduce the cost of inventory and salvaging.

A few popular systems coding and classifications systems relevant to maintenance are explained here.

5.7.1 Part Classification System

Part classification can be determined based on the following three attributes:

- Part manufacturing attributes
- Part design attributes
- Both manufacturing and design attributes

The CODING system for part classification can be of two basic structures:

(a) Hierarchical structure

(b) Chain-type structure

We shall explain the above two systems with an example. Consider a two digit code, such as 05, 15, 25, 35, 45, etc. Let the first digit symbolise general shape of the cross section of the part so that digit 0 means round part, 1 means irregular shape, 2 means flat (**II**) rectangular part, 3 means triangular (Δ), 4 means square (□) or rectangular, and so on.

(a) **Hierarchical Code System**

The interpretation of the second digit, i.e., 5 in this example will depend on the value of the first digit. Therefore, for the code 15, 5 may indicate Length/Diameter ratio whereas, for the second code 25. The digit 5 may indicate the overall length.

(b) **Chain-type Code Structure**

The digit 5 indicates the same meaning irrespective of whether the first digit is 1 or 2.

Example: The second digit '5' will indicate *overall part length*, whether the part is circular, irregular, or rectangular.

The advantages of hierarchical code structure are that more information can be stored in the code. Note that some classification and coding systems use a combination of both hierarchical and chain type structure.

The number of digits required can range from 6 to 30. Coding system that includes only design characteristics needs fewer digits, i.e., 12 or less. Most modern classification and coding system incorporates both design and manufacturing data into the code. For this code numbers with 20 to 30 digits may be needed.

Example: Let us see the coding system in general. Consider a tube manufacturing company where the tubes' internal diameters range from 5 mm to 500 mm. The materials used for the tubes are brass, aluminium alloys, steel (of a particular composition). The length of the tubes varies from 100 cm to 1000 cm.

In this case, we shall use four-digit code:

Column-1	Column-2	Column-3	Column-4
X	X	X	X

- Digit in the first column represents the diameter of the tube.
- Digit in the second column represents length of the tube.
- Digit in the third column represents the material of the tube.
- Digit in the fourth column represents any special treatment on the tube.

Now the range of the dimensions:

Say the diameter of the following values is produced—5 mm, 50 mm.

(i) 100 mm, 150 mm, 200 mm, 300 mm, 350 mm, 400 mm and 500 mm.

(ii) The length 50, 100, 250, 300, 400, 500, 650, 750, 900 mm.

(iii) Material of the tube: brass, aluminium, steel.

Let the diameter of the tube be represented by the following numbers:

Diameter		
5 mm	--	1
50 mm	--	2
100 mm	--	3
150 mm	--	4
⋮		⋮
500 mm	--	9

Let the length be represented in the following way:

Length	*Digit*
0–99 mm	0
100–199 mm	1
200–299 mm	2
300–399 mm	3
400–499 mm	4
500–599 mm	5
⋮ ⋮	⋮
900–999 mm	9

Let the material be represented as follows:

Material	*Digit*
Brass	1
Aluminium	2
Steel	3

Let us take the fourth digit code as

Material with special treatment	1
Material with no special treatment	0

Now, let us consider the example 3621.

The first digit is 3 which means the diameter of the part is 100 mm.

The second digit is 6 which means the length of the part is 650 mm.

The third digit is 2 which means the material of the part is aluminium.

There is special treatment on the material since the fourth digit is 1.

5.7.2 The OPITZ Classification System

This part of coding and classification system is developed by H. Opitz of Germany.

It uses the following digit sequence:

x x x x x	x x x x	x x x x
1 2 3 4 5	6 7 8 9	10 11 12 13

- The *basic code* consists of 9 digits (1...9).
- The first five digits (1, 2, 3, 4, 5) are called *form code*. They describe the design attributes.
- The next four digits (i.e., 6, 7, 8, 9 in the code) are *supplementary codes*. These attribute to manufacturing such as dimensions, work material, starting raw material shape, and accuracy.
- The last four digits (10, 11, 12, 13 positions) are *secondary codes*. They are used to identify the production operation type and sequence.
- The secondary code is designed by individual firm for its specific needs.

ILLUSTRATION 5.1:

The digit in the first column indicates relationship of its dimensions. There will be 0 to 9 digits, i.e., 10 numbers in the first column, as explained above. It may be for rotation component or it may be for non-rotational component.

For rotation components let L be the length of the part and D be the diameter of the part.

For non-rotation components let A be the length of non-rotational component; B be the breadth of non-rotational component; and C be the thickness of non-rotational component.

Then the code digits of the first column will be as given in the following table.

Digit in the first column		*Significance*
	0	$L/D \leq 0.2$
	1	$0.2 < L/D < 1$
Rotational	2	$1 < L/D < 2$
	3	$2 < L/D < 3$
Component	4	$L/D \geq 3$
	5	SPECIAL
	6	$A/B \leq 2$, $A/C \geq 3$
Non-rotational	7	$A/B > 2$
Component	8	$A/B \leq 2$, $A/C <$
	9	SPECIAL

The digit in the second column indicates external shape. The codes may be given as: 1 for plane, 2 for stepped, 3 for tapered, 4 for threaded,

The digit in the third column indicates internal shape. Similar codes as shown may be given as: 1 for plane, 2 for stepped, 3 for tapered, 4 for threaded.

In a similar fashion other digits also can be coded suitably. However, the digits in various columns can be broadly coded to indicate as follows:

- The digit in the fourth column indicates plane surface machining.
- The digit in the fifth column indicates auxiliary holes and gear teeth.
- The digit in the sixth column indicates the absolute dimensions.
- The digit in the seventh column indicates the material of the part.
- The digit in the eighth column indicates the original shape of the raw material.
- The digit in the ninth column indicates accuracy on the part.

ILLUSTRATION 5.2:

Interpret the following code on the basis of Opitz coding and classification system:

320136786.

- The digit in the first column is 3, which means $2 < L/D < 3$.
- The digit in the second column is 2, which denotes stepped on one side.

- The digit in the third column is 0, which says that the spare part is a solid part i.e., no hole or not hollow.
- The digit in the fourth column is 1, which indicates one external groove at one end of the shaft.
- The digit in the fifth column is 3, which indicates spur-gear tooth.
- The digit in the sixth column is 6, which indicates the external diameter of the part (say, it is 100 cm).
- The digit in the seventh column is 7, it indicates the material, say, brass.
- The digit in the eighth column is 8, indicates the original shape of the raw material, say it is a cold-rolled product.
- The digit in the ninth column is 6, which indicates the accuracy desired, say it is ± 0.1 mm.

5.7.3 Multiclass Coding and Classification System

The multiclass coding and classification system is developed by the Organisation for Industrial Research. It has the following features.

- The hierarchical or decision-tree coding structure is used.
- Since the hierarchical coding structure is used, the succeeding digits depend on the values of the previous digits.
- The coding structure contains of up to 30 digits.
- Up to 9 different types of components can be included in a single multi-class software structure.
- It is relatively a flexible system.
- It can be used for a variety of spare parts like bolts, belts, etc., as well as for manufacturing goods such as sheet metal parts, machined parts, tooling, electronics, machine tools, etc.

The significance of first 19 digits on multi-class coding and classification system are as follows:

Column	*Meaning or Significance*
1	Code system prefix
2	Main shape
3, 4	External and internal configuration
5	Machined secondary elements
6, 7	Functional descriptors
8 to 13	Dimensional details such as length, diameter, thickness
14	Tolerance
15, 16	Chemistry of material
17	Shape of raw material
18	Quantity of production
19	Orientation of machined element

5.7.4 General Coding System (For Spare Parts)

This system is exclusively meant for spare parts, and, hence, uses local codes of the organisation. The organisation may use any codes in alpha-numeric form and there is no restriction or limit of digits in the system. For example, a company can code an item as follows:

Character 1 – Code of section or department in which the equipment is used
Character 2 – Code of the function (the equipment performs) in the section
Character 3, 4 – Code or number of the machine in the section
Character 5 – Code of sub-unit or sub-assembly for which the part is used
Character 6, 7 – Number of such spares used for the equipment
Character 8-10 – Serial number of the spare part as per stock list

Now, for example, in a machine shop (say the code number of machine shop is 9) for a lathe (say lathe code is 8), on 48th machine's gear box (say the code of gear box sub-assembly is 3), three gears are used. A spare part used for third gear, say key enumerated as 263 in the spares list (which contains all the details of the specifications and required information such as date of purchase, expiry date, etc.) is to be replaced.

This key should be given a code number as 9 8-48-3-03 263.

Drawbacks of the system

- Every time, whenever a machine is transferred from one department to the other, the code is to be changed.
- If a particular spare part is useful for more than one machine, it gets two code numbers and, thus, leads to confusion.
- When a machine is updated with new technological features or with design out maintenance or new features, it becomes difficult to give the coding.
- This system is not universal; hence a new entrant into the system or department needs lot of time to get familiarised and needs assistance or training.

BUSTLE - 5.3

For the spare parts listed in Bustle 5.1 or 5.2, prepare a codification system.

____

____

____

____

Test Your Understanding 5.3

1. What is codification? What is its significance in spare parts management? What are the advantages of codification?
2. Explain the part codification and classification system with examples. Distinguish between the hierarchical and chain-type coding structures with examples.
3. Describe the Opitz codification and classification system with examples.
4. Discuss the features of multi-class coding and classification system.
5. Explain the general coding system with the help of suitable examples. What are its drawbacks?

5.8 STANDARDISATION

The term 'standard' is multi-valued as it may mean a physical standard, prototype of mass, or a technical document. A standard is defined as a general agreement, a model, a rule, or a method established by authority or consensus or custom, or is created and used by various levels of interest. For example, an industrial standard is the generally accepted statement of the requirement. Standardisation is setting such standard and observing, adhering to and practicing. In other words, standardisation is the orderly and systematic formulation, adoption, application, and review of industrial standard which leads to simplification or variety reduction.

The standardisation implies reducing unnecessary varieties to the most economical sizes, grades, shapes, colours, or types of parts. A common misconception about standardisation must be removed at this stage. There are some people who think that to achieve standardisation, one should limit ones purchases to a single manufacturer's product out of many others in the field. True standardisation does not restrict competition. It focuses on standardizing product (or process or service) but not the producer. One of the objectives of standardisation is cost reduction and, in fact, the standardisation on a single manufacturer's product only defeats the objective. Moreover, by restricting purchase in this fashion one may not have a standard at all.

5.8.1 Need for Standarisation

We need standardisation for the following reasons:

- Suppose you buy a pair of jeans of waist size, say 36. When you try the jeans, it may turn out to be too loose and ill-fitting even though your waist size is exactly 36 inches.
- A person buys an electric plug. On reaching home, he finds that it does not fit the socket.
- A lady purchases a blouse of sky blue shade to match the colour of her sari. But she finds that it does not match or after one or two washes the colour fades out and does not match the saree at all.
- A repairman on a service call goes to the machine along with tools and expected spares but finds the tools are not fitting correctly to open the machine or the spares are not compatible.

All these are simple examples of non-standardisation that usually occur in our day-to-day life. Imagine a situation that there are no standard sizes of garments and shoes, standard sizes of books and magazines, standard traffic signals and road signs, standard forms used by business organisations, ball pen refills, etc. In these scenario our daily life will become very difficult. Thus, standardisation of products, methods, processes, activities, etc., is very much necessary to make life easier and simpler.

Standardisation is applied to two distinct areas in the industry:

1. Standardisation of products—their shape, dimensions, colour, physical and chemical properties, etc. This is industrial standardisation.
2. Standardisation of business practices—forms, procedures, systems, operating practices, etc. This is managerial standardisation.

Standardisation process may take the form of a document containing a set of conditions to be fulfilled, a fundamental unit or physical constant or an object for physical comparison. The apex body of standards in India is the Bureau of Indian Standards (BIS), formerly known as the Indian Standard Institution (ISI) and at the international level, International Organisation for Standardisation (ISO), Geneva.

5.8.2 Objective of Standardisation

The primary objectives or goals of standardisation are as follows:

- Maximising the use of common parts in different product.
- Minimising the number of different types of parts, components, assemblies, and other items.
- Maximising the use of interchangeable and standard or off-the-shelf parts and components.
- Minimising the number of different models and makes of equipment in use.
- Controlling and simplifying inventory and maintenance.
- Reducing storage problems and the effort spent on part coding and numbering.

5.8.3 Factors Leading to High Failure Rate in Non-Standard Items

Factors that contribute to the high failure rate for non-standard items are as follows:

- Poor uniformity of manufacture in small production runs. Consequently, parts are needed to be ordered more frequently, increasing the logistical burden. Also, small production runs can make the parts more difficult to obtain.
- Incorrect use, application, handling, installation, or maintenance of parts and components since they are less familiar to workers.
- Lengthy storage periods of items for which there is low demand and the resulting deterioration of the items during this time.

5.8.4 Interchangeability

This is an important maintainability design factor that is made possible through standardisation. *Interchangeability is defined as an intentional aspect of design, any component, part, or unit can be replaced within a given product or piece of equipment, by any similar component, part, or unit.*

There are two types of interchangeability:

(a) Functional interchangeability, and

(b) Physical interchangeability.

In functional interchangeability, two specified items are interchangeable as they serve the same function. In physical interchangeability, two items can be mounted, connected, and used effectively in the same locations and in the same manner.

Requirements of Interchangeability

Interchangeability requirements are as follows:

- Field use conditions or functionality
- The economy of design and manufacture
- Inspection and checking parameters

Principles of Iinterchangeability

Basic interchangeability principles are as follows:

- Liberal tolerances
- Uniformity
- Dimensional similarity

- In equipment, products, or units that require frequent replacement and servicing of component parts due to wear or damage, each and every part must be totally interchangeable with each and every other similar part.

5.9 METHODS OF STANDARDISATION

From Standard Institutions

The standards are available from sources such as Bureau of Indian Standards (Indian Standards Institution) or any standard institution. The same can be applied. Particularly, this method is more appropriate for an already existing product.

From Frequency Distribution

The frequency distribution of demand for each type and size of an item in terms of dimensions or quantity of materials will be the best starting point for standardisation and variety reduction in an organisation. For example, a company wants to serve soft drink in bottles. It can start with 100, 150, 200, 250, 300, 350 ml sized bottles. The statistics of consumption pattern may be taken. Suppose 200 ml and 250 ml bottles are the most preferred, then these two can be fixed as standard and the remaining can be gradually discontinued if these items are not critical from performance and safety angles.

From Renaud Standard Series

The preferred number series—a series of standard sizes in suitable steps used in bulbs and currency—developed by Renaud and known as R-5, R-10, R-29, R-40, etc. is another way of starting the standardisation process. The values of the series are available in the standards 1076 published by Indian Standards Institute.

5.9.1 How to Form an Indian Standard

The following steps are in practice for standardisation in India:

- If no standard exists for a product or process, the manufacturer can contact ISI (or BIS) for the formation of a standard.
- ISI themselves may initiate action.
- Usually any ISI member or any authoritative body may put forward the subject matter of a proposed standard.
- ISI examines the existing standards, if any, along with the other proposals, and identifies the necessity.
- ISI analyses the matter through its suitable division or council. ISI has number of division councils according to different subjects. Members of the council are drawn from the relevant industry and associations. The technical departments also assist the members.
- If the relevant provisional council accepts the necessity, preparation of the standard is then taken into account.
- While preparing the standard, care is to be taken in the following aspects.
  - Ensure that over specification is avoided.
  - Ensure that adequate attention is paid to convenience in handling and storage.
  - Specify the tests required to carry out at inspection stage.

  - Specify the special marking for the product and packing required to be done.
- A draft standard is prepared and sent to the concerned for their comments.

Despite over 15000 standards published by ISI so far, a lot of work is yet remaining. Many more standards have to be there in the offing. Further, as technology and market conditions change, existing standards also have to be redefined. Standardisation is, thus, a continuous and never ending process.

5.10 LEVELS OF STANDARDS

Various levels of standards are as follows:

- Individual level standards
- Industry-wise standards
- National level standards
- International level standards

(a) **Individual Level Standards:** Examine the following instances:

- A builder, contractor, or a real estate company lays down its own specifications to suit the special needs of a house, a dam, or a bridge.
- A corporate, a body, a government department sets its own rules and regulations (standard procedures).
- A private sector company or a factory produces certain items of spare parts with its own standards. Example, XYZ mosquito repelling machine will have its own standard for which the refill bottle of XYZ make only suits.

The instances described here belong to the individual level standards. Usually, these will have qualitative factors such as fragrance of perfume, colour preferences, finish of surfaces, fashions. Such individual level standards deal essentially with non-repetitive manufacturing or short-term productions or batch production.

(b) **Industry-wise Standards:** The standardisation activity is undertaken at the industry level, particularly when advances are made by an industry in any particular field of knowledge. In India, such work is undertaken by statutory bodies and authorities established by the government. The list of such bodies is given below.

Exclusively for the organisations run by Government:

(i) Directorate General of Supplies and Disposals
(ii) Research Design and Standards Organisation of Railways
(iii) Directorate of Standardisation for Defence Production

Other statutory bodies are as follows:

(i) Directorate of Marketing and Inspection
(ii) The Drug Controllerate
(iii) Chief Inspector of Explosives
(iv) Textile Commissioner
(v) Central Committee for Food Standards
(vi) Director General of Mines Safety

(vii) Export Inspection Council

(viii) Indian Road Congress

(c) **National Level Standards:** In India, after independence, in 1947 Indian Standard Institution (ISI) was formed with the help of Government of India (GOI). The major part of financial needs is met through grants by GOI and the balance needs are met by industry. Thus, this body is jointly managed by government and industry. The standards established by the institution are adopted throughout the country. The institution is headed by the Central Minister for Civil Supplies as its President and industry is represented by two Vice-Presidents. There are other bodies also in specific areas of specialisation to look after the standardisation activity. For example, Department of Weights and Measurements under the Union Ministry of Civil Supplies looks after the entire process of defining, enforcing, and monitoring standardisation in weights and measures. The National Physical Laboratory is entrusted with the responsibility of standardisation of measurement in coordination with International Organisation for Legal Metrology.

(d) **International Level Standards:** The ultimate goal of standardisation should be to achieve international standards in all technical matters related to economic activities between nations. In the year 1947 when ISI was born, International Organisation for Standardisation (ISO) was also instituted. It comprises over 70 national organisations including ISI and is a body responsible for general standardisation work at international level. Some other important bodies which specialise in specific areas are:

(i) **Organisation for Legal Metrology:** To standardise the practices and procedures in measurement.

(ii) **International Conference on Weights and Measures:** To define and standardise the units of measurement of physical quantities, weights, and measures.

(iii) **Codex Alimentarious Commission (CAC):** Set up jointly by Food and Agricultural Organisation (FAO) and World Health Organisation (WHO) to define and compile food standards for universal application.

Important Foreign Standards Admissible In India

1.	API	American Petroleum Institute
2.	ASA	American Standards Association
3.	ASTM	American Society of Test Materials
4.	BSS	British Standard Specification
5.	CSA	Canadian Standards
6.	DIN	German Standards
7.	GOST	USSR Standards
8.	ISO	International Organisation for Standardisation
9.	JIS	Japanese Industrial Standards
10.	NF	French Standards
11.	SAE	Society of Automobile Engineer
12.	UNI	Italian Standards

5.11 ADVANTAGES OF STANDARDISATION

Standardisation offers numerous advantages when considered from different angles. However, the advantages listed here are viewed from the maintenance aspect:

In Manufacturing vis-à-vis Maintenance

(a) **Cost reduction:** The most important advantage is reduction in unit cost. Standardisation limits the manufacturer to a few identified models and, hence, rationalising his product range to a few preferred sizes, shapes, etc.

(b) **Less tool cost and setting time:** If there are few models, obviously there will be fewer tool changes, reduced set up times and longer runs due to higher volume for each item.

(c) **Scope for design out maintenance and LRA:** Standardisation provides a large scope for the maintenance by Least Replaceable Assembly (LRA) units and Design Out Maintenance.

(d) **Increased mechanisation:** Since continuity of each operation for longer period is assured, there is greater justification for the manufacturer to introduce increased mechanisation, better tools, or special purpose plant and machinery.

(e) **Reduced material handling:** Material handling costs are also reduced. All these benefits result in higher productivity.

(f) **In achieving goals of TPM and TQM:** Standardisation assists in achieving the goals of TPM owing to the following reasons:

(i) Standard parameters

(ii) Unified dimensional tolerance

(iiii) Standardised composition of material (chemical/metallurgical)

(iv) Unified performance tests, (hence testing and inspection becomes easy and cheaper)

(v) Training the staff becomes simpler because of standardised repetitive operations

(vi) Since a manufacturer can expect repeated orders, production can be continued in anticipation of orders. This results in employment stabilization in the factories.

(g) **In spare parts purchasing and controlling**

(i) With standardisation, the buyer, the vendor, and the inspector are in common terms. So, disputes and controversies are eliminated and litigations are avoided.

(ii) Quality assurance through standards strikes a balance and guarantees economical buying.

(iii) Buying on brand names limits the choice of suppliers. But with standardisation, the purchase to brands is not necessary, so it extends the sources of supply and cuts across proprietary rights. Arrival of new vendors obviously reduces prices.

(iv) It also helps in improved availability.

(v) Standardisation helps in saving time and effort in screening of vendors.

(vi) Further, it helps the organisation in reduction of negotiation burden before purchase while reduction in inspection costs after purchase.

(vii) Standardisation is the best tool to reduce the variety of spare parts of similar or same type. Lesser spare parts inventory is more conducive to better planning and more accurate forecasting.

(viii) Better control over spare part inventory and a better spare part turnover ratio (annual requirements divided by average inventory).

(ix) Standardisation reduces the number of items, hence less inventory holding leading to less carrying costs. The control is better and easy if the number is less.

(x) Losses due to obsolescence and surplus can be better controlled.

(h) **In stores management of spares**

(i) Lesser number of items required to be stocked in smaller quantities

(ii) Quicker deliveries

(iii) Lower lead time

(iv) Smaller storage accommodation

(v) Capital cost on warehousing less

(i) **Other benefits of standardisation**

(i) Eliminates need for special or close tolerances

(ii) Reduces manufacturing costs, design time, and maintenance time and cost

(iii) Reduces the danger of incorrect usage of parts and, hence, enhances the safety

(iv) Facilitates 'cannibalizing' maintenance approaches

(v) Reduces procurement stocking and training

(vi) Leads to greater reliability

(vii) Reduces errors in wiring and installation by variations in characteristics of similar items

(viii) Reduces the chances of accidents that will stem from wrong or unclear procedure

5.12 BARRIERS TO STANDARDISATION

In spite of many advantages of standardisation listed above, an organisation insisting on a standardisation programme faces numerous difficulties. Some of these are listed below:

- It generally needs a lot of patience to persuade and dogged perseverance to carry out the mandate of standardisation.
- The production (or user) departments are often accustomed to using branded products and their prejudice is so strong that they are averse to trying new, thought standard.
- Similarly, if they have made a design of their own for a product, it is difficult to convince them that a different design can prove equally efficacious and at the same time be economical.
- The likes and dislikes of user departments dominate too immensely to result in lack of participation and cooperation.
- There may even be disagreement or resistance from the management if the company investment in favour of a particular product (or a design) is substantial even though another product or design is found to be more satisfactory from the standardisation point of view.
- A convincing demonstration is necessary that despite the investment, it is essential to switch over to the standard product.

5.13 APPLICATIONS OF COMPUTERS

Computers can be used in the maintenance department in various ways. A few applications are given below:

- Computers can be used for information flow about the breakdown of certain machines through e-mails.
- They are used for maintenance scheduling.
- They help in reliability calculation which is difficult and tedious to calculate manually.
- They are used for calculations of maintenance parameters such as MTBF, MTTF, MTTR, etc.
- They are used for calculations of mean down time (MDT), optimal maintenance interval, etc.
- They enhance maintenance cost analysis.
- They are used for maintenance information system.
- They are used in spare part management and controlling through ABC, VED, or such analyses.
- They are used for preventive maintenance scheduling.
- They are used in purchase of spares, for selection of vendors, cost comparison, stocking, etc.
- They are used for codification and standardisation of the parts.

With the use of computers, the following benefits can be experienced:

- Elimination of human error in maintenance requisition, maintenance scheduling, and maintenance reports.
- Reduction in record keeping of equipment and spare parts, and information regarding breakdowns and maintenance work undertaken.
- Reduction of paper work and clerical work.
- Reduction in inventory time and cost in preparation, identification, searching spares.
- Reduction of ordering cost, lead time of spare parts.
- Reduction in time in selection of vendors for spare parts, etc.

BUSTLE - 5.4

For the spare parts listed in Bustle 5.1 or 5.2, you might have prepared a codification system as asked in Bustle 5.3. Now check whether these parts are standard; if any of them are non-standard, standardise them.

____

____

____

____

Test Your Understanding 5.4

1. What is standardisation? What is its significance in maintenance spare parts management?
2. Discuss various the methods of standardisation.
3. What is inter-changeability? How is it relevant to standardisation of spare parts? Discuss the importance of inter-changeability in standardisation.
4. List the factors leading to a high failure rate in non-standard items.
5. Enumerate the objectives of standardisation.
6. Discuss different levels of standardisation.
7. List the advantages of standardisation.
8. What are the barriers to standardisation? Suggest some remedies to overcome the barriers.
9. Discuss few applications of computers in maintenance department.

5.14 CONCLUSION

Many a times, the downtime or non-availability of the equipment is attributed to the non-availability of the spares in time. At times, this must be a result of not adopting, not maintaining or not adhering to appropriate scientific controlling system. In maintenance department, this is one of the most important functions; particularly, when the company is on the lines of observing a condition monitoring or TPM. A correct policy and strategy adopted to the spare part management leads to improved availability by reducing the losses due to logistics and non-availability of spares. This chapter provides some guide lines in managing the stores of spare parts.

5.15 RECAP

This chapter mainly focuses on the classification and controlling of the spare parts after they are received to the plant engineering stores. The inventory management for procurement, EOQ inventory costs are felt as the pre-requisites to read this unit; the readers are expected to have prior knowledge of them and, therefore, they are not discussed here. However, this chapter discussed three inter-related topics of spare parts management, viz. selective inventory control techniques, codification systems and standardisation of spare parts. Various selective inventory control systems such as ABC, VED, SDE, HML, SOs, GOLF, FSND, VEIN, CIN are exhaustively dealt. Discussions are extended to the maintenance system optimisation and other relevant applications. The codification and classification systems are then explained including parts classification, Opitz classification, multi-class and general classifications. The standardisation, its significance, methods, levels, advantages and barriers are described with a special reference to the maintenance department. Finally, the unit ends with listing of some applications of computers in maintenance department.

5.16 KEY CONCEPTS

1. **Spare Part:** A spare part is the part or component identical to the part (or component) of the equipment that needs replacement due to wear and tear as the life of the part is less than the operating life of the equipment.

2. **ABC Analysis:** Selective control analysis of spare parts based on annual cost consumption, **A Class** (60–70% cost, 10% consumption) spares are most valuable and requires more attention; **B Class** (20–30% cost, 20–30% consumption) spares are intermediate and requires moderate care; **C Class** (10% cost , 70% consumption) spares are least valuable and so do not need much attention.
3. **SDE Analysis:** Selective control analysis of spare parts based on availability; **S** for 'scarce', **D** for 'difficult', **E** for 'easily' available items.
4. **FSND (or FSN) Analysis:** Selective control analysis of spare parts based on consumption, usage, or demand; **F** for 'fast moving', **S** for 'slow-moving', **N** for 'non-moving', **D** for 'dead' items.
5. **VED Analysis:** Selective control analysis of spare parts based on significance; **V** for 'vital', **E** for 'essential', **D** for 'desirable' items.
6. **CIN Analysis:** Selective control analysis of spare parts based on criticality; **C** for 'critical', **I** for 'important', **N** for 'normal' machines.
7. **VEIN Analysis:** Selective control analysis of spare parts based on criticality; **V** for 'vital' items, **E** for 'essential' items, **I** for 'important', **N** for 'normal' items. Normal equipment has low criticality.
8. **SOs Analysis:** Selective control analysis of spare parts based on seasonality; **S** for 'seasonal' items; **Os** for 'off seasonal' items.
9. **GOLF Analysis:** Selective control analysis of spare parts based on the type of market in which the spare part is available; **G** for 'government market'; **O** for 'open market'; **L** for 'local market' and **F** for 'foreign market'.
10. **HML Analysis:** Selective control analysis of spare parts based on the cost of the spare part; **H** for 'high cost', **M** for 'medium cost' and **L** for 'low cost' items.
11. **XYZ Analysis:** Selective control analysis of spare parts based on value of the part; **X** for low value, **Y** for medium value, and Z for high value items.
12. **Two-dimensional Classification:** The inventory classification using any two of the selective control analyses such as ABC–VED and arranging in two dimensions.
13. **Music 3-D Model:** The inventory classification using any three of the selective control analyses, such as ABC–VED–FSN and arranging in three dimensions.
14. **Codification:** A systematic method in alpha-numeric codes given to generalise all the parts for their easy identification and traceability.
15. **Part Classification System:** A kind of coding system for part. It is in two structures, viz. hierarchical and chain-type structure.
16. **Hierarchical Code System:** The interpretation of the second digit will depend on the value of the first digit.
17. **Chain-type Code Structure:** The interpretation of the second digit will not depend on the value of the first digit.
18. **The Opitz classification system:** The coding system developed by H. Opitz, Germany uses 13 digit sequence with Basic code 9 digits (1 to 9) of which the first five digits (first to fifth) are Form Code to describe the design attributes and the next four digits (i.e., sixth to ninth) are supplementary code to describe manufacturing and the last four digits (tenth to

thirteenth positions) are secondary code, used to identify the production operation type and sequence.

19. **The Multi-class Coding and Classification System:** It uses hierarchical or decision-tree coding structure and is developed by the Organisation for Industrial Research.
20. **General Coding System (for spare parts):** The spare parts coding system using local codes of the organisation.
21. **Standardisation:** Standardisation is the orderly and systematic formulation, adoption, application and review of industrial standard which leads to simplification or variety reduction.
22. **Interchangeability:** Interchangeability is defined as an intentional aspect of design, any component, part, or unit that can be replaced within a given product or piece of equipment, by any similar component, part, or unit.
23. **Renaud Standard Series:** The preferred number series of standard sizes in suitable steps developed by Renaud such as R-5, R-10, R-29, R-40, etc. (available in the standards 1076 published by the Indian Standards Institute).
24. **ISI:** Indian Standards Institution.
25. **BIS:** Bureau of Indian Standards.

REVIEW QUESTIONS

1. Enumerate the functions of spare parts management with reference to Plant Engineering. What are its distinguishable features of maintenance spares as compared to general spares?
2. Describe the procedure of ABC analysis taking a hypothetical example with maintenance spares.
3. Compare various selective inventory control techniques with reference to their basis and applications.
4. What is maintenance system optimisation? Discuss the application of selective inventory control techniques to maintenance system optimisation.
5. What are the requirements for maintenance system optimisation? Discuss the implications of improper maintenance mix.
6. Discuss the importance and advantages of the following concepts in maintenance spares.
   (a) Codification
   (b) Standardisation
   (c) Interchangeability
7. Enumerate the barriers to the standardisation of maintenance spares. How do you overcome?

CHAPTER 6

Facility Planning and Plant Layout

PREAMBLE

This chapter emphasises on the concept of facility and determines the situation when the decision of facility planning arises. It is greatly swayed by the type of the industry that they support particularly, in the decisions such as product design, plant layout, buildings and grounds, location planning, size of the facility, expansion, diversification, organisation development, etc. All the factors influencing the facility planning and considerations in facility location planning are explained in detail.

CHAPTER OUTLINE

6.1 INTRODUCTION

Facility planning and plant layout issue is an area of arranging facilities such as equipment on a shop floor, departments or sections, etc., inside the plant or work place. It is one of the most critical strategic decisions for the following reasons:

- Plant layout is generally a one-time activity since it is very difficult to rearrange facilities frequently.
- It requires long-term vision about factory so that minimal dislocations or disturbances occur when the factory expands or goes through minor changes in process, production schedule, or production mix.
- Inappropriate arrangements of facilities result in more travel time between machines, which result in more throughput time, more work-in-process, more material handling, etc.

6.2 OBJECTIVES OF GOOD FACILITY PLANNING

The objectives of facility planning are also multi-dimensional. These are listed below:

(a) **Objectives related to work place:**

- Suitable design or workstations and their proper placement.
- Maintaining the sequence of operation on parts by locating succeeding facilities adjacently.
- Safe working conditions with reference to ventilation, lighting, etc.
- Minimum movement or workers.
- Least chances of accidents, fire, etc.
- Appropriate space for machine, workers, tools, etc.
- Optimum utilisation of space (vertical height) available in the plant.

(b) **Objectives related to material:**

- Fast travel of material inside the factory without congestion or bottleneck.
- Less material handling and minimum transportation cost.
- Reduced waiting time for in-process inventory.

(c) **Performance related objectives:**

- Simpler plant maintenance.
- Increased productivity, better product quality and reduced cost.
- Least set-up cost and minimal changeover.
- Exploitation of buffer capacity, common worker for different machines, etc.

(d) **Objectives related to flexibility:**

- Scope for future expansion.
- Considerations for valid product mix.
- Considerations for alternate routine.

6.3 PRINCIPLES OF FACILITY LAYOUT

Facility layout is a multi-criteria problem faced while taking decisions on numerous factors such as location of work centres, offices, computer centre, design and drawing section, tool-room storage space and many more. However, a good facility layout decision involves

- Least material handling cost
- Worker effectiveness
- High productivity and effectiveness
- Group technology

(a) **Least Material Handling Cost**

For less material handling, the following approach is adopted:

- Placement of facilities as per sequence of part operations.
- Proper location of store and packaging area.
- Smooth and continuous flow of material.
- Utilisation of vertical space in material transport.

(b) **Worker Effectiveness**

For better worker effectiveness, the following approach is adopted:

- Integration of similar facilities at the same location. For example, all milling machines may be placed at the same locations.
- Sufficient space for worker movement.
- Sufficient space for the movement of material between machines.
- Safe working environment.

(c) **High Productivity and Eeffectiveness**

For high productivity and effectiveness, the following approaches are adopted:

- Flexibility to expand, re-routing, varied product mix, etc.
- Scope to use multi-skilled worker.
- Proper utilisation of facilities and worker resources, etc.
- Removal of bottlenecks, roadblocks of any form, under-utilisation (or over-utilisation) of resources, etc.

(d) **Group Technology**

For deriving the group technology benefits, the following approaches are adopted:

- Group similar parts in one family.
- For each part family, design a machine-cell. Machine-cell is a combination of machines, which caters to all the processing requirements of corresponding part family.
- As far as possible inter-cell transfer of parts.

6.3.1 Muther's Principles of Layout

According to Muther, there are six basic principles of best layout. These are:

1. **Principle of Over-all Integration:** According to this principle, the best layout is one which integrates all the resources such as men, materials, machinery, supporting activities, etc., to result in the best compromise.
2. **Principle of Minimum Distance:** According to Muther, the best layout is that in which men and materials have to travel the most minimum distance between operations/operating centres.
3. **Principle of Flow:** Muther said that, other things being equal, the best layout should ensure the arrangement of the work area for each operation or process in the same order or sequence that forms, treats, or assembles the materials.
4. **Principle of Cubic Space:** According to this principle, the layout in which all the available space both vertical and horizontal is most economically and effectively used.
5. **Principle of Satisfaction and Safety:** Other things being equal, the best layout must ensure work satisfaction, pleasant and safe working conditions to workers.
6. **Principle of Flexibility:** The best layout is one that can be adopted and rearranged at a minimum cost and with least inconvenience.

BUSTLE - 6.1

List various facilities that are required for establishing the following. Prepare a report on each and compare the significance of each factor.

(a) An educational institution of 10+2 standard with separate hostels for girls and boys.

(b) A technical educational institution with hostel.

(c) A process industry.

(d) A product industry.

____

____

____

____

Test Your Understanding 6.1

1. Enumerate the objectives of a facility planning.
2. Discuss the role of facility planning in the following areas:
   (a) Material handling
   (b) Worker's efficiency
   (c) Productivity
   (d) Group technology
3. List Muther's principles of layout.

6.4 WHEN DOES FACILITY PLANNING DECISION ARISE?

Facility location decisions are strategic, long-term, and non-repetitive in nature. Without sound and careful location planning in the beginning itself, the new facility will pose continuous operating disadvantages for the future operations. Location decisions are affected by many factors, both internal and external, of the organisation. Internal factors include the technology used, the capacity, political, economic, and social conditions in various localities. Most of the fixed and some of the variable costs are determined by the location decision. The efficiency, effectiveness, productivity, and profitability of the facility are also affected by the location decision. The facilities location problem is concerned primarily with the best (or optimal) location depending on appropriate criteria of effectiveness. Location decisions are based on a host of factors; some subjective, qualitative, and intangible while some others are objective, quantitative, and tangible.

6.4.1 Concept of a Facility

Traditionally, location theorists have dealt with industrial plant/factory location. However, the concept of plant location has now been generalised into that of facility location, since facility includes a production operation or service system. The term 'plant' has been traditionally used synonymous to a factory, manufacturing, or assembly unit. This includes fertilizer, steel, cement, rice milling plants, textile, jute, sugar mills, rubber factories, breweries, refineries, thermal, or hydro-electric nuclear power stations, etc. However, with the enlarged scope of a facility, this term can now be used to refer to banks, hospitals, blood banks, fire stations, police stations, warehouse, godown, depot, recreation centre, central repair workshop, etc. In some cases, machines, equipment, desks, workshop, canteen, emergency room, etc. can also mean a facility. Thus, in fact, this is when a facility planning decision arises.

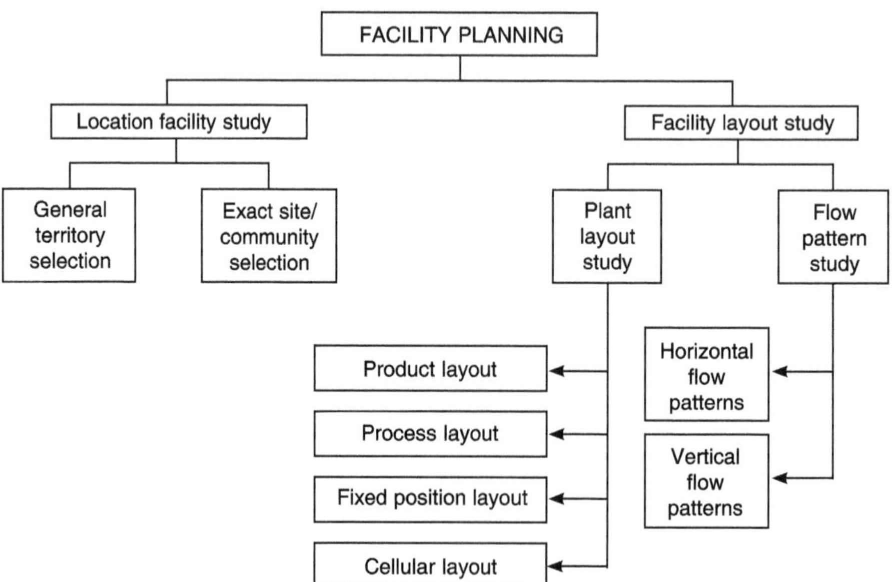

Figure 6.1 Classification of facility planning

In fact, facility planning is composed of two important decisions. Firstly, facility planning involves selection of a place or location, which has maximum facilities. Then, the second decision is to ensure optimising the facilities and converting them into utilities or amenities. In other words, it is to decide how to and where to arrange/position them. In simplest words, it is to layout the available facilities so that greater ease, higher convenience, better comfort, and economic, efficient, and effective usage may be attained. Thus, facility planning deals with—

1. Facility location study
2. Facility layout study

The classification of facility planning is diagrammatically represented in Figure 6.1.

BUSTLE – 6.2

1. *Have you ever come across any situation related to facility planning decision in the organisation where you are working or in the institution where you are studying or a general situation for a ceremony? Brief out the situation and your decisions.*

___

___

2. *Suppose you have been transferred to a new place and are in search of a rented house to live with your family. What sort of facilities would you prefer to have within your vicinity? Explain your answer with an example or situation you have faced in your life hither to. If you have not yet faced such a situation, think of such and answer.*

___

___

3. *Suppose you are working in marketing of consumer durable goods and shortly you are going to set up another go-down-cum marketing centre at another place. The options available are*

   Site A: *A small village 100 km away from Hyderabad (Andhra Pradesh),*

   Site B: *Outskirts of Hyderabad, with pollution-free environment,*

   Site C: *Hyderabad/Secunderabad, reasonably well-planned city with pleasant weather almost throughout the year, and*

   Site D: *Vijayawada, a big town well connected with trains and road transport, a business centre about 300 km away from Hyderabad.*

   Conduct a Strength-Weakness-Opportunity-Threat (SWOT) analysis for each of the sites.

Site	*Strengths*	*Weaknesses*
Site A		
Site B		
Site C		
Site D		
Site	*Opportunities*	*Threats*
Site A		
Site B		
Site C		
Site D		

Test Your Understanding 6.2

1. Discuss certain situations which necessitate the facility planning decision?
2. What do you understand by facility? What is facility planning?

6.5 FACILITY LOCATION STUDY

Location of facilities is a problem associated with the planning phase of a factory or even a service sector. Whether it is small entrepreneurship or a big industrial estate, hospital, or a fashion designer shop, school or a five star hotel, one has to start first with the planning activity with the first and foremost question—

Where should the site be located so that no changes are needed for years to come?

It is a very vital decision, which has long-term implication. This type of decision is called a strategic decision. It is not very easy to answer a location problem. The reasons are

- Uncertainty in future.
- Complexity and conflicting factors associated with the site selection problem.
- Constraints and limitations of resources to procure a site.

Let us consider two sites for location of a new factory. Site A is nearer to market but far from the raw material source; site B is otherwise. Site A is a rural location with easy availability of land and electric power at cheaper cost. Thus, list the factors which will have better advantages at site A while other factors that will be advantageous at site B. This will help in making a comparative analysis and, thus, arriving at a decision on the basis of the requirement of the factors available at the sites. Thus, the decision of site selection depends on several factors and options that are available and required in that context with their relative strengths and weaknesses and on an integrated framework. Various factors that are worth considering while planning location are summarised and tabulated under different categories as shown in Table 6.1.

Table 6.1 Factors to be considered for location planning

Category	*Factors*
Process Input	Raw material, personnel, transportation of raw material, work force availability, availability of water and power, road transport, railways, etc.
Process Output	Market nearness.
Process Characteristics	Environmental factors such as pollution, noise, etc., weather (e.g., knitting industry), level of humidity and season, rainfall.
Personal Preference	Preference of executives and entrepreneur.
Government Policy	Tax exemptions, legal requirements, incentives, availability of loan/land, etc.
Local Conditions	Community, culture and attitude, past history of industry located in the area, incidents of labour unrest in the area, political interference, etc.
Cost Factors	Cost of land, cost of transportation, wages of unskilled labour.
Competition	Location of other industries in the area, market forces for competition, etc.
Intangible Factors	International considerations, possibility for expansion and growth, school, churches, medical, recreation facilities, etc.

6.5.1 Considerations in Plant Location

Perhaps, there is hardly any plant on this globe that is located at a place, which fulfills all the criteria of perfect location. Some factors are compromised to take advantage of the other factors. For example, if the raw material is quite bulky and difficult, the plant is generally considered to locate nearer to raw material source and often this location will be away from its market. Many iron and steel industries are geographically located near the place where iron ore is found. Jamshedpur, Bokaro, and Rurkela are the places where iron ore are found. To run the power plants, coal is needed and, hence, many NTPC power plants are located near the place where coal is found. However, some other factors are compromised such as market, transportation cost, labour availability, and so forth. In general, the following factors are considered in a plant location decision:

- Cheap availability of land.
- Availability of water.
- Nearness to raw material source.
- Nearness to market or consumer.
- Good transportation facility.
- Availability of labour, both skilled and unskilled, at comparative wages.
- Availability of fuel and power.
- Suitable climatic conditions.
- Community receptivity to business.
- Construction cost.
- Quality of life: housing, recreation, school, climate, crèche, etc.
- Taxes and government regulations.
- Site for waste disposal and environmental regulation.
- A vailability of sub-contractors, industrial climate, and organised industrial complexes.

6.5.2 Comparative Study of Rural and Urban Sites

After the general territory for a site has been identified, specific site location is the major issue. The alternative sites may have one of the following locations:

- Urban or city like Bangalore, Chennai, Delhi, Hyderabad, Kolkata, Mumbai, etc.
- Suburban location like Patancheru near Hyderabad; Faridabad near Delhi, etc.
- Country or rural location.

Urban locations have major advantages such as:

- Easy to locate and retain professional managers and engineers.
- Power and water supply is available.
- Easy to rent a building rather than building a new one.
- Supporting industries and ancillaries may be near by.
- Better quality of life.
- Market or consumers are near by.
- Well-connected by rail, road, and communication network, such as telephones, telex, fax, e-mail, Internet, etc.

Table 6.2 Comparison of urban, suburban, and plant locations

Factors	*Urban/City Site*	*Suburban*	*Rural/Country Site*
Availability of land	Less	Moderate	Plenty
Cost of land	High	Moderate	Low
Connected by rail/road, etc.	Well	Moderate	Poor
Availability of labour	Less	Moderate	More
Wages of labour	More	Moderate	Less
Communication network like the Internet, facsimile, telephone, e-mail, etc.	Very good	Moderate	Poor
Power and water availability	Good	Moderate	Poor
Supporting industries and ancillaries units	Near by	Moderate	Far
Quality of life, such as recreation, school, hospital, etc.	Very good	Moderate	Poor
Market and consumer	Near by	Moderate	To be built
Building for site	May be rented or built	Rented	Poor
Availability and retaining potential for professional like manger and engineers	Better	Normal	Not available outside
Training of workers and foreman	In local institutions	May or may not	Less
Security	Better	Moderate	Less
Expansion of site	Difficult	Moderate	Easy
Taxes	More	Moderate	Less
Government support	Less	Normal	More
Union problem and industrial relation	Poor	Normal	Better
Pollution and environmental constraints	More	Normal	Less
Waste disposal	Difficult	Normal	Easy
Location of sub contractors, retailers, etc.	Nearby	Normal	Far
Incentives from financial institutions like banks, IDBI, IFCI, etc.	Less	Normal	More

Thus, the urban and the rural sites have their own strengths and weaknesses. A compromise can be the suburban site. These sites are located near to the cities. For example, Faridabad, Gurgaon, and Ghaziabad are located in close proximity of Delhi. Patancheru, Nacharam, etc. are near to Hyderabad. Many industries have come up in these areas. Maruthi, BTL are located in Gurgaon; ABB, Escort, Goodyear are located in Faridabad. We can find Pennar Industries Ltd (PIL), HBL at Patancheru, ECIL, Leather industries are at Nacharam area. Similarly, near Allahabad city many industries are located in Nani.

6.6 FACILITIES GOVERNING SELECTION OF LOCATION

Some of the industrial needs and characteristics that tend to favour each of these locales are discussed here.

6.6.1 Facilities of a City Location

Facilities governing choice of a city location are

- Availability of adequate supply of labour force.
- High proportion of skilled employees.
- Rapid public transportation and contact with suppliers and customers.
- Small plant site or multi-floor operation.
- Processes heavily dependent on city facilities and utilities.
- Good communication facilities like telephone, telex, post offices.
- Good banking and health care delivery system.

6.6.2 Facilities of a Suburban Location

Facilities governing the choice of a suburban location are

1. Large plant site close to transportation or population centre.
2. Free from some common city building zoning (industrial areas) and other restrictions.
3. Freedom from higher parking and other city taxes, etc.
4. Labour force required resides close to plant.
5. Community close to, but not in, large population centre.
6. Plant expansion is easier than that in a city.

6.6.3 Facilities of a Country/Rural Location

Facilities governing the choice of a country/rural location are

1. Large plant site required for either present demands or expansion.
2. Dangerous production processes.
3. Lesser effort required for anti-pollution measures.
4. Large volume of relatively clean water.
5. Lower property taxes, away from Urban Land Ceiling Act restrictions.
6. Protection against possible sabotage or for a secret process.
7. Balanced growth and development of a developing or underdeveloped area.
8. Unskilled labour force required.
9. Low wages required for meeting competition.

BUSTLE - 6.3

Suppose the institution where you are studying (or the company, where you are working) is going for expansion. Assume that you are deputed as the team manager for location study. Choose any four suitable places and make a thorough study. Write your findings and analysis/results here.

____

____

____

____

Test Your Understanding 6.3

1. What are the considerations needed during facility location planning?
2. Why is facility location planning treated as one of the critical situations in an industry? Discuss in detail.
3. Give a comparative statement in location study of a rural site and an urban site.
4. What facilities would influence (both favourable and/or unfavourable) the location decisions in the following case?
   (i) City/urban sites
   (ii) Suburban sites
   (iii) Rural/countryside sites.

6.7 STEPS IN THE FACILITY LOCATION STUDY

Location studies are usually made in two phases:

1. the general territory selection phase, and
2. the exact site/community selection phase amongst those available in the general locale.

The considerations vary at the two phases , though there is a substantial overlap as shown in Table 6.3.

Table 6.3 Overlap of considerations of factors in the two phases of facility location

SL. No.	*Location Factors*	*Phase I General Territory Selection*	*Phase II Particular Selection of Site and Community*
1.	Market	√	
2.	Raw materials	√	
3.	Power	√	√
4.	Transportation	√	√
5.	Climate and fuel	√	
6.	Labour and wages	√	√
7.	Laws and taxation	√	√
8.	Community services and attitude		√
9.	Water and waste		√
10.	Ecology and pollution		√
11.	Capital availability	√	√
12.	Vulnerability to enemy attack	√	√

A typical team studying location possibilities for a large project might involve economists, accountants, geographers, town planners, lawyers, marketing experts, politicians, executives, industrial engineers, defense analysts, ecologists, etc. It is indeed an inter-disciplinary team that should be set up for undertaking location studies.

6.7.1 Territory Selection

Now in phase (I) for the general territory/region/area selection, the following are some of the important factors that influence the selection decision:

(a) **Markets:** There has to be some customer/market for your product/service. The market growth potential and the location of competitors are important factors that can influence the location. Locating a plant or facility nearer to the market is preferred if promptness of service is required, if the product is fragile, or is susceptible to spoilage. Moreover, if the product is relatively inexpensive and transportation costs add substantially to the cost, a location close to the markets is desirable. Assembly type industries also tend to locate near markets.

(b) **Raw Material and Supplies:** Sometimes accessibility to vendors/suppliers of raw materials, parts supplies, tools, equipment, etc., may be very important. The issue here is promptness and regularity of delivery and inward freight cost minimisation. If the raw material is bulky or low in cost, or if it is greatly reduced in bulk, viz. transformed into various products and by-products or if it is perishable and processing makes it less so, then location near raw materials sources is important. If raw materials come from a variety of locations, the plant/facility may be situated so as to minimise total transportation costs. The costs vary depending upon specific routes, mode of transportation, and specific product classifications.

(c) **Transportation Facilities:** Adequate transportation facilities are essential for the economic operation of a production system. For companies that produce or buy heavy, bulky, and low value per ton commodities, water transportation can be an important factor in locating plants. It can be seen that most civilisations grew along rivers/waterways, etc. Many facilities/plants are located along riverbanks.

(d) **Manpower Supply:** The availability of skilled manpower, the prevailing wage pattern, living costs and the industrial relations situation influence the location.

(e) **Infrastructure:** This factor refers to the availability of common minimum amenities and infrastructural facilities such as schools to children, living conditions, availability of common materials, and commodities.

(f) **Legislation and Taxation:** Factors such as financial and other incentives for new industries in backward areas or no-industry-district centres, exemption from certain state and local taxes, octroi, etc., are important.

(g) **Climate:** Climatic factors can dictate the location of certain type of industries like textile industry which requires high humidity zones.

6.7.2 Site/Community Selection

Having selected the general territory/region, the next step is to determine Phase II, i.e., site/community selection. In addition to the factors in common consideration as shown in the Table 6.3 and discussed in the above paragraphs (under territory selection), there are some more factors that influence the site selection decisions. Let us discuss some factors relevant to this phase:

(a) **Community Facilities:** These involve factors such as quality of life which in turn depends on availability of facilities like schools, places of worship, medical services, police and fire stations, cultural, social and recreation opportunities, housing, good streets and good communication, and transportation facilities.

(b) **Community Attitudes:** These can be difficult to evaluate. Most communities usually welcome setting up of a new industry especially because it will provide opportunities to the local people directly or indirectly. However, in case of polluting or 'dirty' industries, they will try their utmost to locate them as far as possible. Sometimes because of prevailing law and order situation, companies have been forced to relocate their units. The attitude of people as well as the state government has an impact on industrial location.

(c) **Waste Disposal:** The facilities required for the disposal of process waste including solid, liquid, and effluents effluents need to be considered. The plant should be positioned such that prevailing winds carry any fumes away from populated areas and waste may be disposed off properly and at reasonable expense.

(d) **Ecology and Pollution:** These days there is a great deal of awareness towards maintenance of natural ecological balance. There are quite a few agencies propagating the concepts to make the society at large more conscious of the dangers of certain avoidable actions.

(e) **Site Size:** The plot of land must be large enough to hold the proposed plant and parking and access facilities and provide room for future expansion. These days a lot of industrial areas/parks are being earmarked in which certain standard sheds are being provided to entrepreneurs (especially small-scale ones).

(f) **Topography:** The topography, soil structure, and drainage must be suitable. If considerable land improvement is required, low priced land might turn out to be expensive.

(g) **Transportation Facilities:** The site should be accessible by road and rail preferably. The dependability and character of the available transport carriers, frequency of service and freight, and terminal facilities is also worth considering.

(h) **Supporting Industries and Services:** The availability of supporting services such as tool rooms, plant services, etc., need to be considered.

(i) **Land Costs:** These are generally of lesser importance, as they are non-recurring and possibly make up a relatively small proportion of the total cost of locating a new plant. Generally speaking, the site will be in a city, suburb, or country location. In general, the location for large-scale industries should be in rural areas, it helps in regional development also. It is seen that once a large industry is set up (or even if a decision to this effect has been taken), a lot of infrastructure develops around it as a result of the location decision. The location of medium-scale industries can be preferred in the suburban/semi-urban areas where the facilities and advantages of both the urban and rural areas are available. For the small-scale industries, the preferred location will be urban areas where the infrastructure facilities are already available. However, in real life, the situation is somewhat paradoxical as people—rich and affluent are usually based in the cities and will like to locate the units in the city itself.

BUSTLE - 6.4

1. Identify the factors that are affecting you and your way of life in view of the location of the organisation/ institution where you are working/studying. Identify both pros and cons of the location.

Pros:

Cons:

1. In the above question, identify the factors that are merits/demerits to your organisation/institution.

Merits:

Demirits:

Test Your Understanding 6.4

1. List the factors influencing the decisions in exact site selection of location planning. Explain.
2. List the factors influencing the decisions in general territory selection of location planning. Explain.
3. Write a detailed essay on location planning.

6.8 SUBJECTIVE, QUALITATIVE, AND SEMI-QUANTITATIVE TECHNIQUES FOR FACILITY LOCATION PLANNING

Three subjective techniques used for facility location are industry precedence, preferential factor, and dominant factor. Most of us always look for some precedents. So in the industry precedence subjective technique, the basic assumption is that if a location was best for similar firms in the past, it must be the best in the present also. As such, there is no need for conducting a detailed location study and the location choice is, thus, subject to the principle of precedence—good or bad. However, in the case of the preferential factor, the location decision is dictated by a personal factor. It depends on the individual's whims or preferences, e.g., if one belongs to a particular state, he may like to locate his unit only in that state. Such personal factors may override the factors of cost or profit in taking a final decision. This can hardly be called a professional approach though such methods are probably more common in practice than generally recognised. However, in some cases of plant location there can be a certain dominant factor (in contrast to the preferential factor) which influences the location decision. In a true dominant sense, mining or petroleum drilling operations must be located where the mineral resource is available. The decision in this case is simply whether to locate or not at the source.

For evaluating qualitative factors, factor ranking and factor weight rating systems can be used. In the ranking procedure, a location is better or worse than another for the particular factor. By weighing factors and rating locations against these weights, a semi-quantitative comparison of location is possible. Let us now discuss some specific methods.

6.8.1 Equal Weights Method

Equal weights to all factors can be assigned and each location can be evaluated along the factor scale. For example, Bunsen, a manufacturer of fabricated metal products selects three factors by which it has to rate four sites. Each site is assigned a rating of 0 to 10 points for each factor. The sum of the assigned factor points constitute the site rating by which it can be compared to other sites.

Table 6.4 Decision matrix

Factor\Potential sites	*S1*	*S2*	*S3*	*S4*
F1	4	5	9	2
F2	3	4	8	3
F3	2	3	7	6
Site rating (Total)	9	12	24	11

Table 6.4 shows that site S3 has the highest site rating of 24. Hence, this site will be chosen.

6.8.2 Variable Weights Method

The above method can be utilised on account of giving equal weightage to all the factors and evaluating each location site along the factor scale. Hence, factor F1 might be assigned 300 points, factor F2 might be assigned 100 points and factor F3 might be assigned 50 points. Thus, the points scored, out of the maximum assigned to each of the factors, for each possible location site can be obtained and again the site rating can be derived as follows:

Table 6.5 Decision matrix

Factor\Potential sites	*Max. Points*	*S1*	*S2*	*S3*	*S4*
F1	300	220	260	250	50
F2	100	40	60	70	100
F3	50	15	50	20	40
Site rating (Total)		285	370	340	190

Now Table 6.5 shows that site S2 has the highest site rating of 370. Hence, this site will be chosen.

6.8.3 Weight Cum Rating Method

Weight cum rating method is another method of evaluating a potential location site. Variable weights can be assigned to each factor. The location point assignment for the factor is then obtained by multiplying the location rating for each factor by the factor weight. For example, rating weights of one to five can be assigned to the three factors F1 (labour climate), F2 (community facilities), and F3 (power availability and reliability), as 5, 3, 2, respectively. Now for each of the factors, sites S1, S2, S3, or S4 can receive 0 to 10 points as follows. Now each site rating can be obtained.

Table 6.6 Decision matrix

Factor\Potential sites	*Factor rating weights*	*S1*	*S2*	*S3*	*S4*
F1	5	2	5	9	2
F2	3	3	3	8	3
F3	2	6	2	7	3
Site rating (Total)		31	38	83*	25

*Sample calculation: $5 \times 9 + 3 \times 8 + 2 \times 7 = 83$

As shown in Table 6.6, the sample calculation should hopefully suffice to obtain the site ratings. Hence, site S3 with the highest rating of 83 is chosen.

6.8.4 Factor-point Rating Method

In this method, the first step is to establish a subjective scale common to all factors. Then assign points against the subjective scale for each factor and assign the factor points of the subjective rating for each factor. For example, five subjective ratings, poor, fair, adequate, good, and excellent are selected to be used in evaluating each site for each factor. For each of the factors, 'adequate' is assigned a value zero and then negative and positive relative worth weights are then assigned a subjective ratings below and above adequate for each factor as shown in Table 6.7.

Table 6.7 Factor-point ratings sample

Factor	*Desctipton*	*Poor*	*Fair*	*Adequate*	*Good*	*Excellant*
F1	Water supply	– 15	– 12	0	6	10
F2	Appearance of site	– 3	– 1	0	1	2

The range between minimum and maximum weights assigned to a factor is equivalent to the the effect weighs. Each location site S1 to S4 are then rated by selecting the applicable subjective rating for each of the factors for each location and the equivalent points of that subjective factor rating assigned to the factor. Thus, we can now obtain the rating as shown in Table 6.8.

Table 6.8 Decision matrix

Factors	*Site1*	*Site2*	*Site3*	*Site4*
F1	Adequate (0)	Fair (–12)	Good (6)	Adequate (0)
F2	Adequate (0)	Poor (–3)	Excellent (3)	Fair (–1)
F3	Adequate (0)	Adequate (0)	Adequate (0)	Adequate (0)
Total site rating	0	–15	9	–1

Accordingly, Site3 with the highest rating of 9 will be chosen.

In most cases, hardly any attempt is made to establish a direct relationship between the site rating point value and the cost values. Usually, this is left to the management.

The location analyst presents both the cost and the intangible data results to the management. In such cases, management can take a decision based on a simple composite measure method illustrated below with the aid of a numerical example.

6.8.5 Composite Measure Method

Let us list the steps of the composite measure method.

Step-1: Develop a list of all relevant factors.

Step-2: Assign a scale to each factor and designate some minimum.

Step-3: Weigh the factors relative to each other in light of importance towards achievement of system goals.

Step-4: Score each potential location according to the designated scale and multiply the scores by the weights.

Step-5: Total the points for each location and either (a) use them in conjunction with a separate economic analysis, or (b) include an economic factor in the list of factors and choose the location on the basis of maximum points.

Now let us illustrate the composite measure method with a numerical example. There are three potential sites and five relevant factors like transportation costs per week, labour costs per week, finishing material supply, maintenance facilities, and community attitude. The costs are in rupees whereas for the last three factors, points are assigned on 0–100 scale. The data collected is shown in Table 6.9.

Table 6.9 Payoff Matrix

Description	*Factor*	*Site 1*	*Site 2*	*Site 3*
Transportation cost/week (₹)	F_1	800	640	580
Labour cost/week (₹)	F_2	1180	1020	1160
Finishing material supply (Points)	F_3	30	80	70
Maintenance facilities (Points)	F_4	60	20	30
Community attitude (Points)	F_5	50	80	70

The location analyst has pre-established weights for various factors. This includes a standard of 1.0 for each ₹ 10 a week of economic advantage. Other weights applicable are 2.0 on finishing material supply, 0.5 on maintenance facilities and 2.5 on community attitudes. Also the organisation prescribes a minimum acceptable score of 30 for maintenance facilities.

First of all, let us look at the economic factors F_1 and F_2 for which monetary values were possible. If we sum up the costs for each site, we get the costs for sites S_1, S_2, and S_3 as ₹ 1980, ₹ 1660, and ₹ 1740, respectively. Thus, Site 1 will be the worst cost wise. Site 2 will have an economic advantage over Site 1 by a margin of ₹ 320 (1980–1660). Similarly, Site 3 will have an economic advantage over Site 1 to the extent of ₹ (1980–1740), i.e., ₹ 240. Now monetary value in rupees can be converted to a point scale as you would realise that a standard of 1.0 is to be assigned for each ₹ 10 per week of economic advantage. Thus, the following site rating can be obtained as shown in Table 6.10.

Table 6.10 Decision matrix

Factors	*Weightage*	S_1	S_2	S_3
Combined economic advantage ($F_1 + F_2$)	1.0	0	32	24
F_3	2.0	30	80	70
F_4	0.5	60	20	30
F_5	2.5	50	80	70
Composite site rating (Total)		215	402	354

Now, on referring to certain prerequisites for certain factors, because of the nature of the situation, a constraint in the form of a site scoring at least 30 on account of maintenance has been given. It can be observed that this basic requirement is not met by S_2. In fact, any further calculations for S_2 need not have been carried out as soon as one detected this flaw. However, we have deliberately persisted on going through all the calculations. There could have been the possibility of revision of maintenance clause constraint *viz*., perhaps, it might have felt that a bare minimum score of 15 or 20 would suffice. In such cases, it is better to go through all the calculations and when finally taking a decision, do keep the constraints in mind.

BUSTLE - 6.5

1. *On the basis of the information given in Table 6.10, a location analyst selected site S_2 since it has a maximum location site rating of 402. Do you support or oppose? Give reasons.*
2. *Suppose you have to select a site for establishing*
   (a) *A cement factory*
   (b) *A ready-made garments production shop*
   (c) *An ancillary unit for producing cutting tools*
   (d) *A rice mill (to process paddy to rice) or a flour mill (wheat to flour)*

   What factors do you choose for the above cases? Select some conditions or factors (select arbitrarily if not existing) and make a comparative study of any four sites in your vicinity by various methods given above.
3. *Write your analysis and the decision matrices of the following* :
   (a) *A cement factory*
   (b) *A ready-made garments production shop*
   (c) *An ancillary unit for producing cutting tools*
   (d) *A rice mill (to process paddy to rice) or flour mill (wheat to flour)*

Test Your Understanding - 6.5

1. Write notes on the following methods of rating a hypothetical location in a location facility planning.
   (a) Equal weights method
   (b) Variable weights method
   (c) Weight cum rating method
   (d) Factor-point rating method
   (e) Composite measure method
2. Write a detailed note on qualitative methods of rating to select a location with reference to facility planning.

6.9 QUANTITATIVE MODELS FOR FACLITY LOCATION PLANNING

Various types of operations research models (qualitative models) have been used to arrive at best decisions in facility location planning. A few models, which occupy the priority positions, are explained below:

6.9.1 Median Model

This model uses the graphical and statistical techniques to find the best location for a new plant. Before explaining the model, an understanding of the nature of the problem and its basic assumption is necessary.

Suppose a plant is to be established at a site, and now, the decision is to be taken on selection of its location. Obviously, one has to think of various factors, of which quantitative approach recommends two major factors, i.e., nearness to availability of raw materials (say source points) and nearness to markets (say destinations). Mere distance calculations will not suffice the decision because it also depends on the quantity of materials to be transported and the cost of transportation from the proposed location to the markets and from sources to proposed location. To consider all these, first the sources and markets are to be identified and represented in the geographical map.

Now, let us learn the computation in two stages.

Stage-I: *Identification of co-ordinates of sources and destination*

Step 1: Identify the origin: To measure the distances that require to be travelled, a reference point is necessary. When it is represented on a graph, it is required to locate an origin from where the distances are always measured. Take the geographical map that contains all the identified sources and destinations. Locate the origin at convenient position, conventionally it will be at the left-bottom (southwest corner) of the map. It is always convenient if all the sources as well as destinations occupy positions in the first quadrant of the graph.

Step 2: Draw the co-ordinate axes on the map.

Step 3: Measure horizontal (from Y-axis) and vertical (from X-axis) distances of sources and destinations on the map as measured with references to origin. These distances are noted as co-ordinates of the sources and destinations. While measuring distances, it is not necessary to calculate the actual or real distances, but it is sufficient to compute the measured distance on the map in terms of centimeters or millimeters as in any units devised (or scaled) suitably.

After finding the co-ordinate points, we proceed to stage-II, the simple median model.

Stage II: *Simple Median Model*

This is based on the basis of assumption that the mode of interaction or the path of movement, transportation of load is done on a rectilinear/rectangular pattern. In simple words, the movement can be expressed as the movement of rooks on a chessboard (i.e., either horizontal or vertical motions are allowed, but not cross or diagonal motion).

The computation algorithm is as follows:

Step 1: *Computation of median:* Sum up all the loads ($N=\sum L_i$) to be transported, i.e., from sources to new location and new location to destination. If the number of loads to be moved is odd number, then median is (N+1)/2 and if it is even, then there exist two medians viz. N/2 and N/2 + 1.

Step 2: *Graph preparation:* Draw X and Y axes and graduate them with suitable scale.

Step 3: *Locating sources and destinations:* Locate all sources and destinations on the graph paper with the given co-ordinate points.

Step 4: (a) *Horizontal pass:* Locate all S_{nx} and D_{nx} on X-axis and note their distances from the origin.

(b) *Loading calculation:* Put the loads that are to be transported under S_{nx} and D_{nx}. Find the cumulative load and put under each of S_{nx} and D_{nx}.

(c) *Location decision:* Now identify the point on X-axis among all S_{nx} and D_{nx}, where the median is just contained by the cumulative loads. If there are two medians, and each of these fall in different stations (of S_{nx} and D_{nx}), then the middle point of such stations is considered as the X-coordinate of the new location.

Step 5: (a) *Vertical pass:* Repeat as that of 4(a) and locate S_{ny} and D_{ny} on Y-axis.

(b) *Loading calculations:* Repeat as that of 4(b).

(c) *Location decision:* Find Y-coordinate of new location by using similar methods as that of 4(c).

Step 6: *Location of new plant:* Locate the new plant with the coordinate points found by step 4 and step 5.

Step 7: *Cost computations:* Computation of transportation cost can be done in the following tabular form with the formula.

$$\text{Total Cost (TC)} = \sum L_i \left[\, |X - a_i| + |Y - b_i| \,\right]$$

Table 6.11 Table of load distance and cost calculations

The old location $S_i\ (a_i,b_i)$	Load to be transported (L_i)	X-distance from/to new location $\|(X-a_i)\|$	Y-distance from/to new location $\|(Y-b_i)\|$	X-load distance $= L_i \|(X - a_i)\|$	Y-Load distance $= L_i \|(Y - b_i)\|$	Total load distance	Unit cost	Total cost
(1)	(2)	(3)	(4)	(5) = (2) × (3)	(6) = (2) × (4)	(7) = (5) + (6)	(8)	(9) = (7) × (8)

The above method is illustrated with a numerical problem here below:

ILLUSTRATION 6.1:

A particular city is trying to find the best location for a master solid-waste disposal station. At present four sub-stations are located at the following coordinate locations: station1 (4,12), station2 (6.5, 4), station3 (11,9), and station4 (1,13).

The number of loads hauled monthly to the master station will be 300, 200, 350, and 400 from stations1, 2, 3, and 4, respectively. Use simple median model to find the best location.

Solution.

Step 1: Calculate the median: As the number of loads to be moved is four, we get two medians.

Median loads = (300 + 200 + 350 + 400)/2 = 1250/2 = 625 and

{(300 + 200 + 350 + 400)/2} + 1= {1250/2} + 1 = 626

Step 2 and *3* are shown in Figure 6.2:

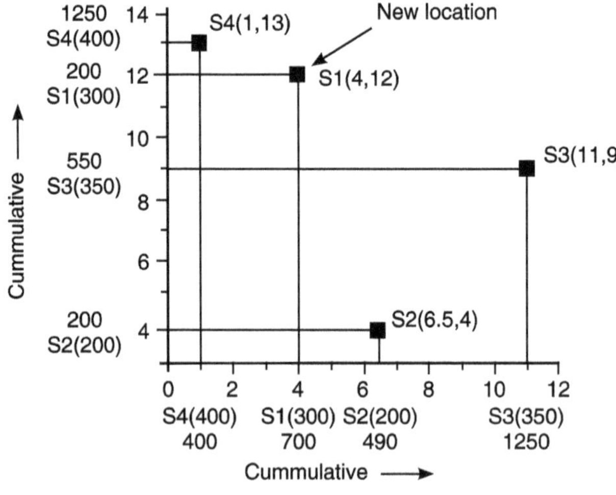

Figure 6.2

Now, according to **step 4**, we can say that both median loads fall under S1(4, 12) along X-axis as the cumulative loads is 700 (both 625 and 626 fall under 700).

Hence, the X-coordinate of the new location should be 4.

Similarly, to find the Y-coordinate, the median loads (625 and 626) fall under 850 which correspond to S1 again.

Hence, the new location will have the Y-coordinate as 12.

Thus, the new location is (4, 12).

ILLUSTRATION 6.2:

Raju & Co. wants to establish a central go-down for their production centres located at $S_1(10, 50)$ and $S_2(30, 10)$ and market centres located as $D_1(40, 60)$ and $D_2(60, 70)$. The loads that are to be transported daily from production centres are 60 and 70 from S_1 and S_2, while that from go-down to the market centres D_1 and D_2 are 40 and 50 units, respectively. Suggest the best location by using simple median method. Also calculate the cost of transportation at the rate of ₹ 1.00 per unit (unit load distance).

After establishing the new location for central go-down, it has become insufficient after one year. Then the top management of Raju & Co. has decided to construct another new go-down and convert the existing go-down as factory sales counter whose estimated demand is 40 units. Suggest the best location for the new go-down and calculate the total cost.

Solution.

First, let us calculate for the first case:

Now, as explained in step 1, let us identify the median. The total load to be moved is 220 (viz. 60 + 70 + 40 + 50 = 220). As there are four (even number of loads) points of loads, two medians will exist viz. 110 and 111 since there are 109 above and below this pair of values.

According to step 2 and 3, the coordinates of the loads is graphically represented in Figure 6.3.

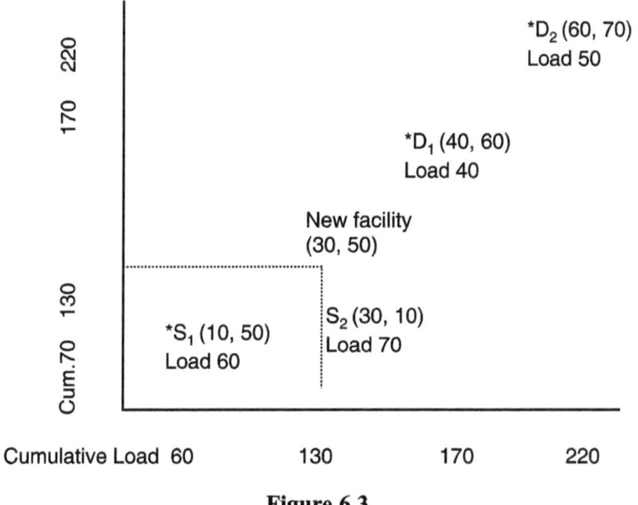

Figure 6.3

As per step 4, let us determine the X-coordinate of the new location (i.e., median load). First place the X-coordinates of the existing facilities in the ascending order, as we go horizontally from left to right; the order of the facilities will be S_1, S_2, D_1, and D_2 having loads to be moved as 60, 70, 40, and 50, respectively. Thus, the loads 1 to 60 are shipped by source S_1 at $X_1 = 10$, loads 61 to 130 are shipped by source S_2 at $X_2 = 30$, from 131 to 170 at D_1 at $X_3 = 40$ and for 171 to 220 at D_2 at $X_4 = 60$. Since both medians fall in the interval 61 to 130, therefore, X = 30 is the best X-coordinate location for the new facility.

Similarly, in step 5, we determine the Y-coordinate by arranging the Y-coordinates of the existing facilities in the ascending order (vertical passing) viz. S_2, S_1, D_1 and D_2, with Y-coordinates as 10, 50, 60, and 70, respectively. It can be observed clearly that the loads are shipped as 1 to 70 by S_2, 71 to 130 by S_1, and so on. The median loads both 110 and 111 fall in the interval 71 to 130. Hence, the best Y-coordinate of the new facility should be located at Y = 50.

Thus, the coordinates of the new location is (30, 50).

Now, let us compute the cost of transportation.

Table 6.12 Table of load distance and cost calculation

The old location $S_i (a_i, b_i)$	*Load to be transported* (L_i)	*X-distance from/to new location* $\|(X - a_i)\|$	*Y-distance from/to new location* $\|(Y - b_i)\|$	*X-load distance* $= L_i \|(X - a_i)\|$	*Y-load distance* $= L_i \|(Y - b_i)\|$	*Total load distance*	*Unit cost*	*Total cost*
(1)	(2)	(3)	(4)	(5) = (2) × (3)	(6) = (2) × (4)	(7) = (5) + (6)	(8)	(9) = (7) × (8)
S_1	60	$\|30 - 10\| = 20$	$\|50 - 50\| = 0$	1200	0	1200	1.00	1200
S_2	70	$\|30 - 30\| = 0$	$\|50 - 10\| = 40$	0	2800	2800	1.00	2800
D_1	40	$\|30 - 40\| = 10$	$\|50 - 60\| = 10$	400	400	800	1.00	800
D_2	50	$\|30 - 60\| = 30$	$\|50 - 70\| = 20$	1500	1000	2500	1.00	2500
Total Cost								7300

Now, we calculate the case after one year.

Step1: Here, we have five (odd number) centers altogether, hence there is only one median. To find median load, the total load to be moved is 260 (viz. 60 + 70 + 40 + 50 + 40 = 260). Median = 260/2 = 130

According to step 2 and 3 we represent the coordinates of the loads on the graph as shown in Figure 6.4.

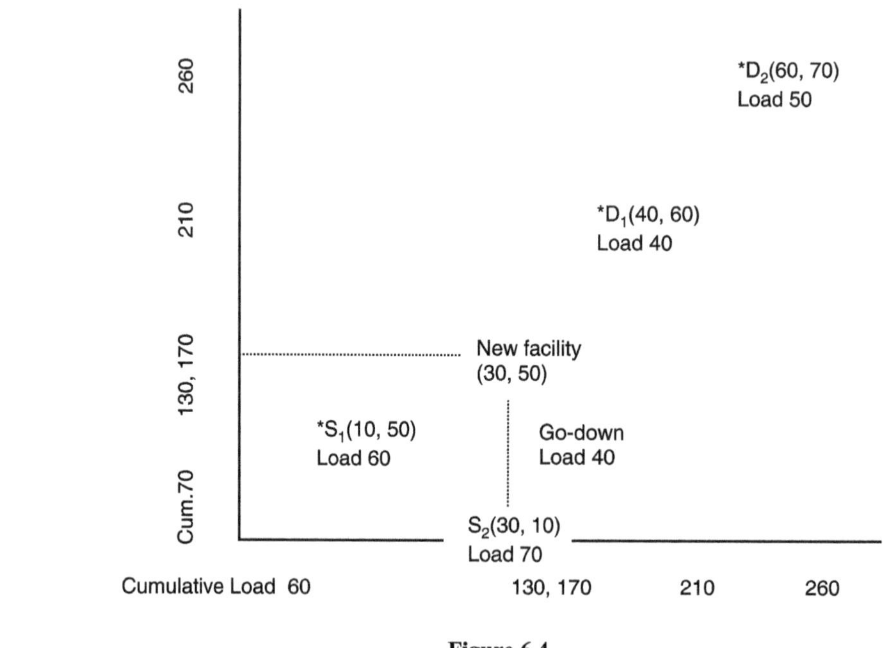

Figure 6.4

As per step 4, let us determine the X-coordinate of the new location (i.e., median load). First place the X-coordinates of the existing facilities in the ascending order, as we go horizontally from left to right the order of the facilities will be S_1, S_2, G, D_1 and D_2 having loads to be moved as 60, 70, 40, 40, and 50, respectively. Thus, we can understand that the loads 1 to 60 are shipped by source S_1 at $X_1 = 10$, loads 61 to 130 are shipped by source S_2, 131 to 170 are shipped at existing go-down at $X_2 = 30$, from 171 to 210 and D_1 at $X_3 = 40$ and for 211 to 260 at D_2 at $x_4 = 60$. Since median falls in the interval 61 to 130, (and 131 – 170 is also in the same point) therefore, X = 30 is the most optimal X-coordinate location for the new facility.

Similarly, in step 5, we determine the Y-coordinate by arranging the Y-coordinates of the existing facilities in the ascending order (vertical passing) viz. S_2, S_1, G, D_1 and D_2 with Y-coordinates as 10, 50, 30, 60, and 70, respectively. It can be observed clearly that the loads are shipped as 1 to 70 by S_2, 71 to 130 by S_1, 131 to 170 at existing go-down, and so on. The median load 130 falls in the interval 71 to 130. Hence, the best Y-coordinate of the new facility should be located at Y= 50.

Thus, the coordinates of the new location is (30, 50), i.e., at the present go-down.

Now, let us compute the cost of the transportation.

Table 6.13 Table of load distance and cost calculations

The old location S_i (a_i, b_i)	Load to be trans-ported (L_i)	X-distance from/to new location $\lvert(X - a_i)\rvert$	Y-distance from/to new location $\lvert(Y - b_i)\rvert$	X-load distance = $L_i \lvert(X - a_i)\rvert$	Y-load distance = $L_i \lvert(Y - b_i)\rvert$	Total load distance	Unit cost	Total cost
(1)	(2)	(3)	(4)	(5) = (2) × (3)	(6) = (2) × (4)	(7) = (5) + (6)	(8)	(9) = (7) × (8)
S_1	60	$\lvert 30 - 10\rvert = 20$	$\lvert 50–50\rvert = 0$	1200	0	1200	1.00	1200
S_2	70	$\lvert 30 - 30\rvert = 0$	$\lvert 50 - 10\rvert = 40$	0	2800	2800	1.00	2800
G	40	$\lvert 30 - 30\rvert = 0$	$\lvert 50 - 50\rvert = 0$	0	000	0	1.00	0
D_1	40	$\lvert 30 - 40\rvert = 10$	$\lvert 50 - 60\rvert = 10$	400	400	800	1.00	800
D_2	50	$\lvert 30 - 60\rvert = 30$	$\lvert 50 - 70\rvert=20$	1500	1000	2500	1.00	2500
Total Cost								7300

Remarks:

If, in a particular problem, there exist two medians and these fall in different intervals then the co-ordinates of the new facility will be average of the two points.

For example, in the first case of the above problem, if the load of S_1 is 40 instead of 60, and that of D_1 is 60 instead of 40, then the median load 110 will fall in the interval 41 to 110, while the median 111 will fall in the interval 111 to 170 along horizontal pass. Thus, the best X-coordinate is the average of 10 and 30, the X-coordinates of S_1 and S_2, respectively. Therefore, the X-coordinate of new location is 20.

Similarly the Y-coordinate of the new location can be calculated as (50 + 10)/2 = 30.

Hence, best location for the new location will be placed at the co-ordinates (20, 30).

6.9.2 Gravity Model

The technique is based on the principle of "center of gravity". It determines the low cost location of new facility with respect to the fixed ancillary existing facilities like source supplies (S_1, S_2, etc.) and distribution points (D_1, D_2, etc.) for which each type of product consumed or sold is known. Let us use the same data of Illustration 6.2 (first case) under the median model given above and, thus, refer to its corresponding Figure 6.4 once again . The only difference is the mode of interaction between the single new facility and the existing facilities. Here, in this case, we assume that all goods move in straight line joining the ancillary facility and the new facility (in Median Model, we assume the movement is as that of rook in a chess board). This is called 'Euclidean" mode of interaction and is in fact the shortest distance between any two reference points.

Thus, D = Euclidean Distance = $\{(x - a_j)^2 + (y - b_j)^2\}^{1/2}$

Thus, the total transportation costs in this case are

TTC (Total transportation cost) (Euclidean case) = Σ (LiDi)

viz. $TTC = \Sigma Li\{(x - a_j)^2 + (y - b_j)^2\}^{1/2}$

The objective, once again is to determine the location of the new facility at (x, y) such that TTC viz. the total transportation costs are minimum. Let us not get into a discussion on certain analytical problems and difficulties in obtaining optimal, solutions at this stage/level, but rather present analogue model and gravity model, which are simple to understand and can be readily applied. The concept underlying the technique is best visualised as a series of strings to which are attached weights corresponding to the loads/weights of raw materials consumed/dispatched at each source and of finished goods sold/received at each distribution point/market. The strings are threaded through holes in a flat plain metallic sheet; the holes correspond to the ancillary facility locations. The other ends of the string are tied together to a small concentric ring. The ring will finally reach an equilibrium based on the principle of equilibrium of coplanar forces. This equilibrium will be the centre of mass or the ton-mile centre. Therefore, it is called the *gravity model.*

$$X = \Sigma L_i (a_i)/\Sigma L_i$$
$$Y = \Sigma L_i (b_i)/\Sigma L_i$$

Thus, for our example under discussion now from supply sources S1, S2 to the new facility

$$\Sigma L_i = 60 + 70 = 130$$
$$\Sigma(L_i a_i) = (60 \times 10) + (70 \times 30) = 600 + 2100 = 2700$$

And from new facility to distribution points D_1 and D_2

$$\Sigma L_i = 40 + 50 = 90$$
$$\Sigma(L_i a_i) = (40 \times 40) + (50 \times 60) = 1600 + 3000 = 4600$$

X = Sum of all load.distantces/sum of all loads

= (2700 + 4600)/(130 + 90) = 33.19

Similarly Y-coordinate can be calculated as follows:

$$Y = \{(60 \times 50) + (70 \times 10) + (40 \times 60) + (50 \times 70)\}/\{60 + 70 + 40 + 50\}$$
$$= 9600/220 = 43.63$$

Thus, the gravity model suggests that the new location is to be situated at a point (33.19, 43.63); the total transportation costs will be minimum.

Only the location decision problems dealing with just a single new facility have been discussed, which is often termed as **minisium** objective of the sum of weighted appropriate distances. There can be cases when the location as determined above turns out to be non-feasible or infeasible, because of the existence of certain restrictions or limitations. There are some methods available to solve such problems such as Composite Location Measure Model suggested by Brown and Gibson, Bridgeman's Dimensional Analysis, etc.

BUSTLE - 6.6

Refer to Bustle-6.4, Question No. 2. Repeat the same exercise with median model and gravity model by assuming any two source points of raw material and two distribution/marketing points for each case. Find the best suitable location.

(a) A cement factory

(b) A ready-made garments production shop

(c) An ancillary unit for producing cutting tools

(d) A rice mill (to process paddy to rice) or flour mill (wheat to flour)

Test Your Understanding 6.6

1. Write a short note on the quantitative models for selection of new location in a location facility planning.
2. Write the algorithmic steps in selecting the new location by using median model for facility planning.
3. Explain the gravity model used in facility planning to select a new location.

NUMERICAL PROBLEMS FOR PRACTICE

1. A company has four factories located at the Cartesian co-ordinates as A(100, 30), B(80, 60), C(50, 50) and D(30, 70). The factories produce 200, 300, 400, and 100 units, respectively. The company is looking forward to establish a storage go-down to stock its products. Use median model to find the coordinates of the go-down location. Also calculate the cost of transportation if it is charged ₹ 2 per unit distance-load.
2. For the same data given above, use the gravity model and find the coordinates of the new location and the cost.
3. P(12, 10), Q(2, 8), R(10, 6), S(5, 3) and T(7, 9) are the proposed markets with the demands 100, 60, 50, 80, and 20, respectively for a company. The company has to establish its production centre now. Find the coordinates of the optimum location using simple median. If the existing transportation tariff is as given here, find the transportation cost.

   From P: ₹ 2 per unit up to 1000 units of load-distance and ₹ 5 above 1000

   From Q: ₹ 1 for first 500 and ₹ 3 for 500 to 1000 and ₹ 5 for above 1000

   From R: ₹ 3 for any load-distance

   From S: ₹ 2 up to 800, ₹ 3 for 800 to 1200 and ₹ 4 above1200

   From T: ₹ 1 for first 200 and ₹ 1 hike for every raise of 200 units.

4. Use gravity model for the above data and calculate.
5. A particular city is trying to find the best location for a master solid-waste disposal station. At present four sub-stations are located at the following coordinate locations: station1 (4, 12), station2 (6.5, 4), station3 (11, 9) and station4 (1, 13).

   The number of loads hauled monthly to the master station will be 300, 200, 350, and 400 from stations1, 2, 3, and 4, respectively. Use gravity model to find the best location.

6.10 PLANT LAYOUT

There are different types of common layouts. Commonly the layouts are of the following types:

- Product or line layout
- Process or functional layout
- Fixed position layout
- Cellular or group technology layout

6.10.1 Product or Line Layout

In a product layout various facilities, such as machines equipment, work force, etc., are located as per the sequence of operation on parts (Figure 6.5). Even a facility is again needed after few other operations, we duplicate the facility at every required sequence. Product layout is preferred when the production is continuous, parts variety is less, production volume is high, and part demand is relatively stable.

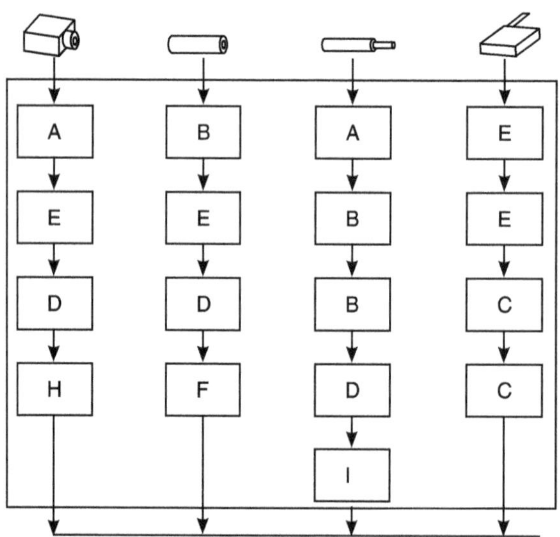

Figure 6.5 Product layout

Advantages of Product Layout:

- Less work-in-process (WIP) inventory as the flow of materials is continuous along a line.
- Compared to process layout it requires less space for the same volume of production.

- Conveyors material handling or automation of material handling is cost effective as the flow of material is well known.
- The throughput time is less as compared to process layout. This is due to less chance of congestion and waiting time on machine.
- Simple production planning and control and better coordination of different activities may be achieved.
- The skill level of workers may be lesser as a particular worker has to do a particular operation, which seldom changes due to standardised production line.
- The flow of material is smooth and continuous.

Limitations of Pproduct Layout:

- Change in product design is difficult to accommodate.
- Product variety is very much limited.
- Breakdown of particular machine in the line halts the production output.
- Capital investment on machinery in product layout may be higher as compared to process that in layout due to duplication of machinery in line.
- The flexibility to increase the production capacity is limited.

Suitability of Product Layout:

- Assembly lines such as automobile factory.
- Low variety high volume production systems.
- Standardised products, which have quite demand in near future.

6.10.2 Process Layout

Process layout is also called functional layout. Similar machines or similar operations are located at one place as per the functions (see Figure 6.6). For example, all milling operations carried out at one place while all the lathes are placed at separate location. Grinding or finishing operation is kept at separate location. This functional grouping of facilities is useful to job production and non-repetitive manufacturing environment.

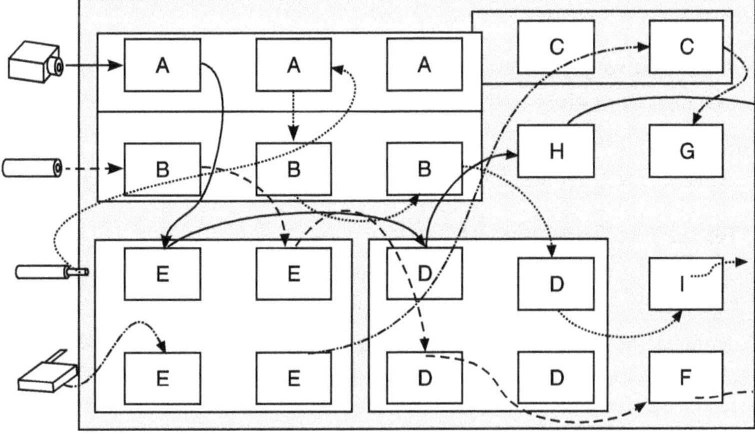

Figure 6.6 Process layout

Advantages of Process Layout:

- Initial investment in the process layout is low.
- High degree of machine utilisation is found in process layout as machine is not dedicated to a single product.
- Greater flexibility and scope of expansions exist in this layout.
- Different product designs and different production volumes can be easily adopted. Therefore, high product variety can be easily handled.
- The overhead costs are low.
- The maintenance of machine is easy. Breakdown of one machine may not result in total stoppage of production.
- Easy effective and specialised and supervision of each functional area is easy to achieve.
- There is low set up and maintenance cost.
- Due to different departments for different processes better teamwork can be achieved in each section.

Limitations:

- There is high degree of material handling. Parts have to backtracked in the same department.
- Large work in process, inventories are common. This may lead to more storage area.
- Workers are more skilled. This is because of variety in product and difference in design. Therefore, labour cost is higher.
- Total cycle time is high. This is due to long wait in different departments and longer martial flow.
- Inspection is more frequent which results in higher supervision cost.
- It is difficult to fix responsibility for a defect or a quality problem. The work moves in different departments in which the machine preference is not fixed. Therefore, the machines or the operators that were faulty during quality lapse are difficult to trace in some cases.
- The production planning and control is difficult.

Suitability of Process Layout:

- Non-standardised product.
- Low volume-high variety manufacturing environment.
- Frequent change in product design.
- Job shop manufacturing.
- Very expensive machine like CNC milling, coordinate measuring machine, etc.

6.10.3 Fixed Position Layout

In fixed position layout, the main product being produced is fixed at a particular location. Resources, such as equipment, labour, and material are brought to that fixed location. This type of layout is useful when the product being processed is very big, heavy or difficult to move. Some examples of fixed position layout are shipbuilding, bus body building (see Figure 6.7), etc.

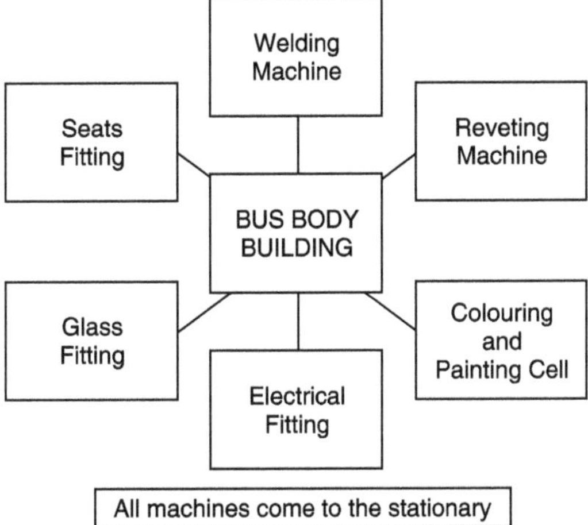

Figure 6.7 Fixed position layout

Advantages of Fixed Position Layout:

- Easy for products which are difficult to move.
- Flexibility for change in design, operation sequence, labour availability, etc., exists in this layout.
- This layout is cost effective when many orders of similar type are in different stages of progress.
- Large project type of jobs such as construction is suited in this layout.

Limitation of Fixed Position Layout:

- High capital investment due to long duration project to complete a product.
- Space requirement for storage of material and equipment is large.
- Careful project planning is required and focussed attention on critical activities otherwise confusion, delay, and conflict may arise.

6.10.4 Cellular or Group Layout

Cellular layout is based on the group technology (GT) principle. Therefore, it is also called group layout. This layout is suitable for a manufacturing environment in which large variety of product is needed in small volumes (or batches). The group technology principle suggests that parts, which are similar in design or manufacturing operations, be grouped into one family, called part family. For each part family (called machine-cell), dedicated cluster of machines are identified. Generally, all the processing requirements of a particular part family are completed in its corresponding machine cell. In other words, the internal transfer of parts should ideally be zero.

The cellular layout is, thus, a combination or process and product layout. Therefore, it possesses the features of both. Cellular manufacturing system (CMS) involves decomposition of manufacturing system into subsystems of similar parts/machines.

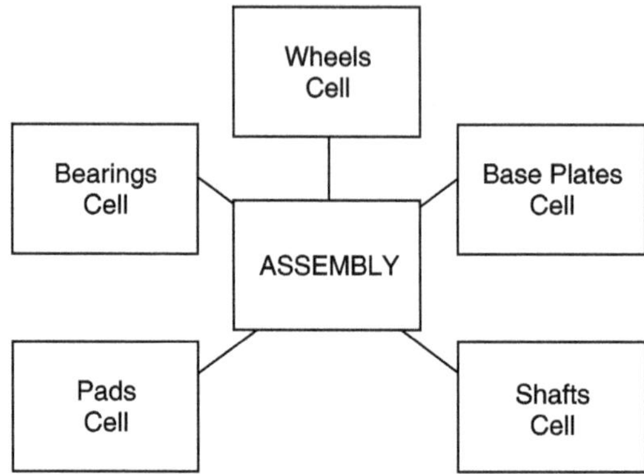

Figure 6.8 Cellular layout

CMS allows batch production to give economical advantages similar to that of mass production with additional advantages of flexibility, normally associated with job shop production systems.

Advantages of Cellular Manufacturing System (CMS):

The advantages of CMS are represented in a tabular form in Table 6.14:

Table 6.14. Advantages of CMS

<table>
<tr><th>Sl.No.</th><th>Feature</th><th>Reason</th></tr>
<tr><td>1.</td><td>Short throughput times</td><td rowspan="6">Group's complete parts, and machines are close together under one foreman in each group.</td></tr>
<tr><td>2.</td><td>Low stock</td></tr>
<tr><td>3.</td><td>Low stock holding cost</td></tr>
<tr><td>4.</td><td>Better customer service</td></tr>
<tr><td>5.</td><td>Better quality</td></tr>
<tr><td>6.</td><td>Lower material handling cost</td></tr>
<tr><td>7.</td><td>Better delegation and accountability</td><td rowspan="3">All parts and machines of the group are close together and under same supervision, which can be made responsible for cost, quality, and due-date.</td></tr>
<tr><td>8.</td><td>Reduced indirect labour cost</td></tr>
<tr><td>9.</td><td>More reliable production</td></tr>
<tr><td>10.</td><td>Training for promotion</td><td>Unlike process organisation, which produces a specialist, CMS gives experience in a much wider range of tasks.</td></tr>
<tr><td>11.</td><td>Stepping stone to automation</td><td>CMS is the first evolutionary step in automation. In general, it is FMS with some manual operations.</td></tr>
<tr><td>12.</td><td>Increased capacity</td><td>Due to easier sequencing and reduction in set up times, some buffer capacity is created.</td></tr>
</table>

13.	Improved job satisfaction	• Team work. • Processing of products in the cell. • Better feedback in the cell. • Identification of quality and performance of job with the cell workers.
14.	Easy retrieval of parts	Standardised product design and tooling and classification.
15.	Shorter lead-time	Due to low setup and throughput times.
16.	Efficient production planning and control	Due to similarity of parts of each cell.
17.	Flexibility to adapt to market fluctuations	• Low stock. • Easy switch over from one part to another inside a cell.
18.	Reduced scrap and wastage	Due to specific machines for each part family.
19.	Easy plant maintenance	Due to decomposition of plant into smaller cells.
20.	Simplified tooling and set-ups	• Specific jigs and fixtures, designed for each part family. • Machine tools, not requiring may change over for any part family due to similarity.
21.	Simplified estimation, accounting, and work measurement	• One supervision for each cell. • Decomposition of plant into smaller but independent cells.
22.	Better utilisation of manufacturing resources and space	• Dedicated machines cell for each part family. • Less material handling. • Less WIP.
23.	No fallout in human factor	Due to teamwork and personal involvement in cell.

Table 6.15 Quantitative estimates of some reported benefits of CMS

Sl.No.	*Feature*	*Value*
1.	Reduction in new parts design.	52%
2.	Reduction in number of drawings through standardisation.	10%
3.	Reduction in new shop drawings.	30%
4.	Reduction in production floor space required.	20%
5.	Reduction in scrap and rework.	15–75%
6.	Reduction in production and quality control costs.	80%
7.	Reduction in production lead-time.	20–88%
8.	Reduction in tooling.	20–30%
9.	Reduction in set up time.	20–69%
10.	Reduction in throughput time.	70%
11.	Reduction in over due orders.	82%

12.	Reduction in raw material inventory.	42%
13.	Reduction in work in process (WIP)	88%
14.	Reduction in finished goods inventory.	60%
15.	Reduction in labour.	15–25%
16.	Increase in employee output per unit time.	33%

Suitability of CMS

The implementation of CMS can significantly increase productivity, which is essential for survival in increasingly competitive industries. Many large and medium size manufacturing firms have experienced quality improvements after adopting CMS.

CMS and Batch Manufacturing

CMS has wide applicability, since it can be implemented in both job-shop and assembly lines. However, CMS is most suitable for organisations that have degree of part standardisation and moderate batch size. The importance of CMS can be well realised, as according to a study more than 75% of all manufacturing units are engaged in the production of a large variety of parts in batches.

CMS and JIT

Cell manufacturing is an important element in the successful implementation of just in time (JIT). It reduces production related wastes, such as inventory (WIP, finished goods, or raw material), production set-up times, insufficient job scheduling, and parts having long queues for a long time at various work centres.

CMS and FMS

CMS is building block of flexible manufacturing system (FMS). It is the first evolutionary step in automation. In general, it is the FMS with some manual operations. The applicability of CMS concept in FMS may be justified on account of four reasons (Kusiak, 1986):

(a) It is easy to process large volume of information with the decomposed manufacturing system, such as CMS.

(b) AGV and robot are the most common type of material handling carriers in FMS.

(c) Often, the services of these material-handling carriers are limited to functional manufacturing. This is due to limitation on cell size.

(d) There may be technological compulsion for grouping some operations like forging machines and heat treatment unit.

Limitations of CMS:

- Difficulty in initial planning and implementation.
- High cost in layout.
- Difficult to bringing too many or too often changes.
- Difficult to produce new part or non-standardised part.
- Too much dependence of GT curtails flexibility of the cellular manufacturing systems.

Table 6.16. Comparison of different layouts

Factors	*Type of layout*			
	Fixed position	*Product (Line)*	*Cellular*	*Process (Functional)*
Type of operation	Ship building, large-scale projects, constructions, or industrial project	Continuous and repetitive	Small to medium batch	Job or small batch
Arrangement of facilities	Facilities move to a fixed product/project	Placed along line of product flow	Similar parts are grouped in part family for each part family one machine cell is formed which contains all facilities needed by corresponding part family	Grouped by specialty
Cost of layout	Moderate to low	Moderate to high	Moderate to high	Moderate to low
Material handling	Moderate	less	Less	High
Material travel	Variable path	Fixed path	Fixed path	Variable path
Utilisation of facilities	Moderate	Very high	High	Low
Operating facilities	General purpose	Special purpose	Special purpose	General purpose
Employee skill	Unskilled/skilled	unskilled	Multi-skilled as one operator may handle more than one operation	Skilled
(Q/P) ratio: Q is production quality, P is number of products or variety	Normally 1 as single product production	Large Q/P	Moderate Q/P	Small Q/P

6.10.5 Computerised Layout Design

With the advent of computerisation, layout design has taken a new dimension and ease. The computerised layout design procedures can be broadly classified into two types. These are:

(a) **Constructive type algorithm**

- ComputerisedRelative Allocation of Facilities Technique (CRAFT)

(b) **Improvement type algorithms**

- Automated Layout Design Process (ALDEP)
- Computerised Relationship Layout Planning (CoReLaP)

Systematic Layout Design Process

The layout planning developed by Muther, often referred to as systematic layout planning (SLP) has received considerable publicity. This procedure is represented in the following flow chart (Figure 6.9).

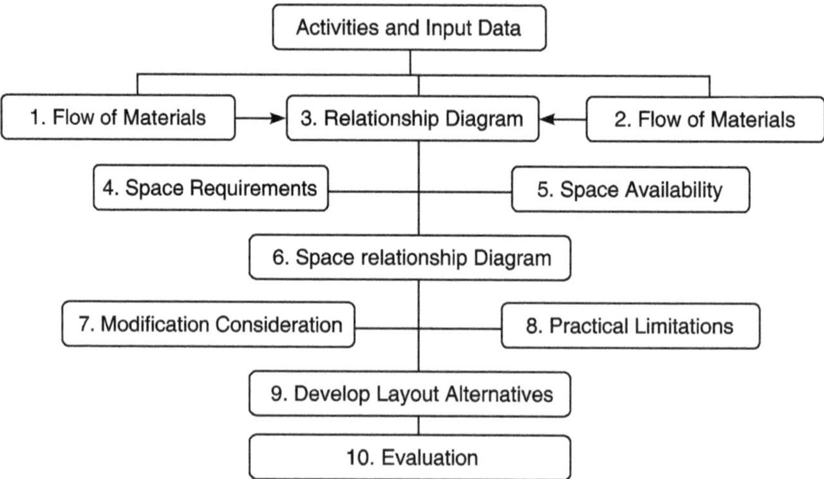

Figure 6.9 Muther's systematic layout planning

6.10.6 Factors Affecting Facility Layout

The following are the factors that usually influence the facility layout studies.

(a) **Type of Industry:** Different types of industry require the facilities in different ways. Knowledge of the different types of industries and their classification is necessary. Industries are classified in many ways, of these the most general ways are according to the nature or their process of manufacture or the scale/volume as given below:

According to nature, industries are divided into three types:

- Process industry
- Product industry
- Service industry

According to process of manufacture, industries can be classified into four types:

- *Synthetic process industry*: It is the industry where two or more materials are mixed or combined or assembled to get a product. The process is known as synthetic process or synthesis. Example of such a process is production of cement by mixing limestone and clay.
- *Analytical process industry*: The opposite of synthetic process, i.e., breaking up of a material into several parts is carried out in such industries. Refining of petroleum is the best example of this case.
- *Conditioning process industry*: In the conditioning process industry, the form of raw materials is converted into the desired product through some transformation processes as done in jute industry, engineering industries such as blade manufacturing, fan manufacturing, etc.

- *Extractive process industry*: In this type, by applying heat, or any other engineering process a desired product is extracted from the original raw material, e.g., manufacture of aluminium from bauxite, manufacture of iron from hematite/limonite/iron pyrites, etc.

According to the scale of manufacture, the industries are

- Large-scale industries
- Small-scale industry

(b) **Type of Product:** Type of product, i.e., whether the product is large or small, heavy or light, liquid or solid or gas will definitely matter while designing the facilities and in plant layout.

(c) **Volume of Production:** Accordingly, while plant layout is being done, volume or type of production is required to be considered. Normally, the following are the types of productions. Any industry will adopt the nearest suitable or the most appropriate one of following.

- Mass production
- Flow production
- Batch production
- Job production
- Project production

(d) **Material Handling:** Last but not least factor to consider, is the material flow line and the quantity and/or distance through which the materials are to handled or transported. Much care should be given to the material-handling problem, position of store room, and tool room, etc.

BUSTLE - 6.7

1. *Observe what type of layout exist in the institution you are studying. Enumerate the demerits or limitations in the layout.*
2. *Design the best suitable layout plan for the following. (Assume the data such as sequence of operations, number of machines, etc., arbitrarily).*

   (a) A cement factory

   (b) A ready-made garments production shop

   (c) An ancillary unit for producing cutting tools

   (d) An educational institute of 10 + 2 level

   (e) A flour mill (to process wheat to flour)

   (f) A cool drink bottling factory.

   (g) Assembly unit of wrist watches.

Test Your Understanding 6.7

1. With a neat sketch explain product layout. What are its advantages and limitations?
2. With examples and a neat sketch explain process layout. What are its merits and demerits?
3. With a neat sketch and some applications, explain fixed position layout. What are its advantages and limitations?
4. With a neat sketch explain Cellular or Group technology layout. What are its advantages and limitations?
5. Give a comparative chart for process, product, fixed position, and cellular layouts with reference to their relative advantages and limitations.
6. Give the suitability of application for the following. Give some evidential supports to your answer.
   (i) Product layout
   (ii) Process layout
   (iii) Fixed position layout
   (iv) Cellular layout
7. Compare the process and product layout. Also compare their areas of applications merits and demerits.
8. How do you fit just in time (JIT), batch manufacturing, part-machine incidence matrix, and flexible manufacturing system (FMS) in cellular manufacturing system (CMS) design?
9. Enumerate the advantages and limitations of CMS.
10. Explain the factors that influence layout decision.

6.11 FLOW PATTERNS TO FACILITATE ASSEMBLY LINES

Study of flow pattern is another significant area of facility planning as it facilitates the assembly/ manufacturing lines and has considerable impact on the production. Flow pattern is the system or path to be adopted for the movement of raw materials from the starting point of manufacturing process to the end of the process where it comes out as the finished or semi-finished or some desired output. The overall objective of the flow pattern study is to plan for the economic and effective movement of raw materials within the plant.

6.11.1 Advantages of a Good Flow Pattern

The following are the advantages of having a good flow pattern system in an industry:

- Length of total production time can be reduced which in turn curtails the production cost.
- Inventory levels particularly work in process can be reduced and controlled effectively.
- Capital and space tied up due to piled up work in process can be easily identified and thence reduced.
- Material handling will be cost effective, safe, and quick.
- Flow pattern helps in achieving high rate of performance.
- Flow pattern provides good coordination among operations.
- Amount of physical and mental strain on operators can be drastically reduced.
- Supervision and control mechanisms are made easy.

6.11.2 Factors Governing Flow Pattern

In fact, a plant layout design starts with the flow system around which services and other facilities are added and accordingly buildings are designed and/or modified. In some occasions, the flow may be adopted to existing buildings. However, the flow pattern design depends on the following factors:

- Number of products to be handled.
- Number of operations on each product.
- Number of units to be processed.
- Number of sub-assemblies made up of assembly line.
- Necessary flow between work areas.
- Size and shape of available land.
- External transport facility.

6.11.3 Types of Flow Lines

The flow system can be classified into horizontal and vertical. The horizontal flow system is adopted on a shop floor while vertical flow is adopted where material has to move in a multi-storey building.

(a) **Horizontal Flow Lines:** There are five basic types of horizontal flow lines.

(i) *I-flow or line flow or straight flow*: It is the simplest form of flow. In this flow system, materials are fed at one end and output comes out at the other end. This type is economical in space and convenient in I-shaped buildings. Linear flow is preferred for building watch assembly, automobile industries, etc.

(ii) *L-flow*: It is similar to the I-flow and is designed where I-line cannot be accommodated in the available space.

(iii) *U-flow*: In this type of flow system, both feeding and output take place at the same end, i.e., both receiving and dispatching of goods are done at one side. In comparison to I or L-flow, this method is easier for supervision. This type of flow can be adopted in the manufacture of Electric Motor Industry.

(iv) *S-flow or serpentine flow*: If the production line is so long that zigzagging on the plant floor is necessary, then S-flow is adopted. This flow system provides efficient utilization of space and is compact enough to allow effective supervision.

(v) *O-flow or circular flow*: This type is employed where operations are performed on a rotary table or a rotary handling system. The components are moved from one working station to the other along O-shaped line. O-flow can be adopted by industries manufacturing electric bulbs.

These above mentioned five basic flow lines are mostly used by industries in various combinations. Examples of combinations of basic horizontal flow systems are shown in Figure 6.10.

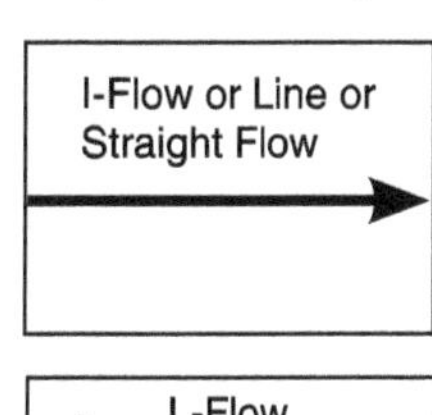

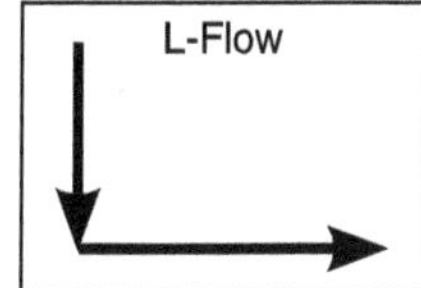

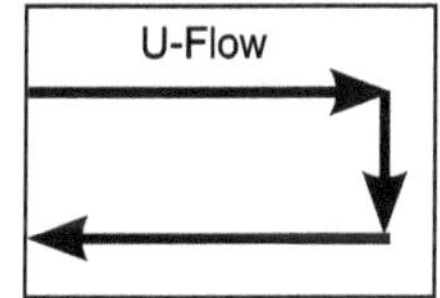

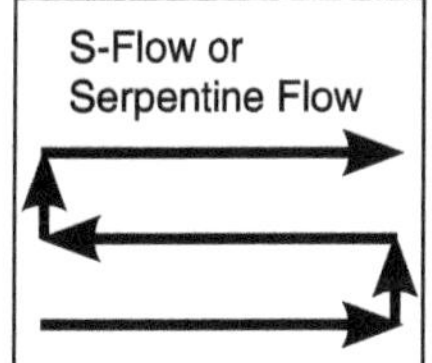

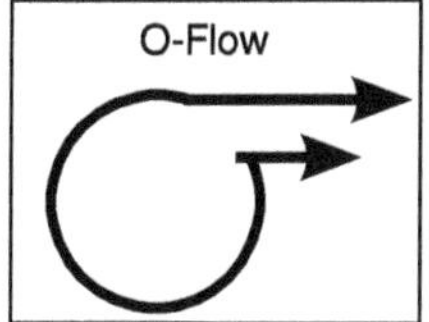

Figure 6.10
Horizontal flow lines

Unidirectional and Retraction Flow

Unidirectional flow: The flow is said to be unidirectional when the material is passed from one to another without passing along the same path. The flow methods explained above are unidirectional flow type.

Retraction flow: This flow is also known as "repeated flow". Here, the flow is repeated, i.e., two or more non-consecutive operations are performed on the same machine. This aspect of flow allows full utilisation of available machine time but schedules have to allow for repeated machine setting due to the intermittent localised halts in the production line, i.e., each time a machine is switched over from one operation to another.

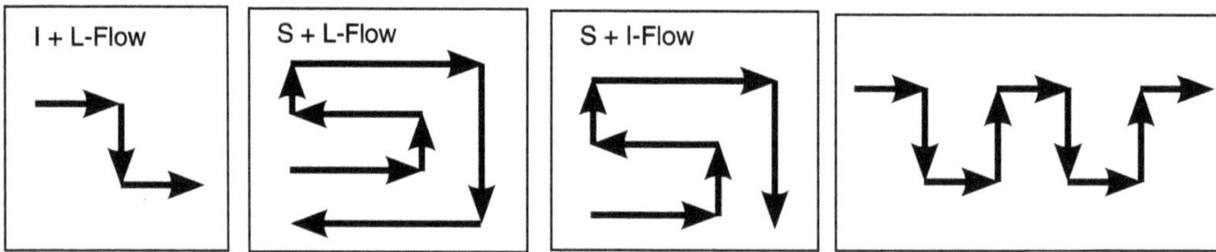

Figure 6.11. Horizontal combination flow

(b) **Vertical Flow Line:**

This type of flow is suitable for multi-storey buildings, so that the materials handling systems and control mechanism operate effectively. The following six basic aspects are to be considered while designing vertical flow lines.

(i) *Processing downwards or upwards*: In downwards processing, the materials are fed at the top floor and the finished product is received at the bottom floor. In upward processing, the materials are fed from the bottom floor while the finished product is received at the top floor.

Downward processing is most suitable in many gravity handling systems such as roller lines, bunkers, chutes, pipes, buckets, hand operated lifts, etc. They are economical in installation, operation, and maintenance, etc. The upward system is quite suitable for processing gaseous or vapour type products.

(ii) *Centralised or decentralised elevation*: All the material handling devices are installed at one common or central place of the buildings in a centralised elevation. Therefore, this system is economical in supervision and maintenance. It sometimes reduces installation costs also. This method is usefully employed when the flow on each floor is a U-flow.

A decentralised elevation method is more costly in installation. Maintenance and space handling on each floor using this method can be greatly reduced and more flexibility in design of the flow lines is possible.

(iii) *Unidirectional or retraction flow*: In unidirectional flow, the material is moved from one place to the other without passing while in retraction flow, material has to come back on the floor, which it had already passed previously. This may be done purposely to achieve better utilisation of available space and machines.

(iv) *Vertical or inclined flow*: The vertical flow is suitable for bulk solid materials processing such as coal processing, extraction of metals in blast furnaces, etc. The inclined type of

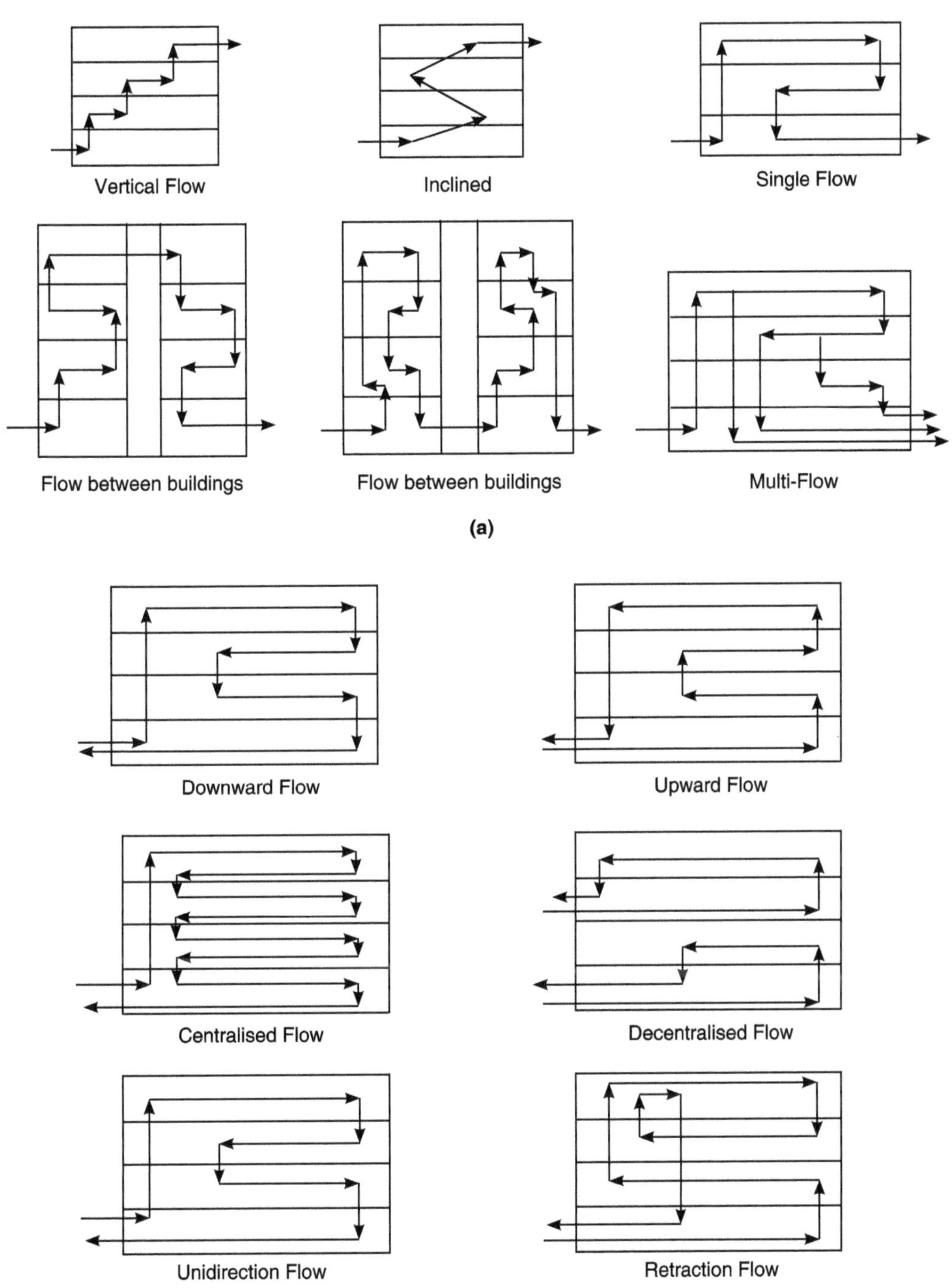

Figure 6.12 Vertical flow lines

flow is more economical and is carried on with material handling devices such as elevators, chutes, and buckets, etc. Inclined flow can also be carried out by conveyor belts used in coal handling plants and chain system is used to move boiler grates, etc.

(v) *Single or multi-flow*: In a single flow, there is only one-flow line of materials while in multi-flow, there will be several flow lines and all others feed one assembly line.

(vi) *Flow between buildings*: When one production is executed in several adjacent buildings the flow of goods achieved may be either on elevated flow or on ground flow. Ground flow is cheaper but requires more handling than that in an elevated flow. Elevated flow allows the ground for traffic and storage purposes.

BUSTLE - 6.8

1. *What type of flow did you observe in the organisation where you are working or studying (both horizontal and vertical, if they exist)? Enumerate the disadvantages or limitations of the flow lines. What modifications do you suggest?*
2. *Design a flow line for the following: (Assume data suitably)*
   (a) Personal Computer assembly unit
   (b) A pharmaceutical company
   (c) Beverage bottling

Test Your Understanding 6.8

1. With neat sketches explain various horizontal flow lines used in assembly layouts.
2. Describe different vertical flow line patterns in a multi-storey building.
3. Discuss how you can improve the productivity with flow patterns.
4. What are the advantages offered to employees and employer through a good flow pattern?
5. Discuss the importance of flow patterns in facility and location planning.
6. Briefly describe the factors governing the flow pattern decisions.

6.12 CONCLUSION

The problem of arranging facilities for an industrial process has been in existence as far as the industrial revolutions. Facility design (i.e., facility layout and location planning) has been the crux of the problem for centuries. However, as scientific management has taken the shape of the industrial engineering, facility design has grown along with activities such as time and motion study, ergonomic study, cost accounting, and production control. Today, we observe that there exists a strong interdisciplinary interest in facility layout and location planning. Economists, operations researchers, urban planners, management scientists, architects, designers, and engineers from all disciplines have realised a common interest in their concern for design of facilities. Facility designing picks up information from many areas, each of which tends to bring to the subject of a different interpretation of the problem and different approaches, i.e., qualitative or analytical for its solution. In analytical/ quantitative approach, mathematical optimisation models are developed to obtain solutions, whereas the traditional qualitative approaches rely heavily upon intuition, logical conclusions, and some engineering analysis. No doubt, the judgement by the analysis serve as an aid in decision making, but in some cases, it is impractical since it may not be possible to include all factors in making models due to complications involved. Therefore, analytical approach yields a solution to the model, but not necessarily the problem and these models should be used as design tools only. Factors not included in the analysis should be considered along with the qualitative approach. Hence, it may be concluded that one may not be developed at the cost of the other. One should not overlook the value of qualitative aspects of the facility design, nor ignore the quantitative approach in favour of a completely qualitative approach.

6.13 RECAP

Facility planning can be described as an overall design of an enterprise. It covers the determination of location, production process, equipment, physical arrangements, provisions for personnel, and all functions that are necessary to the completion of the goods. It is greatly swayed by the type of the industry that they support particularly, in the decisions such as product design, plant layout, buildings and grounds, location planning, size of the facility, expansion, diversification, organisation development, etc. These are discussed in this unit extensively under three major heads, namely,

1. Facility location study,
2. Facility layout, and
3. Flow patterns of materials.

Facility location planning involves a decision on selection of region as well as a specific site within the region. It also covers the criteria of location by which various location alternatives can be evaluated both qualitatively and quantitatively. Layout planning is making available facility to its fullest utility. Flow is subordinate to the overall design of facility and is another important phase of facility design. All these are thoroughly discussed in this chapter.

6.14 KEY CONCEPTS

1. **Facility:** Facility is the provision or availability to fulfil certain need.
2. **Facility Location Study:** Location of facilities is a problem associated with the planning phase of a factory or even a service sector to take a strategic decision where to locate the site so that maximum needs can be fulfilled to a great extent and no change is needed for years to come.
3. **Territory Selection:** A strategic decision involved in identifying and selecting the general territory/region/area based on various factors such as markets, raw material, and supplies, transportation facilities, manpower supply, infrastructure, legislation and taxation, climate, etc., for establishing a plant or firm.

4. **Site/Community Selection:** The decision to identify and select an exact site for establishing a factory or firm or a plant or an organisation based on community attitudes, waste disposal, ecology and pollution, site size, topography, transportation facilities, supporting industries and services, land costs, community facilities such as schools, places of worship, medical services, police and fire stations, cultural, social and recreation opportunities, housing, good streets and good communication, and transportation facilities, etc.
5. **Plant Layout:** The arrangement of certain machinery or equipment or any other facility in such a way that the arrangement provides ease or comfort or high productivity or economy.
6. **Product or Line Layout:** In a product layout, various facilities, such as machines equipment, work force, etc., are located as per the sequence of operation on parts.
7. **Process or Functional Layout:** Process layout or functional layout is that layout in which similar machines or similar operations are located at one place as per the functions.
8. **Fixed Position Layout:** In fixed position layout, the main product being produced is fixed at a particular location. Resources, such as equipment, labour and material are brought to that fixed location.
9. **Cellular or Group Layout:** Cellular layout or group layout is based on the group technology (GT) principle and is a combination of process and product layout. Therefore, it possesses the features of both. Cellular manufacturing system (CMS) involves decomposition of manufacturing system into subsystems of similar parts/machines.
10. **Flow Patterns:** Flow pattern study is a facility planning activity, which facilitates the assembly/manufacturing lines to provide ease or speed or comfort in material movement.
11. **Horizontal Flow Lines:** The path of movement of materials in the same level or floor of a building is known as horizontal flow line. This is commonly found in five shapes (I, L, U, S, O) or their combinations.
12. **Vertical Flow Line:** The mechanism of flow line in a multi-storey building, where the material is moved upward or downward (in vertical direction) is called vertical flow line patterns.
13. **Unidirectional and Retraction Flow:** The flow is said to be unidirectional when the material is passed from one to another without passing along the same path. Retraction flow is repeated flow, i.e., two or more non-consecutive operations are performed on the same machine.
14. **Downward and Upward Flow:** In downward processing, the materials are fed at the top floor and the finished product is received at the bottom floor. In upward processing, the materials are fed from the bottom floor while the finished product is received at the top floor.
15. **Centralised and Decentralised Elevation Flow:** All the material handling devices are installed at one common or central place of the buildings in a centralised elevation. A decentralised elevation method is that in which the handling devices are installed at different floors individually.
16. **Vertical and Inclined Flow:** The vertical type of flow is that flow in which material moves in exactly vertical, i.e., at right angles to horizontal axis and inclined type of flow is flow of material in certain vertical angle (to horizontal axis).
17. **Single or Multi-Flow:** In a single flow, there is only one-flow line of materials while in multi-flow, there will be several flow lines and all others feed one assembly line.

REVIEW QUESTIONS

1. What do you understand by 'facility'? What are the principles of a good facility layout? What the advantages of good facility layout?
2. Compare and contrast between an urban and rural locations for establishing an industry with suitable examples.
3. Briefly discuss various methods of rating a facility location.
4. Explain the following methods of facility location decision:
   (a) Median Method (b) Gravity Method
5. Distinguish between product and process layouts with suitable sketches.
6. What is fixed position layout? What are the situations warranting to fixed position layout? What are its limitations?
7. What is cellular layout? Discuss its merits and demerits.
8. What do you understand by the term 'flow pattern' with reference to facility and location planning? With suitable sketches explain various types of flow patterns.
9. Give the advantages of a good flow pattern in an industry. How does it help in improving productivity?

CASE STUDIES

Case-1 (Solved)

Selection of Site for XYZ Company

Objective: To select a site for XYZ company

For XYZ company, five alternative sites are selected for possible location of its plant. These are city1, city2, city3, city4, and city5. The management hired a consultant group to help on site selection.

The consultant prepared a list of factors that affect the location. They did this by asking the managers and executives of the XYZ company. Nine major issues are identified (see Table-II).

Then, the consultant made a pair-wise comparison of all the nine alternatives. They gave a score of 1, 2 or 3 on every 36 comparisons (it is C3). The scoring scheme in comparison is done as follows (Table I):

Table I

Comparison	*Score*
If factor X scores as a major factor as compared to factor Y	X_1
If factor Y scores as a major factor as compared to factor X	Y_1
If factor X scores as a major factor as compared to factor Y	X_2
If factor Y scores as a major factor as compared to factor X	Y_2
If factor X scores as a major factor as compared to factor Y	X_3
If factor Y scores as a major factor as compared to factor X	Y_3

Table II

Sl.No.	*Factor*	*Description*
1	A	Nearness to raw material source
2	B	Transport facility and logistics support
3	C	Availability of water and power
4	D	Adequacy in labour availability
5	E	Quality of life, like market, school and recreation
6	F	Competition in local market
7	G	Nearness of market
8	H	Incentive from financial institutions
9	I	Cost of land

Each factor is compared pair-wise. For each pair a weight is provided as per the scheme of scoring. The following table for relative score for each factor (Table III) is obtained. The example for F weight in column F and row F is added: (F2 + F3 + F2 + F3) + (F1 + F3) = F14.

Table III

	B	**C**	**D**	**E**	**F**	**G**	**H**	**I**	**Score**	**PERCENTAGE SCORE(WEIGHT)**
A	A3	A1	A3	A3	F2	A2	A2	A3	17	0.218
B		C2	B1	B3	F3	G1	H2	I1	4	0.052
C			C2	C2	F2	G2	H1	C2	8	.0102
D				D3	F3	G2	H2	I2	3	0.038
E					E2	G2	H3	I2	2	0.026
F						F1	F3	I1	14	0.179
G							H2	I1	7	0.090
H								H3	13	0.167
I									10	0.128
Total									78	1.00

All the five cities are given weight (a_{ij}) on a scale of 0 to 100 for each factor A to I. This is done on the basis of extensive field survey, questionnaire and interview of experts in the area. Following observations (see Table IV) are obtained:

Table IV

Sl. No.	Candidate for site	Factors									Weight Score $\Sigma W_i a_{ij}$
		A	B	C	D	E	F	G	H	I	
		0.218	0.052	0.102	0.038	0.026	0.179	0.09	0.167	0.128	
1	City 1	70	75	80	65	25	65	90	90	80	75.445
2	City 2	40	80	80	60	20	90	80	70	80	65.48
3	City 3	100	90	80	70	30	85	60	90	90	85.24*
4	City 4	70	100	45	80	100	40	95	50	70	63.71
5	City 5	55	80	80	70	30	30	30	30	60	48.51

Looking at the last column of the above table city3 scores highest followed by city1, city2, city4 and city5. Therefore, city3 becomes the obvious choice for the site of XYZ company.

Case-2
Selection of Site from Software Company

Objective: To select a site for a software company.

Case details: Samir, non-resident Indian, working with a multinational company in USA is planning to come to India. He is a computer science graduate of IIT, Delhi and PG in software from a Prestigious US Institute. He has now worked for over 10 years in this area. Before coming to India, he asked his friend Rakhesh, to locate a proper site in India. Rakhesh is working with a software industry in Bangalore. Rakhesh scans newspaper and suggests for Bangalore. His recommendation is based on statement of ten software companies based in Bangalore. (See Table A.)

Samir has family links in Delhi. His personnel preference to Delhi is also due to his past association, contact family owned land available for use. He finds it disgusting that Bangalore does not have international airport which Delhi has. He comes to know that IBM has moved to Delhi, of course a small one. He sends e-mail to Rakhesh about his dilemma and asks for more details. Rakhesh sends details (Table A) and recommends for Bangalore.

Table A. Reasons for selecting Bangalore as their site of operations by ten major software companies

Company	*Year of starting*	*Major criteria for selecting Bangalore as their site*	*How they feel now about their decision on site selection*
Novell (fully owned subsidiary of US based Novell.Inc.)	1994	• Fast emerging silicon valley of India. • Required software talents are available. • Easy move from any part of the globe.	• Since the last four years Novell Bangalore has established its creditability as the quality product organisation. • Planning to expand in Bangalore. -Vikram Shah, MD

Tata IBM	Early 90's	• Bangalore is helpful in attracting top IT talent to meet customer satisfaction. • Government of Karnataka is encouraging outlook and supportive means. • Chances of joint venture and Government support in spreading IT message.	• Good decision as the company is continuously growing above the industry rate. -Ravi N. Marwah, MD • Expecting more direct internet connections.
Compaq India	—	• Availability of vast pool of IT skill. • No problem in attracting as well as retaining quality in talent. • Availability of support industries and services at reasonable cost.	• Bangalore continues to the real IT city of India. -Som Mittal, MD
Texas Instrument	1984 (After assess-ment of 9 potential locations)	• Best Hi-Tech climate. • Presence of world class educational institutes. • Unique ability to attract professional talent.	• Today it is the largest R&D centres in Asia and among top three R&D centres outside Asia. • Friendly Government support • Worked closely with Government and DOE in formulating business , friendly software technology park. -Srini Rajaram, MD
Sun micro systems	1995	• Bangalore has the best AIT talent. • Cost of operations, less expensive as compared to Mumbai and New Delhi. • Partners and customers mostly located there.	• Ability of talent and cosmopolitan nature of city, well definitely attract other IT companies. - Bhasker Pramanik, MD
Wiper	Early 80's	• Supportive role of the Government of Karnataka. • Easy land acquisition. • Quick approval of plans. • Incentive for software. • Sales tax exemptions for capital goods and diesel. • Lower power tariff.	• Wiper has emerged as Karnataka's largest software exporter. • Shifted to corporate office from Mumbai to Bangalore. • Investment is over 300 crones. • Generated 4500 employment of IT professionals. • Top ten R&D investors of India. • By year 2000 it plans to hire 1500 IT professionals in Bangalore. -Ashok soota, Group President

Microland	1989	• It is centre of science and educational institutions and electronic industry. • State has maximum number of Engineering colleges. • Better place to mainly due to brilliant weather. -Pradeep Kar, Chairman and MD	• In the long run infrastructure will be a matter of concern, but the climate is so good that it will continue to be here.
Sannata	–	• Choice for single rather than geographical dispersed locations • Better availability of man power • Better quality of living • Infra structure • State Government has IT policy • Better Government interface to support -B.Ramaswamy, President and MD	• No plan had another centre out side Bangalore.
Aditi technologies	–	• Wonderful climate • Good people available locally -Pradeep Singh, CEO	
Infosys technologies	1983	• Excellent reservoir of technical resources • Network of educational institutions • Good climate • Quality of life -Nandan Nilikani Dy., MD	

Based on News Paper (Economic Times) report

REVIEW QUESTIONS

1. In your opinion what problem is likely to come if Delhi is opted as the site for the upcoming software company of Samir?
2. Compile the major pros and cons for Software Company's site in Delhi, and Bangalore. What problems do you perceive if instead a suburban site like Faridabad is opted?
3. Assume that the land in Delhi is not available with Samir. For purchasing, the initial expenses are too high. Do you recommend a rural site, as the cost of land is very cheap there?
4. Enumerate the factors that govern here to take a decision.

   Prepare a comparative chart for the three locations (Delhi, Faridabad and Bangalore) with their advantages and limitations.

CHAPTER 7

Replacement Analysis

PREAMBLE

This chapter deals with the failure pattern of given equipment. It helps distinguish if the equipment is repairable or replaceable (non-repairable). It focuses on assessing replacement age based on the costs. The replacement age of a given machine is discussed in this chapter. It identifies and determines suitable replacement policy, if the equipment is of replaceable type. It helps understand and interpret the failure trend of the equipment if it is repairable type.

CHAPTER OUTLINE

7.1 INTRODUCTION

All those dealing with maintenance and operation of machines will recognise more than an element of truth in *Murphy's* laws pertaining to maintenance stated as follows:

— *If an equipment can fail, it will; and*

— *Failure will usually occur at the most inconvenient time.*

Without any iota of doubt, we all agree that occurrences of failures cannot be avoided completely but failures during operation can be reduced through effective maintenance programmes provided the plant engineer can predict the occurrence of failures. The trend analysis is the best tool for plant engineer in preliminary estimation in this regard. Further, if cost or frequency failures are beyond a certain predetermined level of tolerance it is better to replace partly or completely as the case may be deemed fit.

7.2 REPAIRABLE AND REPLACEABLE SYSTEMS

It is important to first identify the repairable or replaceable features in equipment when its maintainability and reliability characteristics are estimated. Thus, the system is to be first defined as one of the two alternatives, viz. replaceable (non-repairable) system or repairable system. For a non-repairable system such as an electric bulb, invariably we replace, if it fails to function at desired level. The life cycle characteristics can also be used for replacement decisions of such items.

Interestingly, the repairable equipment also needs replacement. This occurs when the equipment is unreliable and/or cost of repair exceeds the cost of the new equipment. That is to say that replacement becomes better than its repair.

In this chapter, the aspect of replacement, i.e., 'when a machine is to be replaced, under various considerations of costs (money value, purchase cost, running cost, scrap value, depreciations, present worth, etc.), reliability (failure probability, group replacement) are discussed.

Thus, the system is considered for replacement based on two important aspects, cost and reliability.

(a) **Cost:** The cost aspect is given preference during replacement decision if the cost of failure is not hazardous or disastrous and that does not lead to catastrophes. The average annual total cost to be spent is compared to its replacement cost for a new one and, thus, a decision is taken on replacement.

(b) **Reliability:** This aspect is considered if the failure is disastrous and can lead to a catastrophe in terms of heavy loss of money, human life, or assets. For example, consider the failures in an aircraft. The cost is immaterial in such cases where its reliable operation is more important than its cost or replacement. In some cases, reliability of the equipment is translated and estimated/measured in terms of cost of failures particularly with reference to cost of damages if it fails when it is under operation.

7.3 COST ANALYSIS OF MACHINE LIFE CYCLE (MLC)—BATHTUB CURVE

Period or time or duration that a machine experiences through various stages from its state of erection and commissioning to the state of discarding or scrapping is often called *machine life cycle* (MLC).

The machine life cycle almost resembles with human life cycle. A human life cycle is considered to have three major stages such as infancy (or childhood), young (or youth), and old age, so also, a machine. The machine life cycle is also assumed to undergo the three phases as shown in the following tabular form (see Table 7.1).

Table 7.1 Stages of machine life cycle

S. No.	*Phase/Stage*	*Other Names Failures*	*Rate of Failure*	*Cause of Failure*
1.	Infancy phase	Early failures or burn-in or teething problems	Decreasing Failure Rate (DFR)	Occur due to design faults or ignorance of user or change of environment
2.	Youth phase	Random failures or rare event failures	Constant Failure Rate (CFR)	Occur once a while, suddenly or slowly develop due to carelessness in maintenance
3.	Old age phase	Wear out failures or age failures	Increasing Failure Rate (IFR)	Occur due to wear/tear, grow slowly and become severe as the usage or age increases

7.3.1 Cost of Early Failures at Infant Stage of Machine

Generally, it is observed that immediately after bringing from the original equipment manufacturer (OEM), a machine shows certain troubles or does not operate efficiently up to the expected satisfaction. This may be due to various reasons. A few reasons commonly claimed are listed below.

- The initial friction between the moving or rotating parts
- Environment is changed and the machine is not adjusted to the environment
- Lack of skill and knowledge of the person working on it
- User is new to the machine or machine is new to the user
- Lapses in design or faulty design
- Errors during the manufacture of the machine
- Erratic operation
- Wrong instructions or guidance by the manufacturer
- Technological misconceptions

An exemplary graphical representation of cost (volume/effect/seriousness/number) of failures with time for a hypothetical machine is shown in Figure 7.1. As shown in the figure, cost of these failures generally start with high volume and are noticed within a short time after putting it under use after bringing it from the manufacturer. However, these costs decrease slowly and become negligible finally after few days or certain time. Since these failures are noticed in the early stage, these failures are called *teething problems or infant problems or early failures or start up problems*. (See chapter 11, section 11.3.1.) These failures and the cost decrease with time, hence, the failure trend can be stated as *decreasing failure rate (DFR)* or the machine is said to be improving

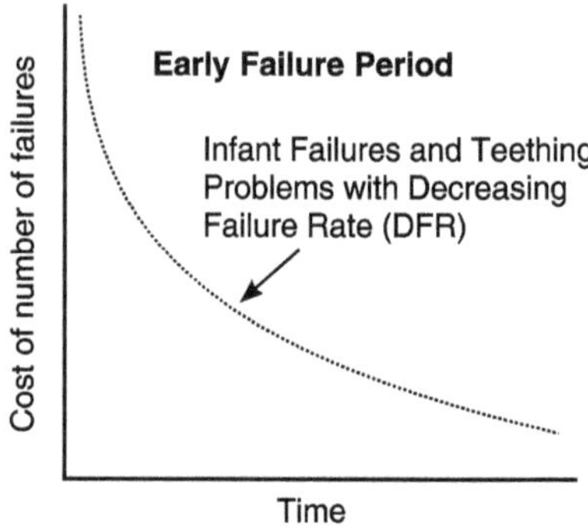

Figure 7.1 Faliure pattern in early failure period

(see Figure 7.1, 7.4 and Table 7.2). During this period, the event of machine failure is generally referred to OEM if the machine is usually covered under guarantee, warrantee, or any contractual maintenance agreement. The costs are usually borne by the OEM. Hence, the costs incurred by the user during this period are considered to be very less or negligible. However, such failures may be disgusting to the user and would cost in terms of lost productive hours.

To control the cost of these failures one has to know thoroughly about the machine, its operation and follow the instructions strictly. By adherence to the instructions given by the OEM strictly, the output may be slow or low or delayed. Yet, it is necessary in the interest of deriving the advantages in the long run in the form of good condition and long life of the machine.

For example, when you buy a bike, the OEM gives you instructions that you have to run the vehicle at not more than 40KMPH up to the first 1500 km. This obviously restricts and slows down the job and does cause inconvenience to the user. But if this is not followed it can lead to a more serious failure by affecting the piston movement in the cylinder or high fuel consumption which will be uneconomical.

Such problems are very common to any new machine. Although there is no considerable direct cost of failures, the cost due to reduced/slow/low production depends on the length of infant failure period which may vary from machine to machine.

7.3.2 Cost of Random or Rare-Event failures

These failures rarely occur and cost cannot be predicted so easily because the machine experiences its youth, i.e., the machine will run with its full efficiency and the user will enjoy its fullest benefit in this period only. The effective utilisation and optimum maintenance can extend this period. Generally, there is a chance that the user may become careless or negligent as he oberves that the machine works or operates efficiently although much care is not taken. This act in fact may or may not result as a failure immediately, but its impact can be seen in the long run which will cost by affecting its life and wearing out or breakae of the parts, etc. However, rarely do some failures appear, known as *random* or *rare-event* or *chance failures* as they occur rarely, random or by chance. (See also Chapter 11, Section 11.3.2.) The reliability of machine will be very high during this stage. Depending on the cost and type of failures, the maintenance strategy is formulated. For low cost failures, usually preventive maintenance and breakdown maintenance are adopted while RCM (reliability-centred maintenance) and CBM (condition based maintenance) techniques may be employed on the machines that are high cost or in case the failure leads to high cost or damage. More or less the occurrence of these failures will be almost constant over a period of time and, hence, the machine is said to be exhibiting constant failure rate (CFR). The average cost or number of failures vs. time graph shows almost straight line parallel to X-axis. (See Figures 7.2 and 7.4 and Table 7.2.)

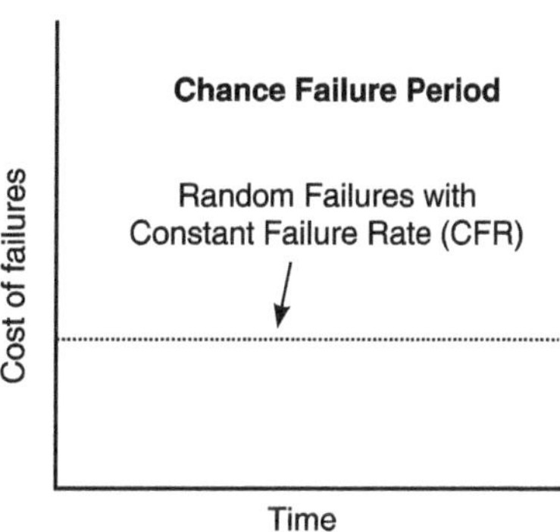

Figure 7.2 Faliure pattern of chance failure period in Machine Life Cycle

7.3.3 Cost of Old Age or Wear Out Failures

This chapter, in fact, lays more emphasis on this stage of machine life cycle. The stage at which the cost of failures of a machine becomes prohibitively high, replacing it or selling it out or discarding is the alternative that can be thought of. The identification of this stage is the prime objective of a maintenance engineer.

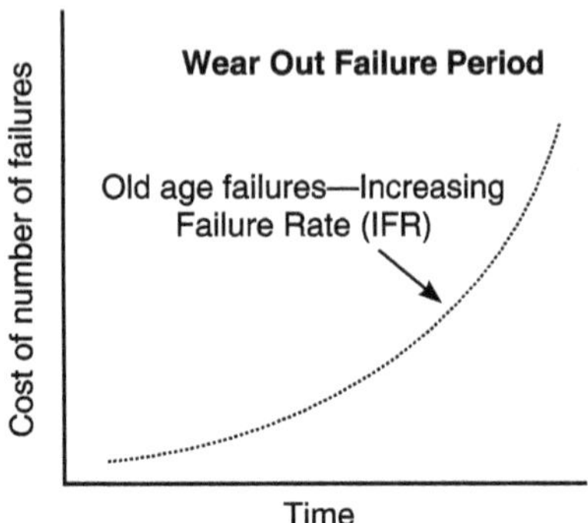

Figure 7.3 Failure pattern of old age faliures in machine life cycle

The efforts put by maintenance during this stage go futile or unnecessary. The costs can go high at this stage if this stage is not detected. Calendar time is not only the scale for detection of this stage. For instance, a machine used sparingly and another machine used continuously may not reach to the old age stage at the same time, though they are bought at the same time. Hence, one should notice that the operational period, conditions maintained while usage, care, and efforts put on it to enhance its life during its infant and young stages, etc., are a few factors governing the old age failures. These failures are mainly because of the worn out parts of the machines. Therefore, these failures are called *wear out failures or old age failures.* (Also see Chapter 11, section 11.3.3.) Since there is tendency of increasing the severity or frequency or volume of these failures, the increasing failure trend or *increasing failure rate (IFR)* will be registered and the machine is said to be deteriorating. (See Table 7.2 and Figures 7.3 and 7.4.). This results in high maintenance and repair costs. Hence, the cost analysis must be done when the maintenance engineer suspects or observes the symptoms. Soon after this stage is noticed in the machine, a plant engineer has to choose one of the following alternative strategies:

- Replacement of the machine with a new one
- Reconditioning of the machine
- Updating the new technological features
- Operating failure and corrective maintenance (as long as its average annual maintenance cost is less than or equal to the interest on the cost of the new machine) or selling in second sale
- Scrapping

It is important to note here that the age of a machine (infant/young/middle/old) is not just decided by the calendar time but by its running time and other factors such as environment, usage, etc. The cost analysis can be a helpful tool for the maintenance engineer in this regard.

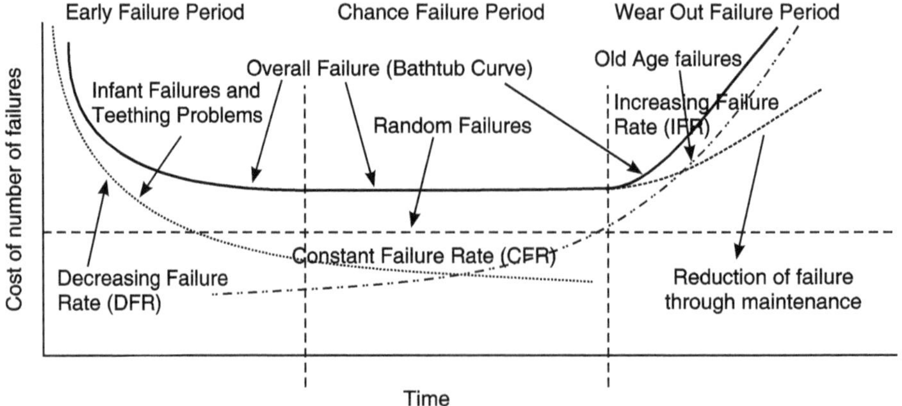

Figure 7.4 Faliure costs in machine life cycle—bathtub curve

Now, consolidating the three types of failure patterns, the overall failure curve is shown in Figure 7.4. The shape of this curve looks like a bathtub and, hence, is named as *bathtub curve (BT curve)*. This curve is also called *machine life cycle (MLC)* curve since it represents all the stages of life of the equipment from its design to discard. This curve can be understood by dividing into three parts, viz. early failure period (or burn in), chance (or random) failure period, and wear out failure period. These periods can also be named as infant, youth, and old age period as analogous to human beings. The summary of failure patterns, failure cause, and appropriate maintenance strategy suitable to the situation are summarised and tabulated in Table 7.2.

Table 7.2 Failure analysis through machine behaviour and machine life cycle

Cost of failure	*Stage in MLC*	*Type of failure*	*Failure rate*	*Probable causes*	*Suitable remedy*
Low cost	Infant	Teething or early failures	DFR	Change in environment, static friction	Refer to OEM for claim warantee
	Youth	Rare events, carelessness	CFR	Operational errors	Close supervision, precautionary measures
	Old age	Age related	IFR	Fatigue, creep	BM/PM
Medium cost	Infant	Installation errors, erratic operation	DFR	Unskilled or untrained operators or supervisors	Refer to OEM and train
	Youth	Random failures	CFR	Over loading, overrun	PM, BM, CBM, RCM
	Old age	Wear and tear	IFR	Overrun	Reconditioning
High cost	Infant	Design faults	–	Technological inefficiency	Refer to OEM to claim guarantee
	Youth	Random failures	CFR	Over loading, overrun	PM/BM
	Old age	Wear and tear	IFR	Weakened parts	Replacement, scrapping, reconditioning

7.4 SITUATIONS WARRANTING REPLACEMENT

The replacement of parts or the entire machine will become necessary in the following cases:

- When average cost of repairs or maintenance or operating goes higher than the costs of the machine or, in other words, the cost of maintenance will increase to such an extent that the average annual repair or maintenance cost is greater than or equal to costs of the new machine.

  $$\text{Annual Maintenance Cost} \geq \text{Cost of New Machine}$$

- When a machine runs with less efficiency and therefore not economical.
- When the machine completely fails to work very frequently by which the production schedules are interrupted.
- If it is expected that the existing model may become obsolete or resale value may drastically come down.

- Modified or new designs in the market may give an edge of advantage such as reduced cost of production or ease and comfort of operation or more functions are available in new design, etc.
- If the item is non-repairable type or use and throw type.

7.5 COSTS INVOLVED IN REPLACEMENT ANALYSIS

While analysing the machine failure, concern with the following costs exist:

- Purchasecost of machine or equipment capital investment.
- Depreciationor scrap value or salvage value.
- Runningcosts including maintenance, repair, and operating (MRO) costs.
- Failure costs and damage costs.
- Cost or loss due to slowdown of production or reduced machine efficiency.

Among the above five, the first three are inevitable, but the failure costs can be prevented by better maintenance policies. On critical examination, it can be noticed that the first three costs depend on age of the machine. These costs vary with the running age as follows:

(a) **Purchase cost** of machine or equipment is considered to be independent of age of machine. However, the interest on the investment is assumed to be lost.

(b) **Resale value or salvage value or scrap value** of the machine decreases with the running age of the machine. As the machine grows old, its resale value comes down. This decrease depends on actual or expected condition of the machine.

(c) **Running cost or operating cost or maintenance cost** of the machine is due to its minor failures or preventive maintenance or operating costs, etc. These will increase on the machine as the age grows. This is assumed to be increasing due to the wear and tear on the moving parts of the machine.

The above costs are shown graphically on a hypothetical machine in Figure 7.5.

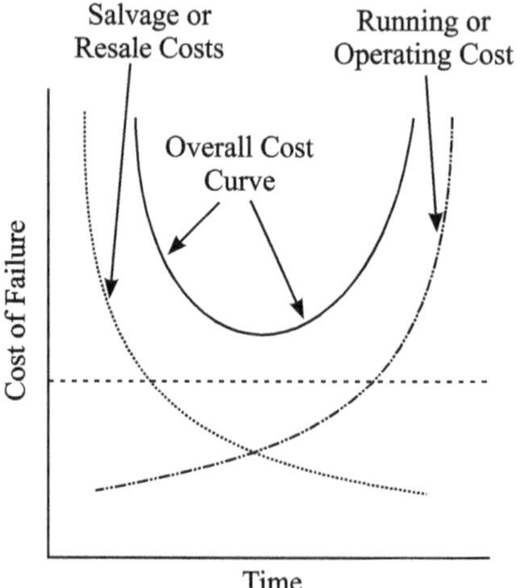

Figure 7.5 Failure costs in replacement analysis

Summing the above three costs, the average total cost decreases for certain period and then increases. The age in the graph that shows its minimum costs will be the optimum age of replacing the machine.

Thus, the concept and the central idea behind this replacement analysis can be encapsulated in the following points:

(a) Collect the data relevant to cost incurred on the machine such as
- Purchase cost
- Resale value
- Running or maintenance cost
- Depreciations or appreciations if any

(b) Cumulate each of the costs.

(c) Sum up to find the total costs directly if the money value is assumed to be not changing.

(d) If the money value is assumed to be changing, then calculate equivalent value by using suitable formula.

(e) Find the mean (average) value for the period. This decreases as the period increases up to certain period and then increases.

(e) Locate the point of the period when this turn is appearing. Till this point the cost incurred is said to be optimum above which it is not advisable to keep the machine because its overall costs will increase then after. Therefore, this will be its replacement age.

BUSTLE - 7.1

List various equipment in the organisation of your choice and identify the stage of each machine in its life cycle. Classify them into three categories as infant, young, and old age machines. Justify your answer as why a particular machine is so classified.

____

____

____

____

Test Your Understanding 7.1

1. Discuss the stages in bathtub curve with suitable examples. Explain various maintenance strategies that are appropriate at each stage.
2. Explain the different costs involved in machine failure and replacement analysis.
3. Describe the nature and behaviour of a hypothetical machine in its infant stage and discuss the maintenance policies befitting to adopt at this stage.
4. Elucidate the nature and behaviour of a hypothetical machine in youth stage and discuss the maintenance policies suitable to adopt at this stage.
5. List the situations warranting the replacement of a machine.
6. Distinguish between repairable and non-repairable systems.
7. Examine the suitability of various maintenance policies for their application in the old age failures.

7.6 REPLACEMENT MODELS

The three models that determine the replacement of machine are

Model 1: Replacement policy for items when money value is assumed to remain unaltered with time.

Model 2: Replacement policy for items when money value changes with time.

Model 3: Items that fail completely—group replacement policy.

7.7 MODEL 1: REPLACEMENT POLICY WHEN MONEY VALUE DOES NOT CHANGE WITH TIME

Let us now find the optimal policy for the case of replacement when money value does not change with time.

Let C = Capital or purchase cost of new item.

S = Scrap or salvage or resale value of the item at the end of 'n' or 't' years.

R_n or $R(t)$ = Running (or operating) cost for the year 'n' or 't'.

n or t = Replacement age of the equipment.

Here two cases arise.

Case I: When time t is continuous variable:

If the equipment is used for 't' years, then the total cost incurred over this period is given by

T_c = Capital (Purchase) Cost – Scrap Value (at the end of t yrs) + Running Cost (for t yrs)

$$= \frac{1}{n}\left\{C - S\int_0^n R(t)dt\right\}$$

Therefore, average cost per unit time incurred over the period of n years is

$$ATC_n = \frac{1}{n}\left\{C - S\int_0^n R(t)dt\right\} \qquad \text{... (1)}$$

To obtain optimal value of n for which ATC_n is minimum, differentiate ATC_n with respect to n and set the first derivative equal to zero, i.e., minimum of ATC_n.

$$\frac{d}{dn}[ATCn] = -\frac{1}{n}[C - S] + \frac{R(n)}{n} - \frac{1}{n}\int_0^n R(t)dt = 0$$

or

$$R(n) = -\frac{1}{n}\left\{C - S + \int_0^n R(t)dt\right\} \cdot n \neq 0$$

$$R(n) = ATC_n \qquad \text{... (2)}$$

Hence, the following replacement policy can be derived with the help of equation (2).

Policy: Replace the equipment when the average annual cost for n years becomes equal to the current/running cost.

i.e., $$R(n) = \frac{1}{n}\left\{C - S + \int_0^n R(t)dt\right\} \qquad \text{... (3)}$$

Case II: When time t is a discrete variable:

The average cost incurred over the period n is given by

$$\text{ATC}_n = \frac{1}{n}\left\{C - S + \sum_{t=0}^{n} R(t)\right\}$$

If C – S and $\sum_{t=0}^{n} R(t)$ are assumed to be monotonically decreasing and increasing respectively, then there exists a value of 'n' for which ATC_n is minimum. Thus, there will be inequalities.

$$\text{ATC}_{n-1} > \text{ATC}_{n+1}$$

$$\text{ATC}_{n-1} - \text{ATC}_n > 0$$

$$\text{ATC}_{n+1} - \text{ATC}_n > 0$$

Rewriting equation (3) for period $n + 1$, we get

$$\text{ATC}_{n+1} = \frac{1}{n+1}\left\{C - S + \sum_{t=1}^{n+1} R(t)\right\}$$

$$= \frac{1}{n+1}\left\{C - S + \sum_{t=1}^{n} R(t) + R(n+1)\right\}$$

$$= \frac{1}{n+1} - \frac{\left\{C - S + \sum_{t=1}^{n} R(t)\right\}}{n} + \frac{R(n+1)}{n+1}$$

$$= \frac{1}{n+1} \cdot \text{ATC}_n + \frac{R(n+1)}{n+1}$$

Therefore, $$\text{ATC}_{n+1} - \text{ATC}_n = \frac{n}{n+1}\text{ATC}_n + \frac{R(n+1)}{n+1} - \text{ATC}_n$$

$$= \frac{R(n+1)}{n+1} + \text{ATC}_n\left[\frac{n}{n+1} - 1\right]$$

$$= \frac{R(n+1)}{n+1} - \frac{\text{ATC}_n}{n+1}$$

Since $\text{ATC}_{n+1} - \text{ATC}_n > 0$, we get

$$\frac{R(n+1)}{n+1} - \frac{\text{ATC}_n}{n+1} > 0$$

i.e., $R(n+1) - \text{ATC}_n > 0$

or $R(n+1) > \text{ATC}_n$ ... (4)

Similarly, $\text{ATC}_{n-1} - \text{ATC}_n > 0$ implies that $R(n) < \text{ATC}_{n-1}$

This provides the following replacement policy.

Policy 1: If, the next year, running cost, $R(n + 1)$ is more than average cost of nth year, ATC_n then it is economical to replace at the end of n years.

i.e.,
$$R(n+1) > \frac{1}{n}\left\{C - S + \sum_{t=0}^{n} R(t)\right\}$$

Policy 2: If the present year's running cost is less than the previous year's average cost, ATC_{n-1} then do not replace.

i.e.,
$$R(n) < \frac{1}{n-1}\left\{C - S + \sum_{t=0}^{n-1} R(t)\right\}$$

The Algorithm (Procedural Steps)

The procedure for obtaining the decision when to replace the equipment in the case of money value not changing with time can be outlined as follows:

Step 1: Draw a table with columns as shown below and enter the values of column 1, 2, 3, and 4 as given in the problem here.

Year of service (n)	*Cost of equipment (C)*	*Salvage value (S)*	*Running cost (R_n)*	*Net value = (C – S)*	*Cumulative running cost ΣR_n*	*Total cost (TC) = (C – S) + ΣR_n*	*Average total cost = ATC = 1/n[(C – S) + ΣR_n]*
(1)	(2)	(3)	(4)	(5) = (2) – (3)	(6) = Σ4	(7) = (5) + (6)	(8) = (7)/(1)

Step 2: Calculate net value by difference of cost and salvage value (C – S), i.e., second column value – third column value and enter it in column 5.

Step 3: Calculate cumulative running cost, i.e., ΣR_n culmination of column 4 and enter it in column 6.

Step 4: Calculate total cost, TC = (C – S) + ΣR_n, i.e., sum of fifth column value and sixth column value to enter in column 7.

Step 5: Calculate average total cost for n years, i.e., divide seventh column value by first column value, i.e., ATC = 1/n [(C – S) + ΣR_n] and enter it in column 8.

Step 6: Observe the values in column 8 and identify the minimum value. The year corresponding to the minimum value is the age of the equipment to be replaced.

ILLUSTRATION 7.1:

A firm is thinking of replacing a particular machine whose cost price is ₹ 12,200. The scrap price of this machine is only ₹ 200. The maintenance costs are estimated to be as follows:

Year	1	2	3	4	5	6	7	8
Maintenance cost	220	500	800	1,200	1,800	2,500	3,200	4,000

Determine when the firm should get the machine replaced.

Solution.

The calculations of average running cost per year during the life of the machine are shown in the following table:

Cost of machine (C) = ₹ 12,200: S = ₹ 200: C – S = ₹ 12,000

Year (n)	*Cost (C)*	*Salvage value (S)*	*Running cost (₹) R(n)*	*Depreciation cost (₹) C-S*	*Cumulative running cost (₹) ∑R(n)*	*Total cost (₹) TC*	*Average cost (₹) per year ATC_n*
(1)	(2)	(3)	(4)	(5) = (2) – (3)	(6) = ∑(4)	(7) = (5) + (6)	(8) = (7)/(1)
1	12,200	200	220	12,000	220	12,220	12,220
2	12,200	200	500	12,000	720	12,720	6,360
3	12,200	200	800	12,000	1,520	13,520	4,506.67
4	12,200	200	1,200	12,000	2,720	14,720	3,680
5	12,200	200	1,800	12,000	4,520	16,520	3,304
6	**12,200**	**200**	**2,500**	**12,000**	**7,020**	**19,020**	3,170 ←
7	12,200	200	3,200	12,000	10,220	22,220	3,174.29
8	12,200	200	4,000	12,000	14,220	26,220	3,277.5

From the above table, it may be noted that the average cost per year, ATC_n is minimum in the sixth year i.e., ₹ 3170. And this average cost increases from the seventh year onwards. Hence, the machine should be replaced after 6 years.

ILLUSTRATION 7.2:

A plant manager is considering replacement policy for a new machine. He estimates the following cost in ₹:

Year	*1*	*2*	*3*	*4*	*5*	*6*
Replacement cost at the beginning of year	100	110	125	140	160	190
Salvage value at the end one year	60	50	40	25	10	0
Operating costs	25	30	40	50	65	80

Find an optimal replacement policy and corresponding minimum cost.

Solution.

The calculations for replacement of the machine are shown in the following table:

Year	*Replacement cost*	*Resale value (S)*	*Net value*	*Operating cost = R_n*	*Cumulative operating cost = ΣR_n*	*Total cost = $C - S + \Sigma R_n$*	*Average Annual Cost ATC_n*
(1)	(2)	(3)	(4) = (2) – (3)	(5)	(6) = Σ(5)	(7) = (4) + (6)	(8) = (7)/(1)
1	100	60	40	25	25	65	65
2	110	50	60	30	55	115	57.5←
3	125	40	85	40	95	180	60
4	140	25	115	50	145	260	65
5	160	10	150	65	210	360	72
6	190	0	190	80	290	480	80

From the above table it may be noticed that the average cost per year, ATC_n is minimum in the second year, i.e., ATC_2 is ₹ 57.50 which is less than ATC_1 (₹ 65) and ATC_3 (₹ 60). Hence, the machine should be replaced at the end of second year.

ILLUSTRATION 7.3:

A fleet owner finds from his past records that the cost per year of running a vehicle whose purchase price is ₹ 50,000 are as follows:

Year	1	2	3	4	5	6	7
Running cost (₹)	5,000	6,000	7,000	9,000	1,500	16,000	18,000
Resale value (₹)	30,000	15,000	7,500	3,750	2,000	2,000	2,000

There after running costs increase by ₹ 2000, but resale value remains constant at ₹ 2000. At what age is a replacement due?

Solution.

The required calculations are shown in the following table:

Year (1)	C (2)	S (3)	C – S (4) = (2) – (3)	R(n) (5)	Cum R(n) (6) = Σ(5)	TC (7) = (4) + (6)	ATC $8 = \frac{(7)}{(1)}$
1	50,000	30,000	20,000	5,000	5,000	25,000	25,000
2	50,000	15,000	35,000	6,000	11,000	46,000	23,000
3	50,000	7,500	42,500	7,000	18,000	60,500	20,167
4	50,000	3,750	46,250	9,000	27,000	73,250	18,313
5	50,000	2,000	48,000	15,000	42,000	90,000	18,000
6	**50,000**	**2,000**	**48,000**	**16,000**	**58,000**	**1,06,000**	**17,667**
7	50,000	2,000	48,000	18,000	76,000	1,24,000	17,714
8	50,000	2,000	48,000	20,000	96,000	1,44,000	18,000
9	50,000	2,000	48,000	22,000	1,18,000	1,66,000	18,444

The average annual total cost is minimum (17,666.67) during the sixth year and, hence, the machine should be replaced at the end of the sixth year.

ILLUSTRATION 7.4:

Machine A costs ₹ 45,000 and the operating costs are estimated at ₹ 1,000 for the first year, increasing by ₹ 10,000 per year in the second and subsequent years. Machine B costs ₹ 50,000 and operating costs are ₹ 2,000 for the first year, increasing by ₹ 4,000 in the second and subsequent years. If we now have a machine of type A, should we replace it with B? If so when? Assume that both machines have no resale value and the future costs are not discounted.

Solution.

The calculations of average costs running per year during the life of machine A and B are shown in the tables given below:

Table A Calculations of average annual total cost for machine A

Year of service(n)	Running cost in ₹ R(n)	Cumulative running cost in ₹ Σ R(n)	Depreciation cost (₹) C – S	Total cost (₹) TC	Average cost (₹) Rs. ATC_n
(1)	(2)	(3)	(4)	(5)	(6)
1	1000	1,000	45,000	46,000	46,000
2	11,000	12,000	45,000	57,000	28,500
3	21,000	33,000	45,000	78,000	26,000
4	31,000	64,000	45,000	1,09,000	27,200
5	41,000	1,05,000	45,000	1,50,000	30,000
6	51,000	1,56,000	45,000	2,01,000	33,500

From the above table it may be noted that the average running cost per year is lowest in the third year, i.e., ₹ 26,000. Hence, machine A should be replaced after every three years of service.

Table B: Calculation of average annual total cost for machine B

Year of service N	*Running cost (₹) R(n)*	*Cumulative running cost (₹) Σ R(n)*	*Depreciation cost (₹) C – S*	*Total cost (₹) TC*	*Average cost (₹) ATC_n*
(1)	(2)	(3)	(4)	(5)	(6)
1	2,000	2,000	50,000	52,000	52,000
2	6,000	8,000	50,000	58,000	29,000
3	10,000	18,000	50,000	68,000	22,667
4	14,000	32,000	50,000	82,000	20,500
5	18,000	50,000	50,000	1,00,000	20,000
6	22,000	72,000	50,000	1,22,000	20,333

From the above table it may be noted that the average running cost per year is lowest in the fifth year, i.e., ₹ 20,000. This cost is less than the average running cost (₹ 26,000) per year for machine A. Hence, machine A should be replaced by machine B.

Now to find the time of replacement of machine A by machine B, the total cost of machine A in the successive years is computed as follows:

Year	1	2	3	4
Total cost incurred (₹)	46,000	57,000 – 46,000 = 11, 0000	78,000 – 57,000 = 21,000	1,90,000 – 78,000 = 31,000

Machine A should be replaced by machine B at the time (age) when its running cost for the next year exceeds the lowest in average running cost (₹ 20,000) per year of machine B.

Calculations show that the running cost (₹ 21,000) of machine A in the third year is more than the lowest in average cost (₹ 20,000) per year of machine B. Hence, machine A should be replaced by machine B after two years.

BUSTLE - 7.2

Explain the type of replacement policy you are following in your organisation. Discuss its merits and demerits that you experience practically.

Test Your Understanding 7.2

1. The cost of a machine is ₹ 6100 and its scrap value is only ₹ 100. The maintenance costs found from experience are given below.

Year	1	2	3	4	5	6	7	8
Maintenance cost (₹)	100	250	400	600	900	1,200	1,600	2,000

When should the machine be replaced?

Answer: At the end of the sixth year = ₹ 1,583.33

2. A truck owner finds from his past experience that the maintenance costs are ₹ 200 for the first year and then increase by ₹ 2,000 every year. The cost of truck type A is ₹ 9,000.
   (i) Determine the best age at which to replace the truck.
   (ii) If the optimum replacement is followed, what will be the average yearly cost of owning and operating the truck?
   (iii) Truck type B costs ₹ 20,000. Annual operating costs are ₹ 400 for the first year and then increase by ₹ 800 every year. The truck owner has now the truck type A which is one year old. Should it be replaced by B type, and if so, when?

Answer: (i) After every three years, truck A is to be replaced (₹ 5,200).

After every six years, truck B (₹ 5,733.3) is replaced.

(ii) Average yearly cost is ₹ 5,200.

(iii) Comparing both costs, truck B is lower than that of truck A from the third year onwards. Hence, after three years, replace truck A by truck B.

3. (i) Machine A costs ₹ 9,000, annual operating costs are ₹ 200 for the first year, and then increase by ₹ 2,000 every year. Determine the best age to replace the machine. If the optimum replacement policy is followed, what will be the average yearly cost of owning and operating the machine?

   (ii) Machine B costs ₹ 10,000, annual operating costs are ₹ 400 for the first year, and then increase by ₹ 800 every year. You now have a machine of type A which is one year old. Should you replace it with B, if so, when?

   Answer: (i) Total cost in the third year is least, i.e., ₹ 5,200, hence it is better to replace it at the end of the third year. (ii) Average annual cost is least at the end of fifth year. The total cost is ₹ 2,200 for one-year-old machine A remains less than the minimum average cost ₹ 400 of machine B until the second year. Therefore, replace machine A before the third year.

4. Running cost and resale value of a small machine whose purchase price is ₹ 6000 are given below.

Year	1	2	3	4	5	6	7
Running Cost (₹)	1,000	1,200	1,400	1,800	2,300	2,800	3,400
Resale Value (₹)	3,000	1,500	750	375	200	200	200

   Determine at what age replacement is due?

   Let us assume that the owner has three of the above type machines, two of which are two years old.

   Now he is considering a new type of equipment with 50% more capacity than one of the old ones at a unit price of ₹ 8000 with the running costs and resale price as follows:

Year	1	2	3	4	5	6	7	8
Running Cost (₹)	1,200	1,500	1,800	2,000	3,100	4,000	5,000	6,100
Resale Price (₹)	4,000	2,000	1,000	500	300	300	300	300

   Assuming the loss of flexibility due to fewer machines is of no importance and that he will continue to have sufficient work for three of the old machine, what should his policy be?

   Answer: The machine is to be replaced at the end of the fifth year. It is observed that the costs of old machine during the first three years are ₹ 4,000, ₹ 2,700 and ₹ 2,150 while during the fourth, fifth, and sixth years the costs are ₹ 2,175, ₹ 2,475 and, ₹ 2,800, respectively.

   Thus, total cost during the second year is $2 \times 2175 + 2150 = 6500$ and during the third year is $2 \times 3475 + 2175 = 7129$ whereas the minimum average cost for two larger machines is $2 \times 3540 = 7080$. Hence, all the three small machines should be replaced after two years before any of them reaches the normal replacement age of 5 years.

5. Fleet of cars have increased their costs as they continue in service due to increased direct operating cost (gas and oil) and increased maintenance (repairs, tyres, batteries, etc.). The initial cost is ₹ 3,50,000 and the trade in value drop as time passes until it reaches a constant value of ₹ 40,000. Given the cost of operating, maintaining, and the trade in value, determine the proper length of service before cars should be replaced.

Years of service	1	2	3	4	5
Yr end trade in value (₹)	2,90,000	2,10,000	1,50,000	1,10,000	40,000
Annual operating cost (₹)	11,500	12,800	13,600	14,000	15,000
Annual maintaining	3000	5000	8000	12,000	15,000

   Answer: $ATC_3(84.43) > ATC_4(79.97) < ATC_5(83.98)$, therefore, the machine is to be replaced at the end of the fourth year.

7.8 MODEL 2: REPLACEMENT POLICY FOR ITEMS WHEN MONEY VALUE CHANGES WITH TIME

Money value changes with time. Suppose you keep ₹ 100 in a bank and you get an interest at the rate of 10% p.a. After one year you will receive ₹ 110. Thus, it can be concluded that "if we have ₹ 100 today, it is worth having ₹ 110 after one year". Similarly, a rupee possessed today is worth 1.1 after one year at 10%.

In the other way, if ₹ 1.10 after one year is ₹ 1.00 today, ₹ 1.00 after one year will be 1/1.1 or $(1.1)^{-1}$ today.

In a similar way, one rupee today will be $(1.1)^2$ two years hence because it is 1.1 after one year and 1.1 x 1.1 in the next i.e., second year. Thus, a rupee after two years is worth $(1.1)^{-2}$, when money value is changing at 10%.

Now, if the money value is supposed to be changing at the rate of r percent, the present worth of one rupee, n years hence will be $(1 + r)^{-n}$ or $1/(1 + r)^n$.

Present Worth Factor (PWF): It is the factor that converts a rupee 'n' years hence changing (money grows but value decreases) at the rate of 'r' percent is worth today. This is denoted by $v = (1 + r)^{-n}$. This is also called "discount rate" or "depreciation rate" or "present worth factor".

Now let C = Initial cost of the equipment

And if R_1 = Operating cost in first year, (present worth is also R_1)

If R_2 = Operating cost in second year present worth of

$R_2 = (1 + r)^{-1} R_2$ or $v\,R_2$

If R_3 = Operating cost in third year, and present worth of

$R_3 = (1 + r)^{-2} R_3$ or $v^2 R_3$ and so on

If R_n = Operating cost in n^th^ year present worth of

$R_n = v^{n-1} R_n$

Thus, the present value of all future discounted cost in 'n' year assuming scrap value of the equipment be zero, is

$$P_n = C + R_1 + vR_2 + v^2R_3 + \ldots + v^{n-1} R_n$$

Thus, P_n is the amount of money required to pay all future costs of purchasing the equipment and operating it assuming that it is to be replaced after 'n' years.

Now if we assume that the manufacturer invests the amount P_n by borrowing at the rate of interest 'r' and repays it in 'n' years with a fixed annual installment 'x' on diminishing balance, then

$$P_n = x + vx + v^2x + \ldots + v^{n-1}x$$

$$= [1 + v + v^2 + \ldots + v^{n-1}]x$$

$$= \left[\frac{1 - v_n}{1 - v}\right]x$$

$$= \left[\frac{1 - v}{1 - vn}\right]P_n$$

Hence, the best period to replace the machine is the period n which minimises $\frac{1-v}{1-vn} \cdot P_n$. But $1v$ = A positive constant quantity and so we can write $F_n = \frac{Pn}{1-vn}$ and find out the value of n the period at which to replace the machine that minimises F_n. Since n can assume only discrete values [1, 2, 3, …] we can use the method of finite differences to calculate its optimal values.

F_n will be minimum if $\Delta F_{n-1} < 0 < \Delta F_n$ ... (5)

Now $\Delta F_n = F_{n+1} - Fn$ ... (6)

$$= \frac{P_{1+1}}{1-V^{n+1}} - \frac{P_n}{1-V^n} = \frac{(1-V^n)P_{n+1} - (1-V^{n+1})P_n}{(1-V^{n+1})(1-V^n)}$$

$$= \frac{1}{(1-V^{n+1})(1-V^n)}\left[(P_{n+1} - P_n) + (V^{n+1}P_n - V^n P_{n+1})\right] \quad \text{... (7)}$$

We have $P_{n+1} = (C + R_1 + vR_2 + ... + v^{n-1} R_n) + v^n R_{n+1}$

$= P_n + v^n . R_{n+1}$ or $v^n R_{n+1} = P_{n+1} - Pn$

Hence, $$\Delta F_n = \frac{1}{(1-V^{n+1})(1-V^n)}\left[V^n R_{n+1} + V^{n+1}P_n - V^n[P_n + V^n R_{n+1}]\right]$$

$$= \frac{1}{(1-V^{n+1})(1-V^n)}\left[V^n R_{n+1}(1-V^n) - V^n P_n(1-V)\right]$$

$$= \frac{V^n(1-V)}{(1-V^{n+1})(1-V^n)}\left[\frac{1-V^n}{1-V}R_{n+1} - P_n\right] \quad \text{... (8)}$$

$$= \text{A positive constant } x \left[\frac{1-V^n}{1-V}R_{n+1} - P_n\right]$$

Hence, F_n has always the same sign as $\left[\frac{1-V^n}{1-V}R_{n+1} - P_n\right]$

∴ From equation (5), n will be optimal if

$$\left[1 - \frac{V^{n+1}}{1-V}R_n - P_{n-1}\right] < 0 < \left[\frac{1-V^n}{1-V}R_{n+1} - P_n\right] \quad \text{... (9)}$$

From equation (9), we have $\frac{1-V^n}{1-V}R_{n+1} - P_n > 0$

or $$R_{n+1} > P_n\left[\frac{1-V}{1-V^n}\right] \text{ or } R_{n+1} > \frac{P}{\left[(1-V^n)/(1-V)\right]}$$

or $$R_{n+1} > \frac{C + R_1 + VR_2 + V^2R_3 + ... + V^{n-1}R_n}{1 + V + V^2 + ... + V^{n-1}}$$

or $$R_{n+1} > \frac{C + \sum_{r=1}^{n} R_r V^{r-1}}{\sum_{r=1}^{n} V^{r-1}} \quad ...(10)$$

The right hand expression is the weighted average (denoted by w_r) of all costs up to and including period (n – 1). The weights are 1, v, v^2. . v^{n-1} are the discount factors applied to the costs for each period.

From equation (9), the left hand side of expression can be expressed as

$$R_{n+1} > \frac{c + \sum_{r=1}^{n} R_n V^{r-1}}{\sum_{r=1}^{n} V^{r-1}} \quad ...(11)$$

From equations (10) and (11) we conclude that

- The machine should be replaced if the next period cost is greater than the weighted average of previous costs.
- The machines should not be replaced if the next period's cost is less than the weighted average of previous costs.

Note: When money value is not considered, v = 1

$$R_{n+1} > \frac{C + R_1 + R_2 + ...\ R_n}{1 + 1 + ...\text{ to } 'n' \text{ terms}} \text{ or } R_{n+1} > \frac{P_n}{n} \text{ where } \frac{P_n}{n}$$

Average yearly cost with no resale value.

This is identical to the previous case when money value was ignored. In real practice, replacement policy is greatly influenced by complicated tax laws prevailing. In actual dealings, the influence of tax has got to be taken into consideration. However, this aspect has not been included as it is not in the scope of this book.

Steps to find the policy when money value changes with time

***Step 1*:** Note the values of capital cost of machine, salvage value, rate of depreciation PWF, etc.

***Step 2*:** Construct the tabular form as given below, and enter the first 2 columns as per the given data.

Year	*Running cost (R_n)*	*PWF or discount factor v^{n-1}*	$R_n\ v^{n-1}$	$C + \Sigma R_n\ v^{n-1}$	$\Sigma\ v^{n-1}$	*Average annual total cost ATC_n*
(1)	(2)	(3)	(4) = (2) – (3)	(5) = C + Σ(4)	(6) = Σ(3)	(7) = (5)/(6)

***Step 3*:** Calculate the present worth factor for each year by the formula $\frac{1}{(1+r)^{n-1}}$ and put in column 3.

***Step 4*:** Calculate $R_n v^{n-1}$, i.e., the product of columns 2 and 3, values, and enter them in column 4.

***Step 5*:** Column 5 is summation of cumulative running cost and capital cost including (salvage if any), i.e., $C + \Sigma R_n\, v^{n-1}$.

***Step 6*:** Find cumulative of column 3 and enter in column 6, i.e., [$\Sigma\, v^{n-1}$]

***Step 7*:** Calculate the value of column 5 divided by column 6, i.e., $(C + \Sigma R_n v^{n-1})/\ \Sigma v^{n-1}$ and enter in column 7.

***Step 8*:** The lowest value in column 7 corresponds to the year in which the machine is to be replaced.

***Note*:** *Readers are advised to check the problem thoroughly and note whether depreciated amount is given or discount rate is given, etc.*

ILLUSTRATION 7.5:

The initial cost of equipment is ₹ 5000. The running cost varies as given here.

Year	1	2	3	4	5	6	7
Running Cost (₹)	400	500	700	1,000	1,300	1,700	2,100

Allowing a discount rate of 10% (or discounted at 0.90), find optimal replacement interval.

Solution.

The required calculations are shown in the tabular form as follows:

Year	*Running cost (R_n)*	*Discount factor v^{n-1}*	$R_n\, v^{n-1}$	$C + \Sigma R_n\, v^{n-1}$	$\Sigma\, v^{n-1}$	*Average annual total cost ATC_n*
(1)	(2)	(3)	(4) = (2) – (3)	(5) = C + Σ(4)	(6) = Σ(3)	(7) = (5)/(6)
1	400	1.000	400	5400	1.000	5,400
2	500	0.900	450	5850	1.900	3,079
3	700	0.810	567	6417	2.710	2,368
4	1,000	0.729	729	7146	3.439	2,077
5	1,300	0.656	853	7999	4.095	1,953
6	**1,700**	**0.599**	**1016**	**9015**	**4.694**	**1,921**
7	2,100	0.531	1115	10117	5.225	1,936

From the above table, we observe that 1953 > 1921 < 1936, i.e., $ATC_5 > ATC_6 < ATC_7$

Therefore, the optimum replacement interval is 6 years.

ILLUSTRATION 7.6:

A manufacturer is offered two machines A and B. A is priced at ₹ 5000 and running costs are estimated at ₹ 800 for each of the first five years, increasing by ₹ 200 per year in the sixth and subsequent years. Machine B, which has the same capacity as A, costs ₹ 2,500 but will have running costs of ₹ 1,200 per year for six years, increasing by ₹ 2,000 per year thereafter. If the money is worth 10% per year, which machine should be purchased assuming that both machines will eventually be sold for a scrap at a negligible value?

Solution.

Table for A: R = 10% $C_A = 5000$

Year	R_i	v^{n-1}	$R_i v^{n-1}$	$C + \sum_{I=1}^{N} R_i v^{n-1}$	$\sum_{I=1}^{n} v^{n-1}$	*ATC*
(1)	(2)	(3)	(4) = (2) × (1)	(5) = C + ∑(4)	(6) = ∑ (3)	(7) = (5)/(6)
1	800	1.0000	800	5800	1.0000	5,800.00
2	800	0.9091	727	6527	1.9092	3,418.88
3	800	0.8264	661	7188	2.7355	2,627.67
4	800	0.7513	601	7789	3.4868	2,233.85
5	800	0.6830	546	8335	4.1698	1,998.89
6	1000	0.6209	621	8956	4.7907	1,869.45
7	1200	0.5645	677	9633	5.3552	1,798.81
8	1400	0.5132	718	10351	5.8684	1,763.85
9	**1600**	**0.4665**	**746**	**11097**	**6.3349**	**1,751.72**
10	1800	0.4241	763	11860	6.7590	1,754.70

Table for B: R = 10% $C_B = 2500$

Year	R_i	v^{n-1}	$R_i v^{n-1}$	$C + \sum_{I=1}^{N} R_i v^{n-1}$	$\sum_{I=1}^{n} v^{n-1}$	*ATC*
(1)	(2)	(3)	(4) = (2) × (1)	(5) = C + ∑(4)	(6) = ∑ (3)	(7) = (5)/(6)
1	1200	1.0000	1200.00	3700	1.0000	3,700
2	1200	0.9091	1090.91	4790.91	1.9092	2,509.51
3	1200	0.8264	991.98	5782.59	2.7355	2,113.91
4	1200	0.7513	901.56	6684.5	3.4868	1,916.99
5	1200	0.6830	819.60	7503.75	4.1698	1,799.55
6	1200	0.6209	745.08	8248.83	4.7907	1,721.84
7	1400	0.5645	790.30	9039.13	5.3552	1,687.92
8	**1600**	**0.5132**	**821.12**	**9860.25**	**5.8684**	**1,680.23**
9	1800	0.4665	839.70	10699.95	6.3349	1,689.25
10	2000	0.4241	848.20	11548.15	6.7590	1,708.56

From Table for A

$$ATC_8\,(1763.85) > ATC_9\,(1751.72) < ATC_{10}\,(1754.70)$$

Therefore, replacement period for A = 9th year

From Table for B

$$ATC_7\,(1687.92) > ATC_8\,(1680.23) < ATC_9\,(1689.25)$$

Therefore, replacement period for B = 8th year

The fixed annual payment for $A = x_A = \dfrac{1-v}{1-v^2}\ P(9)$

$$\frac{1-0.9091}{1-(0.9091)9}\times 11097 = 1752$$

Similarly, $x_B = \dfrac{1-0.9091}{1-(0.9091)8}\times 9860.25 = 1680 < x_A$

Hence, purchase machine B instead of A.

Alternatively, the weighted damage cost in 9 years for A = 1781.72 and that for B in 8 years = 1680.23 which is the lowest.

Therefore, purchase B.

Test Your Understanding 7.3

1. A machine has been purchased at a cost of ₹ 1,60,000. The value of the machine is depreciated in the first three years by ₹ 20,000 each year and ₹ 16,000 per year thereafter. Its maintenance and operating costs for the first three years are ₹ 16,000, ₹ 18,000 and ₹ 20,000 in that order and increase by ₹ 4,000 every year. Assuming an interest rate of 10%, find the economic life of the machine.

   Answer: The machine should be replaced after 2 years.

2. Let the value of the money be assumed to be 10% per year and suppose that machine 'A' is replaced after every three years where as machine B is replaced after every six years. The yearly costs of both the machines are given as:

Year	1	2	3	4	5	6
Machine A (₹)	1,000	200	400	1,000	200	400
Machine B (₹)	1,700	100	200	300	400	500

   Determine which machine should be purchased?

   Answer: Considering the life periods of the machines A and B, i.e., 3 and 6 years, respectively, the average years' costs will be ₹ 1,512.4/3 = ₹ 504.13 and ₹ 2,765.3/6 = ₹ 460.9. The cost of the machine A (₹ 2,648 is less than that of machine B (₹ 2,765.3). Machine A is to be purchased.

3. A manual stamper currently valued at ₹ 1,000 is expected to last 2 years and costs ₹ 4,000 per year to operate. An automatic stamper which can be purchased at ₹ 3,000 will last for 4 years and can be operated at an annual cost of ₹ 3,000. If money carries the rate of interest 10% per annum, determine which stamper should be purchased.

   Answer: Since the present worth of future costs of the automatic stamper (₹ 13,460) is less than that of manual stamper (₹ 15,773), it is better and profitable to purchase an automatic stamper.

4. A machine has initial investment of ₹ 30,000 and its salvage value at the end of 'i' years of its use is estimated as ₹ 30,000/('i'+1). The annual operating and maintenance cost in the first year is ₹ 15,000 and increases by ₹ 1,000 in each subsequent years for the first five years and increases by ₹ 5,000 in each year thereafter. Replacement policy is to be planned over a period of seven years. During this period, the cost of capital may be taken as 10% per year. Solve the problem for optimal replacement.

   Answer: Since the lowest annual average cost occurs during the fifth year, it is better to replace at the end of every 5 years.

5. The cost of a new machine is ₹ 5,000 and the maintenance cost of nth year is given by Rn = $500(n-1)$, n = 1,2,3... Suppose the discount rate per year is 0.5, after how many years will it be economical to replace the machine by a new one?

   Answer: 5 years

7.9 MODEL 3: GROUP REPLACEMENT POLICY

Group Replacement of Items that Fail Completely

This policy is concerned with the items that either work perfectly or work partially or inefficiently or fail completely. This situation generally happens when the system consists of a large number of identical low cost items that are increasingly liable to fail with age. In such cases, the replacement of individual items will incur a set of costs, which is independent of the number replaced. However, it may be advantageous to replace all the items at a time at a fixed interval. This policy is known as *group replacement policy* and is very attractive, particularly when

- the value of any individual item is so small.
- the cost of keeping records of individual ages is high that cannot be justified.
- the purchase of such identical items in bulk can be had at discounted rate.
- average individual replacement would be costlier than the average group replacement.
- sufficient number of standby machines are available.
- new designs of the equipment considerably increase the production rate.

In all the above cases, two types of replacement policies considered are:

(a) **Individual Replacement:** Under this policy, an item is replaced immediately after its failure.

(b) **Group Replacement:** Under this policy, a decision will be taken so as to replace all the items irrespective of the fact that the items have failed, not failed, provided if any item fails, before the optimal time it may be replaced individually.

Algorithm for Deciding Group Replacement Policy

Step 1: Find the probability of failure of items at the end of each period.

Step 2: Find the number of replacements made at the end of each period with reference to probability of failures at the end of each period considering its pervious replacements.

Step 3: Calculate the cost of individual replacement at the end of each period.

Step 4: Calculate the cost of group replacement at the end of each period.

Step 5: Calculate the total cost of group replacement including individual replacements by adding step 3 and step 4.

Step 6: Calculate the average cost per period by dividing the result in step 5 with period number.

Step 7: Identify the least among the average cost per period as the period of group replacement policy.

ILLUSTRATION 7.7:

1000 bulbs are in use and it costs ₹ 10 to replace an individual bulb which has burnt out. If all bulbs were replaced simultaneously it would cost ₹ 4 per bulb. It is proposed to replace all bulbs at fixed intervals of time, whether or not they are burnt out and to continue replacing burnt out bulbs as and when they fail.

The failure rates that have been observed for certain type of light bulbs are as follows:

Week	1	2	3	4	5
Percent failing by the end of week	10	25	50	80	100

At what intervals should all the bulbs replaced? At what group replacement price per bulb would a policy of strictly individual replacement become preferable to the adopted policy?

Solution.

Step 1: To find out the probability of failure of items at the end of each week.

The probability of failure of light bulbs in first week = P_1 = 10/100 = 0.10

The probability of failure of light bulbs in second week = P_2 = (25 – 10)/100 = 0.15

The probability of failure of light bulbs in third week = P_3 = (50 – 25)/100 = 0.25

The probability of failure of light bulbs in fourth week = P_4 = (80 – 50)/100 = 0.30

The probability of failure of light bulbs in fifth week = P_5 = (100 – 80)/100 = 0.20

Sum of all probabilities is 1, i.e., $P_1 + P_2 + P_3 + P_4 + P_5 = 1$

Therefore, all further probabilities P_6, P_7, P_8 and so on will be zero.

Step 2: Calculation of number of replacements made considering previous replacements.

Let N_i be the number of replacements made at the end of ith week, if all 1000 bulbs are new initially.

Thus, $N_0 = N_0 = 1000$

$N_1 = N_0P_1 = 1000 \text{ x } 0.1 = 100$

$N_2 = N_0P_2 + N_1P_1 = 1000 \times 0.15 + 100 \text{ x } 0.10 = 160$

$N_3 = N_0P_3 + N_1P_2 + N_2P_1 = 1000 \times 0.25 + 100 \times 0.15 + 160 \times 0.10 = 281$

$N_4 = N_0P_4 + N_1P_3 + N_2P_2 + N_3P_1 = 377$

$N_5 = N_0P_5 + N_1P_4 + N_2P_3 + N_3P_2 + N_4P_1 = 350$

$N_6 = 0 + N_1P_5 + N_2P_4 + N_3P_3 + N_4P_2 + N_5P_1 = 230$

$N_7 = 0 + 0 + N_2P_5 + N_3P_4 + N_4P_3 + N_5P_2 + N_6P_1 = 286$

From the above results, it is clear that number of bulbs burnt out increases up to fourth week and decreases up to sixth week and again starts increasing. The whole system comes to a steady state where the proportion of bulbs failing in each week is the reciprocal of their average life.

As the mean age of bulbs

$= 1 \text{ x } P_1 + 2 \times P_2 + 3 \times P_3 + 4 \times P_4 + 5 \times P_5$

$= 1 \times 0.1 + 2 \times 0.15 + 3 \times 0.25 + 4 \times 0.30 + 5 \times 0.20 =$ **3.55 week**

Therefore, number of failures in each week in steady state becomes 1000/3.35 = 229.

So, cost of replacing bulbs individually only on failure: 10 × 299 = ₹ 2,990.

Step 3: Calculating the cost of individual replacement at the end of each period cost of individual replacement at

End of first week = 100 × 10 = 1000

End of second week = 160 × 10 = 1600

End of third week = 281 × 10 = 2810

End of fourth week = 377 × 10 = 3770

Step 4: Calculating the cost of group replacement at the end of each period.

End of first week = 1000 × 4 = 4000

End of second week = 4000 + 1000 = 5000

End of third week = 5000 + 1600 = 6600

End of fourth week = 6600 + 2810 = 4410

Step 5: Calculating the cost of group replacement including individual replacement, i.e., adding values of step 3 and step 4.

End of first week = 4000 + 1000 = 5000

End of second week = 5000 + 1600 = 6600

End of third week = 6600 + 2810 = 9410

End of fourth week = 9410 + 3770 = 13180

Step 6: Calculating average cost per week.

End of first week = 5000/1 = 5000

End of second week = 6600/2 = 3300

End of third week = 9410/3 = 3136.67

End of fourth week = 13180/4 = 3295

Step 7: Identifying the least among average cost per period.

It is identified as third week, i.e., 3136.67, so it will be optimal to replace all the bulbs after every 3 weeks, otherwise the average cost will be increasing.

ILLUSTRATION 7.8:

A factory has a large number of bulbs, all of which must be in working condition. The durability of bulbs is given in the following table:

Week	Proportion of bulbs failing during the week
1	0.10
2	0.15
3	0.25
4	0.35
5	0.12
6	0.03

If a bulb fails in service, it costs ₹ 3.50 to replace but if all bulbs are replaced at a time, it costs ₹ 1.20 each. Find the optimum group replacement policy. (Assume 1000 bulbs as available in the beginning).

Solution.

$N_0 = 1000$

$N_1 = N_0P_1$

$= 1000 \times 0.1 = 100$ bulbs

$N_2 = N_1 P_1 + N_0 P_2$
$= 100 \times 0.10 + 1000 \times 0.15 = 160$ bulbs

$N_3 = N_2 P_1 + N_1 P_2 + N_0 P_3$
$= 160 \times 0.10 + 100 \times 0.15 + 1000 \times 0.25 = 281$ bulbs

$N_4 = N_3 P_1 + N_2 P_2 + N_1 P_3 + N_0 P_4$
$= 281 \times 0.1 + 160 \times 0.15 + 100$ x $0.25 + 1000 \times 0.35 = 427$ bulbs

$N_5 = N_4 P_1 + N_3 P_2 + N_2 P_3 + N_1 P_4 + N_0 P_5$
$= 427 \times 0.1 + 281 \times 0.15 + 160 \times 0.25 + 100 \times 0.35 + 1000 \times 0.12 = 279$ bulbs

$N_6 = N_5 P_1 + N_4 P_2 + N_3 P_3 + N_2 P_4 + N_1 P_5 + N_0 P_6$
$= 279 \times 0.01 + 427 \times 0.15 + 281 \times 0.25 + 160 \times 0.35 + 100 \times 0.12 + 1000 \times 0.03$
$= 260$ bulbs

End of the week	*Total number of bulbs failed*	*Cumulative number of failure*	*Cost of replacement (₹ 3.50)*	*Cost of group replacement (₹ 1.20)*	*Total cost*	*Average cost per week*
1	100	100	350	1,200	1,550	1,550
2	160	260	910	1,200	2,110	1,055
3	**281**	**540**	**1,893.5**	**1,200**	**3,093.5**	**1,031.1**
4	427	968	3,388.5	1,200	4,588	1,147
5	279	1247	4,364.5	1,200	564.5	1,113
6	260	1507	5,274.5	1,200	6,474.5	1,079.08

Bulbs should be replaced by the end of every third week.

Test Your Understanding – 7.4

1. There is a special light bulb that never lasts longer than two weeks. There is a chance of 0.3 that a bulb will fail at the end of first week. There are 100 new bulbs initially. The cost for individual replacements is ₹ 1.25 and the cost per bulb for group replacement is ₹ 0.50. Is it cheaper to replace all the bulbs?
   (i) individually
   (ii) every week
   (iii) every second week
   Answer: Replace after every one week [87.5 (1st week) > 93.125 (2nd week)].
2. The following mortality has been observed for a certain type of IC's used in a digital computer:

Week	1	2	3	4	5
Percent failing by the end of week	10	25	50	80	100

   Group replacement of IC's costs ₹ 0.30 per transistor, where as individual replacement costs ₹ 1.25. What is the best interval between group replacements?
   At what group replacement price per transistor would a policy of strictly individual replacement become preferable to the adopted policy?
   Answer: It is optimal to adopt group replacement after every two weeks [1st week (425) < 2nd Week (312.5) > 3rd week (325.4)].

3. A large hospital complex has several operation theatres. Each operation table has special light bulb attachments. The bulb is prone to failure. There are 200 bulbs installed in all. Considering 500 hours as period, the failure of similar bulb has been given:

   Out of 100 bulbs, 9 failed by the end of first period; 20 failed by the end of second period; 33 failed by the end of third period; 61 failed by the end of fourth period; 77 failed by the end of fifth period; 90 failed by the end of sixth period; 100 failed by the end of seventh period.

   The management considers making it a practice to replace all in a group at one time, then replace the individual bulb as and when it fails and after a fixed interval of time again replace entire group of 200 bulbs. If the bulbs are replaced in group, it costs ₹ 5 per bulb and when replaced individually it costs ₹ 20 per bulb. What should be the replacement policy of the hospital?

   Answer: ATC_2 (920) > ATC_3 (813.33) < ATC_4 (930); Therefore, it is better to go for group replacement after every third period.

4. Find the cost per period of individual replacement policy of an installation of 300 bulbs given in the following:

   (i) Cost of replacing individual bulb is ₹ 3

   (ii) Conditional probability of failure is given below:

Week number	0	1	2	3	4
Conditional probability of failure	0	1/10	1/3	2/3	1

   Answer: After every 3 weeks.

5. A decision has to be made for group replacement versus individual replacement policy for 1000 fluorescent tubes of a particular make in the university campus. The survival rate for the tubes was recorded as under:

End of month	0	1	2	3	4
Bulbs survived	1000	850	500	200	100

   Cost of replacing an individual tube is ₹ 0.50 and when replaced as group it is ₹ 100. Find out whether group replacement policy is economical or not. If economical at the end of which month should the tubes be replaced as a group?

   Answer: One week group replacement policy; Average cost = ₹ 175 per week; Cost of individual replacement per week = ₹ 196.

7.10 CONCLUSION

Maintenance department is often concerned with the replacement decision for which the plant engineer is required to be aware of and keep on tracking machine health. The trend analysis and cost analysis of failures can help the engineer in taking right decisions easily and timely. The cost analysis for replacement depends on the money value which is dynamic in nature. In addition to the individual replacement decisions, the plant engineer has to examine the group replacement decisions for optimal maintenance.

7.11 RECAP

This chapter focuses on assessing replacement age based on the costs. The situations warranting replacement and relevant costs involved in machine failure analysis are explained. Replacement models for individual replacement, i.e., replacement policy when money value does not change with time (Model 1) and replacement policy for items when money value changes with time (Model – 2) are discussed with numerical examples. Group replacement policy is also narrated. Sufficient number of numerical examples is given.

7.12 KEY CONCEPTS

1. **Purchase Cost:** The cost of machine or equipment at the time of its purchase. It is considered to be independent of age of machine.
2. **Resale Value or Salvage Value or Scrap Value:** It is the value or cost that can be realised when it is sold. It decreases with the running age of the machine.
3. **Running Cost or Operating Cost or Maintenance Cost:** It is cost incurred during the useful life of the machine to maintain it. These will increase on the machine as the age grows due to the wear and tear on the moving parts of the machine.
4. **Present Worth Factor (PWF) or Discount Factor:** It is the factor that converts a rupee 'n' years hence changing (money grows but value decreases) at the rate of 'r' percent worth today. $v = (1 + r)^{-n}$.
5. **Group Replacement:** Replacement in case of large number of items and replacement of all items at once, provided the items are alike and have similar failure characteristics.

REVIEW QUESTIONS

1. Explain various stages that a machine will experience in its life. Give examples. Discuss what maintenance strategies could be more appropriate to adopt.
2. When does machine replacement decision arise? Explain various possible maintenance strategies applicable at old age of the machine.
3. Describe the stages of machine life cycle (MLC) with reference to failure rate and failure cost.
4. Derive the expression for finding optimal replacement age of a machine under the condition that money value is not changing.
5. Derive the expression for finding optimal replacement age of a machine under the condition that money value is changing.
6. What are the situations warranting to adopt group replacement policy?
7. Write short notes on
   (a) Present Worth Factor (PWF)
   (b) Machine Life Cycle
   (c) Salvage Value
   (d) Reliability

CHAPTER 8

Failure Mode, Effect, and Criticality Analysis (FMECA)

PREAMBLE

This chapter focuses on the approaches of failure analysis—failure mode and effect analysis (FMEA) and failure mode, effects, and criticality analysis (FMECA). The elements of FMECA, the basic information required for FMECA, and the basic analysis procedure of FMEA/FMECA are discussed. The different analysis approaches such as hardware approach and functional approach are also covered in this chapter. The requirement or utility of block diagram in such failure analysis has been explained. Statistical process control uses sampling and other statistical methods to monitor the quality of an ongoing process. Probability sampling methods and non-probability sampling methods are frequently used for cost and time advantages. A few probability sampling methods available are discussed. FMECA can be applied to any stage of design, development, production/operation or usage, but since its main aim is to prevent failure, it is most suitably applied at the design stage to identify and eliminate causes. Design FMEA and process FMEA are discussed and a brief comparative study has been presented in this chapter.

CHAPTER OUTLINE

8.1 INTRODUCTION

Every machinery, equipment, buildings undergo deterioration due to their use and exposure to environmental conditions. This deterioration is to be detected well in advance so as to forestall loss and damage. Industries, therefore, address such issues time to time through repairs, renovations, rejuvenations, recondition, etc., so as to enlarge their useful life to a maximum possible extent. In this context, the maintenance assumes importance as an engineering function and is made responsible for provision of the condition of these machines, equipment, buildings, and services that will permit uninterrupted implementation of plans requiring their use. To achieve this, the maintenance department needs to improve its judicial knowledge on how and when the failure can occur. This means that one should be able to estimate the failure mode, failure effect, and the failure criticality to maintain the machine in a good condition. This chapter particularly deals with the recently developed approaches of failure analysis in this regard, in short, originally named as FMEA and later developed as FMECA.

8.2 FMEA/FMECA—AN OUTLINE

Failure Mode and Effects Analysis, in short, known as **FMEA**, is later developed as **Failure Modes, Effects and Criticality Analysis**, in short known as **FMECA**. The objective of FMECA is to identify all failure modes in a system design. Its purpose is to identify all catastrophic and critical failure probabilities so that they can be minimised at the earliest. Hence, FMECA should be started as soon as preliminary design information is available and investigation is extended as more information is available in suspected problem areas.

FMEA and FMECA are methodologies designed to identify potential failure modes for a product or process, to assess the risk associated with those failure modes, to rank the issues in terms of importance and to identify and address the most serious concerns and carry out corrective actions.

Although the purpose, terminology, and other details can vary according to type (e.g., process FMEA, design FMEA, etc.), the basic methodology is similar for all. Before going into details, let us first familiarise ourselves with its terminology.

Brief general outline of FMEA/FMECA analysis, techniques, and requirements are presented in this chapter.

8.3 THE FMEA/FMECA JARGON

The definitions of the terms given below can be considered as the fundamentals of FMEA/FMECAs. These definitions will enable the reader to understand the differences between part function, part failure mechanism, failure effect, failure detection, and failure modes. (These definitions are taken from MIL-STD-1629.)

Terms	*Definition*
Failure	It is the termination of the ability of an item to perform its required function.
Failure Mode	It is the manner by which a failure is observed. Generally, it is the way the failure occurs and its impact on equipment operation.
Failure Effect	It is the consequence(s) of a failure mode on the operation, function, or status of an item. Failure effects are usually classified according to how the entire system is impacted.
Failure Cause	The physical or chemical process, design defects, part misapplication, quality defects, or other processes that are the basic reason for failure or which initiate the physical process by which deterioration processes to failure.
Detection Mechanism	The means or method by which a failure can be discovered by (i) An operator under normal system operation or (ii) The maintenance crew by some diagnostic action.
Severity	Severity is the worst potential consequence of a failure, determined by the degree of injury, property damage, or system damage that can ultimately occur.
Criticality	It is a relative measure of the consequences of a failure mode and its frequency of occurrence.
Criticality Analysis (CA)	It is a procedure by which each potential failure mode is ranked according to the combined influence of severity and probability of occurrence.
Environments	The conditions, circumstances, influences, stresses, and combinations thereof, surrounding and affecting systems or equipment during storage, handling, transportation, testing, installation, and use in standby status and operations.
FMECA – Maintainability Information	It is the procedure by which each potential failure is analysed to determine how the failure is detected and the actions to be taken to repair the failure.
Single Failure Point	The failure of an item that will result in failure of the system and is not compensated for by redundancy or alternative operational procedure.
Undetectable Failure	It is a postulated failure mode in the FMEA for which there is no failure detection method by which the operator is made aware of the failure.
Failure Definition	This is a general statement of what constitutes a failure of the item in terms of performance parameters and allowable limits for each specified output.

8.4 ELEMENTS OF FMECA

The three elements of a complete F**MEC**A are

(a) Failure **M**ode (b) Failure **E**ffect (c) Failure **C**riticality

These are briefed out here:

(a) **Failure Mode:** *According to John Moubray (1991), "A failure mode has safety consequences if it possesses loss of function or other damage which could lead to the breach of any known environmental standard or regulation or which could hurt or ill someone or is hazardous for usage in anyway".*

The anticipated conditions of operation are used as the background to study the most probable failure mode, locations, and mechanism of the product or system and its components.

(b) **Failure Effect:** The potential failures are studied to determine their probable effects on the performance of the whole product, process or service, and the effects of the various components on each other.

(c) **Failure Criticality:** The potential failures of the various parts of the product or service system are examined to determine the severity of each failure effect in terms of lowering of performance, safety hazard, total loss of function, etc.

8.5 BASIC INFORMATION REQUIRED FOR FMECA

In general, FMEA/FMECA requires the identification of the following basic information:

- Items(s)
- Function(s)
- Failure(s)
- Effect(s)of failure
- Cause(s) of failure
- Currentcontrol(s)
- Other relevant details

Most analyses of this type also include some method to assess the risk associated with the issues identified during the analysis and to prioritise corrective actions. There are two common methods for this risk analysis. These are:

- RiskPriority Numbers (RPNs)
- Criticality Analysis (FMEA + Criticality Analysis = FMECA)

8.6 BASIC ANALYSIS PROCEDURE OF FMEA/FMECA

The basic steps for performing an FMEA/FMECA are

1. Assemble the team.
2. Institute the basic rules.
3. Collect and review relevant information.
4. Identify the items(s) or process(es) to be analysed.
5. Identify the function(s), cause(s), effect(s), and control(s) for each item or process to be analysed.
6. Assess the risk associated in its tangible evaluation with the issues identified.
7. Prioritise and assign corrective actions.
8. Perform corrective actions and reevaluate risk.
9. Distribute, review, and update the analysis, as appropriate.

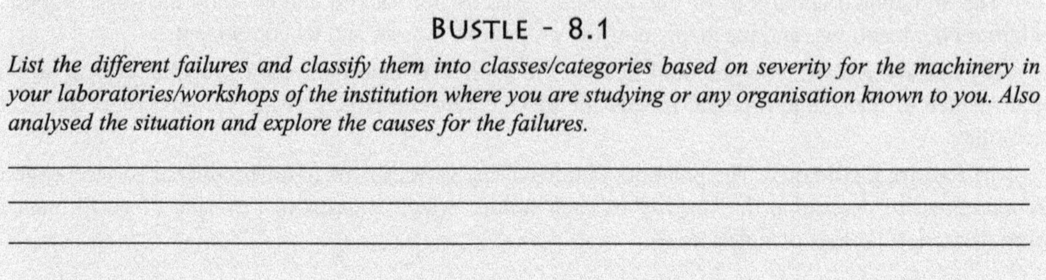

Test Your Understanding 8.1

1. Define the following terms with reference to FMEA/FMECA.
   (a) Failure (b) Failure effect
   (c) Failure cause (d) Failure mode
2. Define the following terms with reference to FMEA/FMECA.
   (a) Single Failure Point (b) Unpredictable Failure
   (c) Environments (d) Maintainability Information
3. What is the basic information required for FMEA/FMECA?
4. Explain analysis procedure of FMEA/FMECA.
5. Define and explain failure, level of analysis, trade-off study units.
6. Explain the elements of FMECA.

8.7 ENVIRONMENTAL INFLUENCES—STANDARDS AND GUIDELINES

The overall design must consider how the environment influences the system during testing, storage, installing, and operating. While considering the failure mode effects, think about the impact on society in general if failure occurs. Society may, in some case, be considered as a customer because if the product is not accepted due to environmental effects, it may not be successful in general, at all.

The FMECA process will address all the problems concerning design, manufacturing, process, safety, and environment. The consequences of these problems are addressed at the stage of studying the failure mode during FMECA. Preventive measures for failures should always be given top priority where an individual's well-being in the work place is concerned.

Standards and Guidelines

There are numerous standards and guidelines for the requirements and recommended reporting format of FMEAs and FMECAs. Some of the main published standards for this type of analysis are found in SAE J1739, AIAG FMEA-3, and MIL-STD-1629A. However, many industries and companies have developed their own procedures to meet the specific requirements of their products/processes. Table 8.4 shows a sample process FMEA in the Automotive Industry Action Group (AIAG) - FMEA-3 format.

8.8 FMECA PLANNING

Planning the FMECA work involves the contractors' procedures for implementing their specified requirements. Planning should include updating to reflect design changes and analysis results. Worksheet formats, ground rules, assumptions, identification of the level of analysis, failure definitions, and identification of coincident use of the FMECA by the contractor, and other organisational elements should also be considered.

8.8.1 Ground Rules and Assumptions

The ground rules identify the FMECA approach (e.g., hardware, functional, or combination), the lowest level to be analysed, and include statements of what will constitute a failure in terms of performance criteria. Every effort should be made to identify and record all ground rules and analysis assumptions prior to initiation of the analysis; however, ground rules and analysis assumptions can be adjusted as requirements change.

8.8.2 Analysis Approach

Variations in design complexity and available data will generally dictate the analysis approach to be used. There are two primary approaches for FMECA. One is the hardware approach that lists individual hardware items and analyses their possible failure modes. The other is the functional approach that recognises that every item is designed to perform a number of outputs. The outputs are listed and their failures analysed. For more complex systems, a combination of the functional and hardware approaches may be considered.

Hardware Approach: The hardware approach is normally used when hardware items can be uniquely identified from schematics, drawings, and other engineering and design data. This approach is recommended for use in a part level up approach often referred to as the bottom-up approach.

Functional Approach: The functional approach is normally used when hardware items cannot be uniquely identified or when system complexity requires analysis from the top down.

8.8.3 Level of Analysis

The level of analysis applies to the system hardware or functional level at which failures are postulated. In other words, it is how the system being analysed is segregated (e.g., a section of the system, component, sub-component, etc.).

8.8.4 Trade-off Study Reports

These reports should identify areas of marginal and state-of-the-art design and explain any design compromises and operating restraints agreed upon. This information will aid in determining the possible and most probable failure modes and causes in the system.

8.8.5 Design Data and Drawings

Design data and drawings identify each item and the item configuration that perform each of the system functions. System design data and drawings will usually describe the system's internal and interface functions beginning at system level and progressing to the lowest indenture level of the system. Design data will usually include either functional block diagrams or schematics that will facilitate construction of reliability block diagrams.

8.9 BLOCK DIAGRAMS

Block diagrams that illustrate the operation, interrelationships, and interdependencies of the functions of a system are required to show the sequence and the series dependence or independence of functions and operations. Block diagrams can be constructed in conjunction with, or after defining the system and will present the system breakdown of its major functions. More than one block diagram is sometimes required to represent alternative modes of operations, depending upon the definition established for the system.

8.9.1 Functional Block Diagrams

Functional block diagrams illustrate the operation and interrelationships between functional entities of a system as defined in engineering data and schematics. An example of a functional block diagram, taken from MIL-STD-1629, is shown in Figure. 8.1.

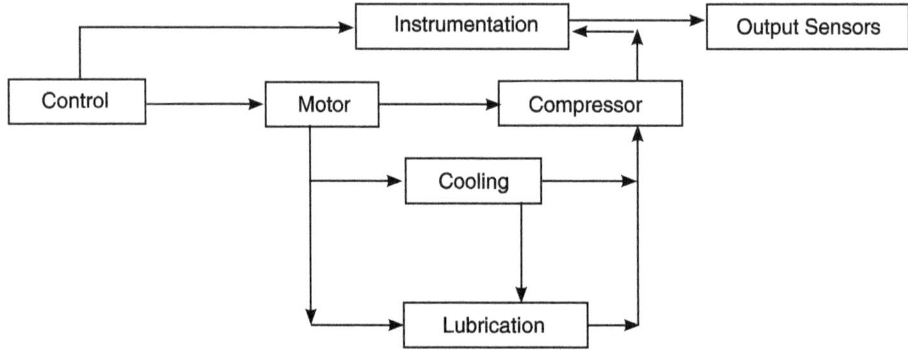

Figure 8.1 Functional block diagram

8.9.2 Reliability Block Diagram

Reliability block diagrams define the series dependence, or independence, of all functions of a system or functional group for each life-cycle event. An example of a reliability block diagram, taken from MIL-STD-1629, is shown in Figure. 8.2.

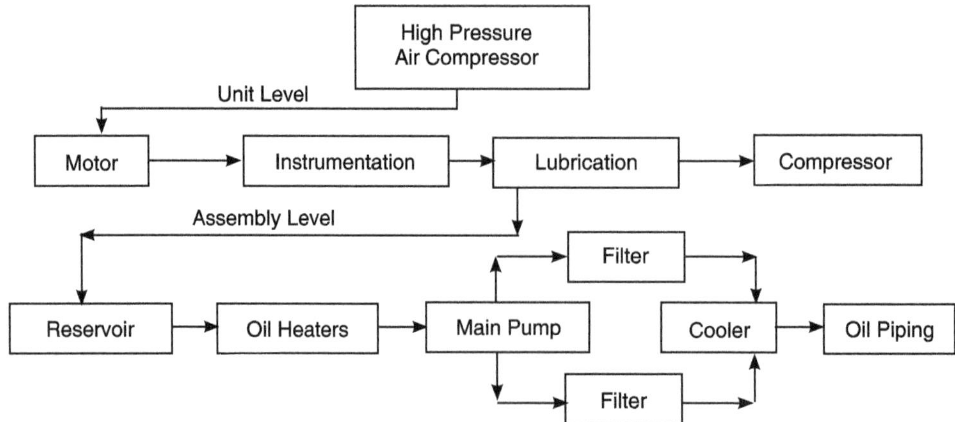

Figure 8.2 Reliability block diagram

8.10 SEVERITY CLASSIFICATION

Severity classification is assigned to provide a qualitative measure of the worst potential consequences resulting from design error or item failure. Classifications should be assigned to each identified failure mode and each item analysed in accordance with the loss statements below. It may not always be possible to identify an item or a failure mode according to the loss statements in the four categories given below, but similar loss statements based on various inputs and outputs can be developed and included in the ground rules for the FMECA activity. Severity classification categories that are consistent with MIL-STD-882 are defined as follows:

- **Category I – Catastrophic:** A failure that can cause injury or death.
- **Category II – Critical:** A failure which can cause severe injury, major property damage, or major system damage that will result in major downtime or production loss.
- **Category III – Marginal:** A failure which can cause minor injury, minor property damage, or minor system damage which will result in delay or loss of system availability or degradation.
- **Category IV – Minor:** A failure not serious enough to cause injury, property, damage or system damage, but will result in unscheduled maintenance or repair.

These categories can be observed through cost or any other factor, but when used in the established criteria, should be consistent throughout the analysis.

FMECA Report

The results of the FMECA and other related analyses should be included in a report that identifies the level of the analysis, documents the data sources and techniques used in performing the analysis, and gives the system definition.

8.11 APPLICATION OF STATISTICAL PROCESS CONTROL (SPC) IN FMEA/FMECA

Statistical process control uses sampling and other statistical methods to monitor the quality of an ongoing process. The graphical displays such as control charts, run chart, Pareto analysis provide a basis for deciding whether the variation in the output of a process is due to common cause (randomly occurring variations) or out-of-the-ordinary assignable causes. Wherever assignable causes are identified, a decision can be made to adjust the process in order to bring the output back to acceptable quality levels.

A goal in the design of sample surveys is to obtain a sample that is representative of the population so that precise inferences can be made. Sampling error is the difference between a population parameter and a sample statistic used to estimate it. For instance, the difference between a population mean and a sample mean is sampling error. Probability sampling methods (where the probability of each unit appearing in the sampling error), non-probability sampling methods (which are based on convenience or judgment rather than on probability) are frequently used for cost and time advantages. Sample, whether or not is representative, is dependent on the judgment of the individuals designing and conducting the survey and not on sound statistical principles. In addition, there is no objective basis for establishing bounds on the sampling error when non-probability sample has been used.

Most governmental and professional polling surveys use probability sampling. It can generally be assumed that any survey that reports a plus or minus margin of error has been conducted using probability sampling. Statisticians prefer probability sampling methods and recommend that they be used whenever possible. A variety of probability sampling methods are available. A few of the more common ones are discussed here.

(a) **Simple Random Sampling:** Simple random sampling provides the basis for many probability sampling methods. With simple random sampling, every possible sample of size '*n*' has the probability of being selected.

(b) **Stratified Sampling:** Stratified random sampling is a variation of random sampling in which the population is partitioned into relatively homogeneous groups called strata and a simple random sample is selected from each stratum. The results from the strata are then aggregated to make inferences about the population. Another benefit of this method is that inferences about the subpopulation represented by each stratum can also be made.

(c) **Cluster Sampling:** Cluster sampling involves partitioning the population into separate groups called clusters. Unlike the case of stratified random sampling, it is desirable for the clusters to be composed of heterogeneous units. In single-stage cluster sampling, the sampled clusters. In two-stage cluster sampling, a simple random sample of clusters is selected and then a simple random sample is selected from the units in each sampled cluster. One of the primary applications of cluster sampling is called *area sampling*, where the clusters are townships, city blocks, or other well-defined geographic sections of the population.

In many segments of industry, the application of statistical process control (SPC) has become a common practice. It is more so in failure data analysis of equipment in the mass production industries such as automobile, automobile component, or consumer electronics manufacture. SPC is extensively applied within the organisations for machine process capability and machine efficiency estimations.

The application of process control charts is only one element of the use of SPC. Many of the benefits of using a systematic and statistical approach to analysing why a particular process varies are derived well before control charts are produced. Indeed, very often improved process understanding lead to improved process capability and, thus, indicates the machine efficiency and also can reduce or even eliminate the need for process control chart.

8.11.1 The 3-Stage Application of SPC in FMECA

The application of SPC is done generally in three stages. These are briefed out here:

Stage-1: The application of SPC commences with developing and understanding of the process, identifying the main process parameters, and defining the control mechanism available. The technique that is widely used during this initial diagnostic plan is failure mode, effect, and criticality analysis (FMECA). This technique helps to identify the critical characteristics of the process, which require further more detailed analysis. A 'criticality' column is added to the analysis, to allow the inclusion of an index related to the criticality of the failure effect. Sometimes as the criticality rating is omitted, the analysis is then FMEA.

Stage-2: The second stage involves the collection and analysis of the process data to determine what is normal variation. The process has to be bounded to include the inherent variability and to exclude the special causes of variation. Determination of normal variation requires a fundamental understanding of the process. The process data is also evaluated during this analysis phase to check whether the distribution is normal.

Stage-3: The third stage in the application of SPC involves the analysis of the critical characteristics to determine the relationship between the natural variation of the process and the allowable specification limits. This relationship, termed the process variability is acceptable. Finally, the ongoing control of the process can be monitored using control charts.

Thus, the use of FMEA represents the initial stage in the application of SPC. Therefore, to apply SPC, it is necessary to have complete understanding on the concept of FMEA/FMECA.

8.11.2 The Concept of FMEA

FMEA primarily plays a role as prevention oriented technique to improve the inherent reliability of a product or service. Each item of a design is considered in relation to its likely mode of failure, the probability of occurrence, and the effects of the failures.

There are two forms of FMEA.

(a) **Design FMEA**: It involves the analysis of the potential failure of a new product or service, and

(b) **Process FMEA**: It involves failure analysis of the process or processes in the manufacture or provision of the product or service.

A comprehensive FMEA for a complete design is a sizable task. However, this method is a very effective formal reliability design review technique. The method is particularly suited to analog system (whether electronic, electrical, or mechanical), for purely digital electronic systems are very complex. Also, the designer has less control over individual elements of the system, since these are contained within the integrated circuit packages. FMEA provides a valuable spin-off, as the analysis is an excellent starting point for producing block diagram, reliability analysis, and diagnostic charts for repair purposes. A considerable amount of work in generating diagnostic routine and preparing user maintenance handbooks can be saved if FMEA worksheets are used as the basis for this work.

The detailed level to which FMEA is taken must be based upon the purpose for which it is performed. For example, consideration of all possible failure modes within a hydraulic sequence valve or an operational amplifier will not usually be included in the analysis of a system of which they were components. Instead, emphasis will be on the failure modes of their outputs. The designer of the valve or operational amplifier will, however, have to analyse the effects of all modes of failure within the item. The level of analysis can also be affected by the stage reached by the project. For large systems, feasibility studies can only generate functional subsystems, with much of the detailed design still to be performed. The initial FMEA will then be related to the failures of these subsystems or blocks. This forms the basis for updated versions as design proceeds, and it is useful in its own right for aiding the assessment of options and identification of critical features.

BUSTLE - 8.2

Did you observe any variation due to abnormal cause in the application of statistical process control (SPC) in FMEA/FMECA. Recall such occasions and analyse why it so happened.

____

____

____

____

Test Your Understanding 8.2

1. Explain quantitative approach of FMECA.
2. Explain the application of statistical process control in FMECA.
3. Discuss the following terms in the backdrop of FMECA:
   (a) Block diagrams
   (b) Severity classifications

8.12 FAILURE MODELS BY FMECA

8.12.1 Model Inputs

The key inputs used in failure modelling using FMEA/FMECA are as follows:

(a) **Severity (S):** Severity (S) is a numerical measure of how serious is the effect of the failure to the customer. It is to assess the failure result on an assumed scale with questioning, if the component or system failure results in a mere nuisance or if it can result in serious injury. The degree of severity is generally measured on a scale of 1 to 10 where 10 is the most severe.

(b) **Occurrence (O):** Occurrence (O) is a measure of probability that a particular mode will actually happen. The degree of occurrence is measured on a scale of 1 to 10, where 10 signifies the highest probability of occurrence.

(c) **Detection (D):** Detection (D) is a measure of probability that a particular mode will be detected in the manufacturer's own operation before reaching the customer. The level of detection is measured on a scale of 1 to 10, where 10 signifies virtually no ability to detect the fault.

8.12.2 Risk Evaluation Methods

A typical FMEA incorporates some methods to evaluate the risk associated with potential problems identified through the analysis. There are two most common methods. These are risk priority numbers and criticality analysis.

Risk Priority Numbers

It is a numerical and relative "measure of overall risk" corresponding to a particular failure mechanism and is computed by multiplying the severity, occurrence, and detection numbers. To use the risk priority number (RPN) method to assess risk, the analysis team must:

- Rate the **Severity** of each effect of failure,
- Rate the likelihood of **Occurrence** for each cause of failure,
- Rate the likelihood of prior **Detection** for each cause of failure (i.e., the likelihood of detecting the problem before it reaches the end user or customer), and
- Calculate the RPN by obtaining the product of the three ratings:

$$\text{RPN} = \text{Severity} \times \text{Occurrence} \times \text{Detection} = \text{S.O.D}$$

The RPN method can then be used to compare issues within the analysis and to prioritise problems for corrective action.

Criticality Analysis

The MIL-STD-1629A document describes two types of criticality analyses: quantitative and qualitative. To use the quantitative criticality analysis method, the analysis team must:

- Define the reliability/unreliability for each item, at a given operating time.
- Identify the portion of the item's unreliability that can be attributed to each potential failure mode.
- Rate the probability of loss (or severity) that will result from each failure mode that can occur.
- Calculate the criticality for each potential failure mode by obtaining the product of the three factors:

  Mode Criticality = Item Unreliability × Mode Ratio of Unreliability × Probability of Loss
- Calculate the criticality for each item by obtaining the sum of the criticalities for each failure mode that has been identified for the item.

  Item Criticality = Sum of Mode Criticalities

To use the qualitative criticality analysis method to evaluate risk and prioritise corrective actions, the analysis team must:

- Rate the severity of the potential effects of failure,
- Rate the likelihood of occurrence for each potential failure mode, and
- Compare failure modes via a criticality matrix, which identifies severity on the horizontal axis and occurrence on the vertical axis.

Critical Characteristics

A failure mode which can conceivably result in personal injury, loss of life or violate a government (e.g., automobile emission norms) mandate is considered critical. Such a failure mode will be assigned a severity level value(s) of 6 or more irrespective of the level of occurrence or detection.

Significant Characteristics

A failure mode which is not deemed to be critical but needs stringent process control (e.g., SPC, 100% inspection, etc.) to ensure product quality is considered significant. Such a failure mode will be assigned a severity level value(s) of 6 or more irrespective of the level of occurrence or detection.

8.13 PROCESS FMEA

As discussed earlier, the process FMEA is used primarily to identify areas of criticality to control and emphasise the design of inherently more reliable process. Usually, the application of SPC in FMEA starts with the preparation of a process FMEA. FMEA helps in understanding the process and in projecting the critical areas to control.

8.13.1 What is Process FMEA?

- It is an analytical technique.
- It identifies potential product related to process failure modes.

- Customer normally is end user but can be the next process too.
- It assesses the potential effects of the failures on the customer.
- It identifies the process causes and process variables on which to focus controls for:
  - occurrence reduction.
  - detection of failure conditions.
- It develops ranked list of potential failure modes leading to prioritisation of corrective actions.
- It documents the results.
- It catalyses to stimulate interchange of ideas between functions affected and promote team approach.
- It is a living document and should be initiated—
  - Before or at the feasibility state.
  - Prior to tooling for production.
  - Taking into account all manufacturing operations.
  - Before introducing a new or a modified process.
  - During manufacturing planning stage of a new model or component.

8.13.2 Objectives of Process FMEA

The objectives of process FMEA are as follows:

- **Process Improvement:** FMEA drives process improvements as the primary objective, with an emphasis on error/mistake proofing solution.
- **High-Risk Failure Modes:** FMEA addresses all high-risk failure modes, as identified by FMEA team with executable action plans. All other failure modes are considered.
- **Control Plans:** The pre-launch and production control plans consider the failure modes from the process FMEA.
- **Integration:** FMEA is integrated and consistent with the process flow diagram and the process control plan. The process FMEA considers the design FMEA, if available as part of its analysis.
- **Lessons Learned:** FMEA considers all major "lesson learned" (such as warranty, campaigns, non-conforming product, customer complaint is input to failure mode identification).
- **Special or Key Characteristics:** FMEA identifies appropriate key characteristics candidate as input to the key characteristics selection process, if applicable due to company policy.
- **Timing:** FMEA is completed during the "window of opportunity" where it can most efficiently impact the design of product or process.
- **Team:** The right people participate as part of the FMEA team throughout the analysis and are adequately trained in FMEA methods. A facilitator should be used, if required.
- **Documentation:** The FMEA document is completely filled out "by the book", including "action taken" and new RPN values.
- **Time Usage:** Time spent by the FMEA team as early as possible is an effective and efficient use of time, with a value-added result.

8.13.3 The Approach to Process FMECA

The failure modes identified by the FMECA process are assessed by their probability of occurrence. To establish qualitative measures of occurrence, severity, and detection, suitable criteria must be established that subjectively relate to the overall effect on the process. Examples are offered in Table 8.1, Table 8.2, and Table 8.3 to serve as guidelines in establishing the measures. The product of the measures of occurrence, severity, and detection is called the risk priority number (RPN). The numbers or criteria assigned to any particular ranking system are at the discretion of the user.

Table 8.1 Severity probabilities

Rank	*Severity Criteria*	*Description*	*Probability*
1	Minor	No noticeable effect. Unable to realise that a failure has occurred.	0.0 – 0.2
2	Marginal	Annoying. No system degradation.	0.2 – 0.4
3	Moderate	Causing dissatisfaction. Some systems degradation.	0.4 – 0.6
4	Critical	Causing a high degree of dissatisfaction. Loss of system function.	0.6 – 0.8
5	Catastrophic	A failure, which may cause death or injury. Extended repair outages.	0.8 – 1.0

Note: The ranking criteria selected must to be consistent throughout FMECA.

Table 8.2 Occurrence probabilities

Rank	*Occurrence Criteria*	*Description*	*Occurrence Rates (Cycles, hrs, etc)*
1	Unlikely	Unreasonable to expect this failure mode to occur.	1/100000
2	Isolated	Based on similar designs having a low number of failures.	1/10000
3	Sporadic	Based on similar designs that have experienced occasional failures.	1/1000
4	Conceivable	Based on similar designs that have caused problems.	1/100
5	Recurrent	Certain that failures will ensue.	1/10

Note: The ranking criteria selected must be consistent throughout FMECA.

Table 8.3 Detection probabilities

Rank	*Detection Criteria*	*Description*	*Probability*
1	Very high	Very high probability of detecting the failure before it occurs. Almost always preceded by a warning.	0.8 – 1.0
2	High	High probability of detecting the failure before it occurs. Proceeded by a warning most of the time.	0.6 – 0.8
3	Moderate	Moderate probability of detecting the failure before it occurs. About a 50% chance of getting a warning.	0.4 – 0.6
4	Low	Low probability of detecting the failure before it occurs. Always comes with little or no warning.	0.2 – 0.4
5	Remote	Remote probability of detecting the failure before it occurs. Always without a warning.	0.0 – 0.2

Note: The ranking criteria selected must be consistent throughout FMECA.

[The method of calculating the product of the above values read from each of the three tables (8.1 to 8.3) to find RPN outlines, the approach used for the FMECA process as prescribed in MIL-STD-1629].

The failure mode and criticality number (C_r) is the portion of the criticality number for the item due to a particular failure mode. This criticality number replaces the RPN used in the previous section. The C_m for failure mode is determined by the expression:

$$C_r = FALT$$

where F = conditional probability of loss of function,

A = failure mode ratio,

L = part failure rate, and

T = duration or operating time.

Table 8.4 Values of 'F' for the analyst's judgment

Failure effect	*Probability of loss of function*
Actual loss	$F = 1$
Probable loss	$0.10 < F < 1.00$
Possible loss	$0 < F < 0.10$
No effect	$F = 0$

The failure mode ratio, A, is the probability that the part or item will fail, if all potential failure modes of a particular part or item are listed, the sum of all the probabilities for that part or item will be equal to one. Individual failure mode multipliers can be derived from failure rate source data or from test and operational data. If failure mode data are not available, the value should represent the analyst's judgment based on an analysis of the item's functions.

Part failure rates, 'L', are derived from appropriate reliability prediction methods using mean time between failure (MTBF) data or possibly other data obtained from handbooks or reference material. Manufacturers often supply failure data; however, it is important that the environment the item will be subjected to is similar to the environment the manufacturer used when obtaining the failure data.

The operating time, 'T', is usually expressed in hours or the number of operating cycles of the item being analysed. The 'A' and 'F' values are often subjective, thus, making the supposed quantitative method somewhat qualitative. All things considered, it is generally understood that the FMECA process is a qualitative method of analysis.

8.13.4 Quality Control Plan and PROCESS FMEA

After the identification of potential risk in the manufacturing process through process FMEA, it is essential that appropriate measures and actions be incorporated within the process quality control plan. For example, a process has identified an X-ray inspection to be performed after a given process step to reduce the RPN to an acceptable level. It now becomes imperative that the quality engineer ensures that the X-ray inspection be installed and carried out at the desired location as well as on the quality control plan. Thus, the process FMEA will always drive the quality control plan and the two documents will be closely linked to each other. With the passage of time as manufacturing engineers gain experience, they will incorporate changes in the process to make it more risk free and less prone to producing defects. Accordingly, both the process FMEA and the control plan get modified simultaneously to reflect the changes and improvements made. Both the process FMEA and the quality control plan are living documents for reference on the factory floor. They should be kept on the factory floor readily accessible to factory personnel when required. The continual updating of these documents contributes significantly in maintaining a healthy manufacturing process flow.

8.14 DESIGN FMEA

Design FMEA is performed to identify and address areas of potential risk within the component or system design. The objective is to review the product design and to establish the following:

- What are the main functional elements of the design?
- Which are the potential failures?
- What are the effects of failure?
- Which are the main causes of failure?
- Which are the controls used to prevent failure?

From this analysis, an assessment of the failure potential of the design is made and this is expressed as the risk priority number (RPN). The reliability of the component or the sub-system is used to calculate the occurrence value. The higher value of the RPN (typically greater than 90), indicates the need for design corrective action to prevent the likelihood of reliability problems. So the design FMEA may recommend changes to be made to a product design so as to reduce risk of failure in the field or of the manufacturing process.

8.14.1 List of Documents

The following is the list of recommended documents which will facilitate the efforts of design FMEA team:

- Customer specifications
- Engineering drawings
- Warrantyinformation of similar project
- Previously performed design FMEA on similar product
- Process flow diagram
- Processcapability data
- Reliability test results (experimental as well as batch analysis reports or prototype evaluation)
- Prototype build failure data and list of problems experienced
- Design of experience data

8.14.2 Steps for Carrying Out Design FMEA

The following are the steps for carrying out Design FMEA:

1. Select a system, sub-system, or part and identify its functions.
2. List potential failure modes.
3. List potential effects of the failure.
4. Estimate the severity number (S).
5. List potential causes or mechanisms of failure.

6. Estimate the occurrence number (O).
7. Assess the design verification measures taken (e.g., prototype testing, pilot run, etc.).
8. Estimate the detection number (D).
9. If D ≥ 5, take a closer look at measures for detection (D > 5 implies high risk).
10. Compute the RPN.
11. Check if the system can fail because of customer abuse or other in-use condition. If so, re-estimate, S, O, D and re-compute RPN.
12. List recommended actions where applicable to lower RPN.
13. List individuals (or departments) responsible for completing recommended corrective actions.
14. List actual actions taken.
15. Re-compute RPN after corrective actions are implemented.
16. Repeat all the steps above for all systems, sub-systems, and components which are deployed in the product.

8.14.3 Severity Level

The guidelines for severity level and severity numbers for design FMEA are given in Table 8.5.

Table 8.5 Severity level and severity number for Design FMEA

Severity Level	*Description*	*Example*	*Severity Number*
High	It can result in injury or loss of life or in significant loss of revenue or business for manufacturer.	Brake failure in automobiles	10
Medium High	It causes serious inconvenience to customer or result in moderate loss of revenue or business for manufacturer.	Automobile does not start	7
Medium Low	Product is functional but causes moderate inconvenience to customer.	Automobile speed pick up is not proper	5
Low	Product functions but causes some annoyance to customer.	Automobile is functional but there is some vibration on door at high speed	2

8.14.4 Occurrence and Detection Rating Level for Design FMEA

The likelihood of occurrence and detection of defects will drive the corresponding occurrence and detection numbers within 1 to 10 range with the higher number corresponding to higher risk. These numbers will have to be assigned by the participating FMEA team members as shown in Table 8.6.

Table 8.6 Occurrence and detection rating level for design FMEA

High probability of occurrence at manufacturers' location	–10
Low probability of occurrence at manufacturers' location	–1
Low probability of detection within manufacturers' location	–10
High probability of detection within manufacturers' location	–1

8.15 COMPARISON BETWEEN DESIGN FMEA AND PROCESS FMEA

The comparison between design FMEA and process FMEA is given in a tabular representation here.

	Design FMEA	*Process FMEA*	*Working variant FMEA*
Purpose	Meeting specification	Production in compliance with drawing	Meets the expressed and implied expectations in working
Responsibility	Design and development	Production	Exploitation/usage/work implementation
Objective	Avoiding design failure	Avoiding failures during process planning, production implementation	Avoiding failures during working (due to design/process/materials failures)
Timings for commencement	After design selection	With production release	Work phase after completion of sales
Period of review	Prior to first trial run	Prior to any investment	Prior to first sale
FMEA first completion	Prior to production release	Prior to bulk production	After first market feedback to produce

Design FMEA should be initiated before or at design concept finalisation at component level, sub-system level, and system level whereas process FMEA should be initiated before or at feasibility stage, prior to tooling for production.

8.16 APPLICATION AND MERITS OF FMEA/FMECA

The FMEA/FMECA analysis procedure is a tool that has been adapted in many different ways for many different purposes.

(a) It can contribute to improved designs for products and processes, resulting in
- higher reliability,
- better quality,
- increased safety,
- enhanced customer satisfaction, and
- reduced costs.

(b) The tool can also be used for repairable systems to
- establish maintenance policy and
- optimise maintenance plans.

(c) The tool contributes to control plans and other quality assurance procedures.

(d) It provides
- a knowledge base of failure mode and
- corrective action information.

(e) It can be used as a resource in
- future troubleshooting efforts and
- training for new engineers.

(f) An FMEA or FMECA is often required to comply with safety and quality requirements, such as ISO9001, QS 9000, ISO/TS 16949, Six Sigma, FDA Good Manufacturing Practices (GMPs), Process Safety Management Act (PSM), etc.

Failure mode, effects, and criticality analysis (FMECA) is a powerful design analysis tool that is used to increase system reliability. It can be applied during the initial design phase or to the existing equipment. To be more effective, FMECA should relate to the nature of the design process itself. In either case, it considers overall design, operating, and service problems, while at the same time addressing process and safety problems.

The preparation of FMEA should always be undertaken by a team of people drawn from all the relevant areas of the organisation bringing the appropriate information and insight.

The advantages of a cross functional team as shown in the people input are as follows:

- Earlier possible, parallel working (not serial) – concurrent engineering
- Benefits of more extensive knowledge and experience
- Increasing creativity
- Increasing acceptance
- Simultaneous and not successive decision process

If used as a design tool, the benefit of FMECA depends upon the timeliness in which information is communicated in early design phase. Timeliness is probably the most important factor in differentiating between effective and ineffective implementation of FMECA. The efforts and sophistication of the approach used depend greatly on the requirements of each individual programme. In any case, FMECA should contribute to the overall programme decision.

BUSTLE - 8.3

1. Brief out the documents which would facilitate the efforts of FMEA in your organisation.

____

____

____

____

Test Your Understanding 8.3

1. Explain quantitative approach of FMECA.
2. List the steps for carrying out design FMECA.
3. Compare design FMEA and process FMEA.
4. Explain applications and merits of FMEA/FMECA.

8.17 CONCLUSION

Failure mode and effect analysis (FMEA) is the study of partial failures to determine their effects, if the results of FMEA are ranked in order of seriousness, then the word criticality is added to give

FMECA. The primary objective of FMECA is to determine the features of product design, production or operation and distribution that are critical to the various modes of failure, in order to reduce failure. It uses all the available experience, from design, technology, purchasing, production/operation, to distribution, marketing, service, etc., to identify the importance levels or criticality of potential problems and simulate action to reduce these levels. FMECA should be a major consideration at the design state of a product or service.

8.18 RECAP

It is possible to analyse products, services, and process to determine possible modes of failure and their effects on the performance of the product or operation of the process or service system. This is another approach for achieving CBM studied under FMEA/FMECA.

FMECA can be applied to any stage of design, development, production/operation or usage, but since its main aim is to prevent failure, it is most suitably applied at the design stage to identify and eliminate causes. Design FMEA and Process FMEA are discussed and a brief comparative study has been made. However, with more complex product or service systems, it can be appropriate to consider these as smaller units or subsystems, each one being the subject of a separate FMECA.

8.19 KEY CONCEPTS

1. **Failure:** The termination of the ability of an item to perform its required function.
2. **Failure Mode:** It is the manner by which a failure is observed. Generally it is the way failure occurs and its impact on equipment operation.
3. **Failure Effect:** It is the consequence of a failure mode on the operation, function, or status of an item.
4. **Failure Cause:** The physical or chemical process, design defects, part misapplication, quality defects, or other processes that are the basic reason for failure.
5. **Criticality:** It is a relative measure of the consequences of a failure mode and its frequency of occurrence.
6. **Criticality Analysis (CA):** It is a procedure by which each potential failure mode is ranked according to the combined influence of severity and probability of occurrence.
7. **Single Failure Point:** The failure of an item that will result in failure of the system and is not compensated for by redundancy or alternative operational procedure.
8. **Undetectable Failure:** It is a postulated failure mode in FMEA for which there is no failure detection method by which the operator is made aware of the failure.
9. **Failure Definition:** This is a general statement of what constitutes a failure of the item in terms of performance parameters and allowable limits for each specified output.
10. **Trade-off Study Reports:** These reports should identify areas of marginal and state-of-the-art design and explain any design compromises and operating restraints agreed upon.
11. **Functional Block Diagrams:** Illustration of the operation and interrelationships between functional entities of a system as defined in engineering data and schematics.
12. **Reliability Block Diagram:** Reliability block diagrams define the series dependence, or independence, of all functions of a system or functional group for each life-cycle event.

13. **Severity (S):** Severity (S) is a numerical measure of how serious is the effect of the failure to the customer.
14. **Occurrence (O):** Occurrence (O) is a measure of probability that a particular mode will actually happen.
15. **Detection (D):** Detection (D) is a measure of probability that a particular mode will be detected in the manufacturer's own operation before reaching the customer.
16. **FMEA/FMECA:** Failure mode and effects (and criticality) analysis, a procedure by which each potential failure mode in a system is analysed to determine the results, or effects thereof, on the system and to clarify each potential failure mode according to its severity.

REVIEW QUESTIONS

1. Define the following terms with reference to FMEA/FMECA.
   (a) Severity
   (b) Criticality
   (c) Criticality analysis
   (d) Detection mechanism
2. Define and explain the three elements of FMECA.
3. Explain the following terms used in FMECA.
   (a) Severity Probability (S)
   (b) Occurrence Probability (O)
   (c) Detection Probability (D)
   (d) Risk Priority Number (RPN)
4. Discuss the use of statistical process control (SPC) in FMECA.
5. Distinguish between process FMEA and design FMEA. Discuss the merits and demerits of FMEA.

CHAPTER 9

Diagnostic Maintenance

PREAMBLE

This chapter deals with the various maintenance philosophies. Before adopting a particular philosophy, it must be thoroughly examined. The implementation of diagnostic maintenance by using fault diagnosis techniques is discussed. The philosophies such as failure-based, time-based, condition-based, reliability-centered, etc., are discussed. The condition-based maintenance techniques and approaches are discussed in this chapter.

CHAPTER OUTLINE

9.1 INTRODUCTION

The industrial infrastructural facilities in the form of machinery, equipment, buildings, and other service facilities are subject to deterioration due to their use and exposure to environmental conditions. This deterioration, if undetected, culminates in rendering these facilities unserviceable and makes them standstill. Industries, therefore, have no other choice but to attend to them time to time, to repair, renovate, recondition, and restore them so as to prolong their useful life to the maximum possible extent. In this context, maintenance assumes importance as an engineering function and is made responsible for provision of the condition of these machines, equipment, buildings, and services that will permit uninterrupted implementation of plans requiring their use. To achieve this, various strategies/philosophies and policies have been developed particularly in recent past. These philosophies, their approaches, and relevant techniques are highlighted in this book. This chapter particularly deals with fault diagnosis, and diagnostic maintenance.

9.2 PHILOSOPHIES OF MAINTENANCE: THE EVOLUTION

It is needless to mention the importance of higher powers and speeds of machinery, utilisation of complex process and equipment, and desirability of minimum downtime and penalty costs, when industries all over the world have entered the era of high technology maintenance management. Particularly, these are practically true in areas with high volume oil production/offshore drilling, mass rapid transport, mining and geotechnical applications, power plants, and other continuous process industries. Downtime in many of these industries causes enormous losses in earnings. A way of reducing plant breakdowns is to select the best maintenance strategy. The various maintenance strategies are already discussed in Chapter 3. However, as a ready reference, the evolution and the significance of all these philosophies/strategies in general and condition-based maintenance (CBM) in particular are discussed in detail in this chapter.

9.2.1 Failure-based Maintenance

The classical and oldest concept of maintenance is carrying out *failure-based maintenance*. This maintenance comes into action when the machine fails either suddenly or gradually due to wear and tear (*corrective or breakdown maintenance*). This philosophy is strongly, but of course wrongly believed by those industrialists who believe that maintenance time is an obstruction for the production and the time utilised for the maintenance is wastage. They feel that their productive hours are lost due to preventive maintenance and that when the machine is running, it is not necessary. And at the same time they do not accept the fact that proper maintenance will improve the life of their machines.

It is no surprise when they say, *"Why should we lose today's productive hours for want of some unknown hours that we may get tomorrow?"* They are just concerned with the present profit-making and production without giving much thought to future consequences that may arise from poor machine condition. Their approach is short-term and they want to make full use of the machine in its working condition. However, in the long run, the machine can break down irreparably without maintenance, causing greater loss.

Perhaps, the pressure to meet the production targets can be another reason for their feelings and views. This philosophy, thus, emphasises on the short range maintenance planning such as breakdown maintenance, operate to failure and corrective maintenance. Clearly, this did give room for neither stopping future failures nor reducing the probability.

Owing to the above reason and knowing the need for prolonging the life of the equipment, a trial has been made to contract the maintenance repairs with the original equipment manufacturer (OEM) after the sale. In this connection, some mediocre agencies have been developed to provide the repair, service, and maintenance services *(contractual maintenance)*.

9.2.2 Time-based Maintenance

When this long range maintenance planning has gained the importance to prevent the failures or reducing the chance of failures, the plant engineers developed and started focusing on the *time-based maintenance* philosophy. In this policy, the machine is overhauled at regular intervals and if any parts are worn out, they will be replaced; otherwise they will be cleaned and reassembled, technically known as *preventive maintenance (PM)*. And a schedule is maintained and adhered to which is named as *planned maintenance*. To facilitate those managers who bother about their lost productive hours due to shutdowns for this type of preventive maintenance, the plant engineers have scheduled in such a way that the preventive maintenance is carried out on holidays or non-productive days or on the days when the production is not or less required. For some equipment that has standbys can be planned in the regular days only without disturbing the regular production. Thus, the PM is sub-divided into two categories as *planned maintenance* for regular days and *scheduled maintenance* for off days. With the impressive results obtained by this philosophy, every industry started adopting these policies so as to increase the life of their equipment and reduce the probability of failures under various heads like *maintenance prevention, predictive maintenance, scheduled maintenance, shutdown maintenance and fixed time maintenance*, etc., with a little conceptual difference as suitable to their requirements. However, all these are derived from the preventive maintenance. The time between two consecutive maintenance and duration of maintenance has become basic parameters for measuring the maintenance effectiveness.

With growing technology and facilities, one has to concentrate on updating their equipment so as to match with the speed, flexibility, and trend. The thought of *design out maintenance* emerged in this connection.

9.2.3 Reliability-centred Maintenance

No doubt, the PM has tremendously added the advantages of preventing or reducing failures and adding the life for the equipment, but at what cost? When this question has come into the minds of plant engineers, some more questions have sequentially developed such as

- Is it necessary to carry out PM at regular intervals?
- Is it economical to perform PM on all machines irrespective of high cost or low cost?
- Is there any relation between the age of the machine and the failures?
- How long can we rely upon the availability of the machine after carrying out the PM?
- Does this reliability show any impact on the maintenance interval?

Yes, that was the origin for development of a new philosophy, *reliability-centered maintenance (RCM)* which made use of the statistics of past failure data and the probability concepts. The terms reliability, maintainability, and availability have taken a shape and the quantification of these terms helped plant engineers to carry out preventive maintenance most economically.

9.2.4 Condition-based Maintenance

It is not a wrong belief that the maintenance carried out at irregular intervals will give more benefits than carrying out at regular intervals (like in preventive maintenance) owing to following major reasons:

- Put a simple question to yourselves whether you carry out this PM according to the calendar time or time of utilisation. When the machine is not utilised as specified in a given period, how far is it justified to conduct PM?
- Though the time of preventive maintenance is reached as per schedule, it is not necessary to conduct the PM when the machine is running without any problem. Since it does not help much improvement, rather simply adding the cost of maintenance in some occasions and disturbing the productive hours in few other cases.
- While conducting the PM as per the schedule but not need, and when reassembling certain machine parts, there is often a possibility of mismatching of the components. It may be because of worn-out parts, change of models, spurious parts, lack of standardisation, lack of skill and knowledge, etc. It will obviously lead to loss.
- Should the PM schedule be same for all the machines uniformly? Is it not different for different machines? Certainly, it should be different for different machines. The high cost machines, the machines which are very critical (though low cost) in the production chain, the machines whose spares are scarcely available require more care.
- Do you care equally for the health of an infant, a child, a teenager, a middle aged, and an old person? The answer must be "No". It is similar in the case of machines also. A newly bought machine and an old machine cannot be treated equally. And an industry will have the machines of different ages. Again simply age will not be the criterion to decide the PM schedule but the hours of run is another factor.
- Are you taking the machine condition into account? The condition of machine plays a greater role as it sometimes depends directly on the age and/or hours utilised, but indirectly affected by environment, the usage, and the complexity of design and amusingly on the PM also.

Now, one must accept the need of conducting the PM at irregular intervals depending on the condition of the machine which includes all the factors pointed out above such as age, hours utilised, cost of machine, cost and availability of spares, its criticality in the production chain, etc. These factors are also not constant but are subjected to dynamic changes. Hence, in such a complex, complicated, and dynamically changing situation, a more appropriate maintenance philosophy one can think of, is condition-based maintenance (CBM). The condition-based maintenance, in simple words, can be described as a preventive maintenance carried out at irregular intervals based on the actual condition of the machine.

9.2.5 Total Productive Maintenance, Physical Assets Management and Tero-technology

Well, a single defect can cost about ₹ 100 in diagnosis and repair if it is detected early in production, whereas the same defect in the field may cost ₹ 1000 to rectify otherwise. If it transpires that the failure is a design fault then the cost of the redesign, documentation, and retest may well be in tens or even hundreds of thousands of rupees. Therefore, an eye should be kept upon the design of the machine. This close vision is to be continued till the end of life of the machine. Hence, the availability can be enhanced and high reliability for any machine can be set so as to facilitate in maximising the production. Thus, the maintenance from "design to discard" concept with a little difference is introduced through the *Tero-technology*, developed by United States, the *Physical Assets Management (PAM)* developed by Europeans and the *Total Productive Maintenance (TPM)*, a brain child of Japanese.

BUSTLE - 9.1

1. *Observe various philosophies of maintenance with reference to their principles, areas of applications, advantages and limitations and prepare a master chart in the tabular form as given below.*

Philosophy	*Principles*	*Areas suitable to apply*	*Advantages*	*Limitations*

2. *Which philosophy is in practice in the organisation/institution where you are working/studying or the one with which you are familiar. Bring out various merits and demerits you are practically facing with the present policy. Which of the above policies do you feel is most suitable to your organisation/institution and why?*

___

___

___

___

Test Your Understanding 9.1

1. Explicate the philosophy of condition-based maintenance.
2. List the various maintenance strategies in practice. Explain.
3. Discuss various philosophies adaptable to maintenance management. Which one of them would you prescribe for (a) project production machinery such as aircraft building, (b) machinery of mass production unit, (c) job production machinery? Justify your answer.

9.3 SIGNIFICANCE OF CONDITION-BASED MAINTENANCE

Consider the following case:

In an office, there are three friends working namely, Ms. Bala Mani, Ms. Priya Mani, and Ms. Chintha Mani. They have different tastes and strategies in life. (Let us refer to them with their abbreviated names as BM, PM, and CM). For example, the health care strategy of these friends is interestingly different. Ms. BM does not care about her health and she never takes any precautionary measures at all for keeping up of good health. Further, she eats any roadside eatables also. And she goes to doctor if and only if it is so necessary even when she is ill. But unfortunately most of the times, it becomes compulsory for her to spend on her ailments; those become chronic due to mere negligence and ignorance. Unlike BM, Ms. PM is so careful about her health. She never takes outside food and water unless she feels it is hygienic. She follows all the precautionary measures for her health and exercises regularly. She goes to her family doctor in regular intervals and gets her blood pressure, blood sugar, etc., checked. Ms. CM finds that BM takes leave from her office quite often on health reasons, sometimes even weeks together. Interestingly, she finds that PM also takes leave for her health check up frequently though not really required. She further notices that PM is much worried about her health and spends lot of money unnecessarily and her regular work is disturbed by herself. Therefore, she finds another good policy that neither too much care like PM nor carelessness like BM is required. She wants to take care of her health in such a fashion that she visits her doctor when she just needs or finds some symptoms of the ailment and, thus, prevent the illness. This strategy has saved most of her leaves and money for CM. It also has given a way for her to work without any tensions.

The above exemplary case clearly presents the significance of condition monitoring. If Bala Mani (BM), Priya Mani(PM), and Chintha Mani(CM) are considered as three machines, the strategy adopted by Bala Mani is *operate to failure* and *corrective maintenance* or *breakdown maintenance* while *preventive maintenance* is followed by Priya Mani. The third kind is *condition-based maintenance* followed by Chintha Mani in which the downtime and unnecessary costs can be controlled most effectively. In other words, the first strategy is a failure-based maintenance strategy, the second one is time-based maintenance and the third is condition-based.

1. Failure-based Maintenance
   (Repair it when it fails)

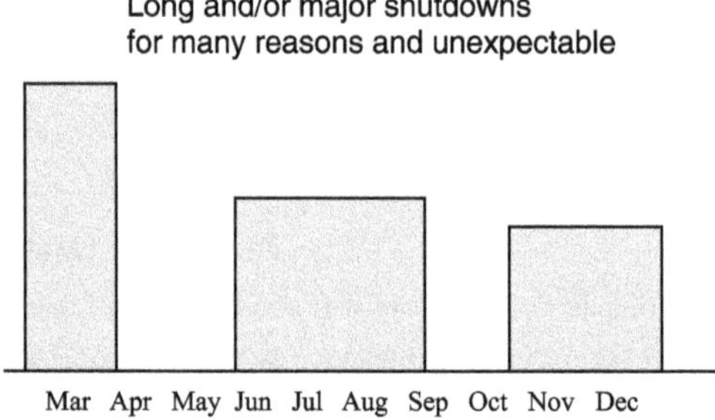

2. Time-based Maintenance
   (Comprehensive maintenance in advance to reduce chance of trouble)

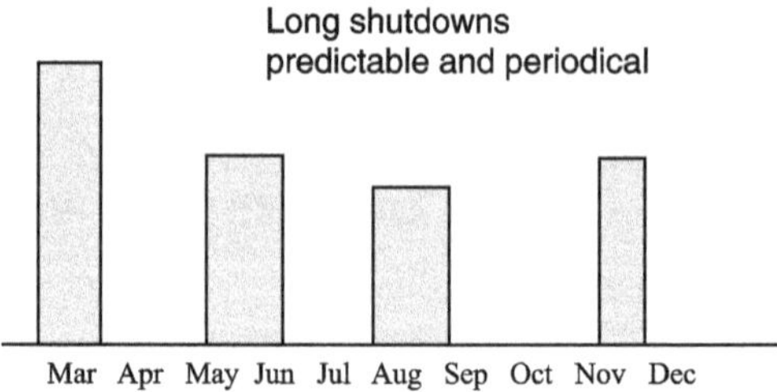

3. Condition-based Maintenance
   (Maintenance as it is needed just before the problem has arisen)

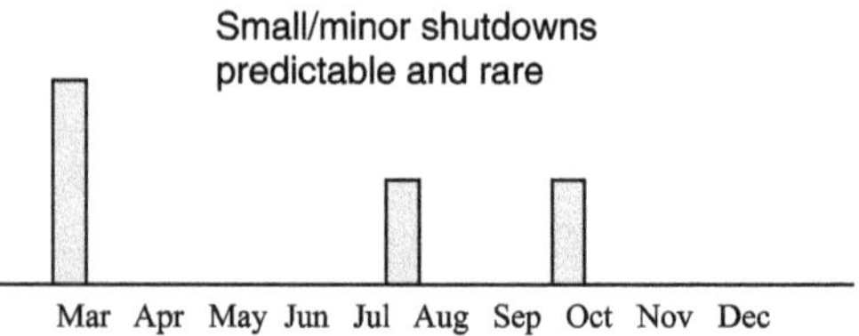

Figure 9.1 Significance of condition-based maintenance

Figure 9.1 shows what can happen to a typical plant if the method of operation of machine is to run until they fail and then repair them in order to restore the acceptable condition. This policy can prove to be expensive in the form of maintenance cost, lost productive hours, and hazards to the personnel. Therefore, it is now well recognized especially in the case of large and expensive plant in order to avoid unscheduled shutdown, that it is much more economical to carryout preventive maintenance (PM) so that major breakdowns are avoided as shown in the figure. The problem in preventive maintenance lies in the choice of an appropriate maintenance interval for the machine because the actual running time before maintenance really needed is not constant. It varies from one occasion to another and from one machine to another due to the difference in the operation of the machine and in the behaviour of its components and many more factors as discussed earlier. If safe overall interval is chosen as shown in Figure 9.1. There will be occasions when the machinery will be overhauled long before it is really necessary, while it could have run on for much longer without failing. This situation wastes production output and increases the incidence of human errors on reassemble of the machine. Too long an interval results in an unacceptable number of machine failures during operations. A compromise between these two extremes can be established by experience but machine failures will continue to occur. Some machine failures are the result of sudden events in which the machine condition deteriorates in few minutes. The greater number of machine failures however is the apex of long slow deterioration during which the machine gets progressively worse over a period of months or even years.

Therefore, a more satisfactory strategy is to carryout preventive maintenance at irregular intervals and to determine these intervals by the actual condition of the machine. In this method of condition-based maintenance (also known as dynamic predictive maintenance or diagnostic maintenance), the plant is maintained just after some problem has arisen, but much before the possible breakdown. The maintenance activity is directed towards not only detection and prevention but also cure of the particular problem. Although condition-based maintenance (CBM) as shown in Figure 9.2 is applicable to the random failure mode, the characteristic of many plant items, it also has an obvious application to age related failure of high cost plant such as where the manufacturer runs a number of old machines and equipment and he needs to be aware of the advent of deteriorating working conditions.

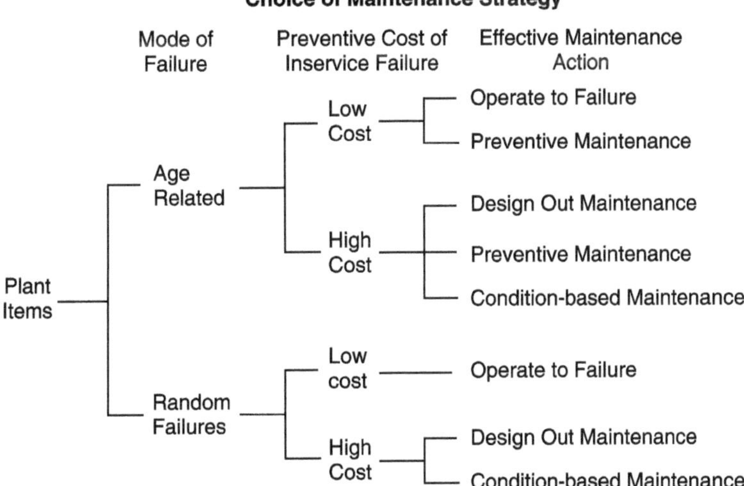

Figure 9.2 Selection of maintenance strategy based on type of failure and cost

The savings by reducing downtime through the implementation of CBM strategy can be considerable. However, for the whole strategy to work there must be some means of knowing the condition of the equipment and getting a warning sign of problems sufficiently well in advance to arrange a schedule of shutdown, make available necessary skilled and intelligent maintenance staff and standby spare parts. The main function of condition monitoring is to provide the knowledge of machine condition and of its failure rate, obtained through selection of a suitable parameter such as vibration for measuring deterioration and reading its values at intervals. In recent years, improved diagnostic techniques have become available to monitor the condition of machinery and equipment with sufficient accuracy and consistency to enable practising CBM.

9.4 APPROACHES TO CONDITION-BASED MAINTENANCE

There are several approaches to implementing condition-based maintenance of which the following five are popular in general practice. These are interlinked but conceptually differ in certain aspects.

(a) Failure Modes, Effects, and Criticality Analysis (FMECA)
(b) Fault Diagnosis and Diagnostic Maintenance
(c) Condition Checking
(d) Trend Monitoring
(e) Trend Analysis

[Among the above five approaches, the concept of FMECA has already been discussed in Chapter 8, while the second one, i.e., fault diagnosis and diagnostic maintenance is described in this chapter. The third and fourth (condition checking and trend monitoring) will be discussed in Chapter 10. Trend analysis, which will discussed in Chapter 11, though is an approach to condition monitoring, is more used with a focus on reliability-centered maintenance (RCM)].

9.5 FAULT DIAGNOSIS AND DIAGNOSTIC MAINTENANCE

This maintenance method is viewed as a part of CBM by most of the plant managers probably due to the reason that the techniques employed are almost similar. The choice of the monitoring techniques to be applied in particular plant is difficult, affected by factors such as skills available, effect of the environment, reliability of the machines, and the monitoring methods. This decision tends to be made on an ad-hoc basis of personal experience or advice to be developed. But the diagnostic maintenance uses the scientific methods and analysis of failures to detect and rectify the faults.

Most of us have experienced the situation of going to a pathological laboratory for a blood check up or urine examination or any related tests referred by our doctor when we visit him with some ailment. Although the doctor knows or suspects the disease, he/she does not diagnose and treat immediately. Do you know why? Because he/she wants to confirm it through scientific test, to know the severity of the problem and to check any scope of development of other linked problems. A plant engineer is none less than the doctor for the machine and thence it necessitates him/her to completely diagnose the fault before the treatment starts. Various tests are available for the plant engineer, which can be categorised into two main groups:

(a) Destructive Tests (DT), and
(b) Non-destructive Tests (NDT).

There are numerous destructive tests to know if the material of the machine part is subjected to any deterioration due to stress, impact, creep, fatigue, and thermal effects. These tests will be resourceful to the plant engineer in ensuring whether to continue or to replace the machine parts. In fact, these tests are performed occasionally by the plant engineer to confirm when there is any doubt on the machine part but not regularly. A plant engineer shows much interest and gives utmost importance to the non-destructive tests rather than destructive tests as the latter involve the destroying of the part while the former can identify, detect, and indicate the severity of the problem without actually destroying the part and, thus, facilitates the engineer to use the same part if he feels that replacement is not immediately necessary. (These tests are dealt in the course metallurgy/material science and strength of materials. The readers are advised to go through these subjects for more details of destructive tests. However, a few most important tests are briefed out here for ready reference).

9.6 DESTRUCTIVE TESTS

Several destructive tests are available to determine the various mechanical properties. Based on the nature of loading, they are classified as follows:

(a) **Static Test** – load is applied gradually.

- Tension Test
- Hardness Tests
- Creep Test

(b) **Dynamic (Impact) Test** – specimen is subjected to shock loads.

- Charpy Test
- Izod Test

(c) **Cyclic (Fatigue) Test** – load is repeatedly varied in magnitude and direction during the test.

9.6.1 Static Test

These tests are used to evaluate the ability of the material to withstand the static loads, i.e., the loading is done gradually in certain predetermined increments. The tension test, hardness test, and creep test are the common important tests performed to evaluate the mechanical properties of materials used for the equipment or components or parts. Often these tests are performed on the part after it is worn out, to know the reason of its worn out (like postmortem). If same has to be performed on a part of a new or working machine, then a small piece of identical material specimen is taken and tested.

Tension Test

In this experiment, Young's modulus, the yield strength, ultimate strength, fracture stress, elongation, and other properties of several tensile shafts or bars using Universal Testing Machine (UTM) will be determined. After testing, the graph is drawn from the data. For this, various measurements of sample geometry are required to calculate engineering stress versus engineering strain to obtain the material properties. To construct the test, one end of standard specimen is attached to movable crosshead with the other end fixed to a stationary member. The cross head is driven such that it stretches the specimen apart, i.e., the specimen is stretched under load until it breaks. Thus, the tensile strength can be measured. The test is suitable to test the load capacity of shafts or bars trusses and some machine parts. However, such tests will be performed on specimen materials of the machine parts by original equipment manufacturer (OEM) of the equipment during or before manufacturing of the machine or during the design stage.

Hardness Tests

Hardness can be defined as the resistance of a material to deformation or indentation. Hardness is measured in hardness number and has no units. There are three methods in practice to measure hardness number. Thus, the hardness tests are classified in the following three ways:

(a) Brinell Hardness Test

(b) Rockwell Hardness Test

(c) Vicker's Hardness Test

A brief description of these tests is given below:

(a) **Brinell Hardness Test:** In the Brinnell hardness test, a hardened steel ball of diameter 10 mm is forced into the surface and an optical measuring device is used to measure the diameter of resulting indentation, d. Then the Brinell hardness number (BHN) can be calculated by the relation,

$$BHN = \text{Load Applied/Surface Area of Indentation}$$

$$= \frac{P}{\frac{\pi D}{2}\left[D - \sqrt{D^2 - d^2}\right]}$$

where P = load applied, kg
= 3000 kg for steel
= 1000 kg for copper
= 500 kg for aluminium
D = diameter of steel ball, mm
d = diameter of indentation, mm

(b) **Rockwell Hardness Test:** Rockwell hardness tests use different loads and penetrators (indenters), and depth of indentation is calibrated on a dial to a hardness number. There are two scales here, viz. B-scale and C-scale. Hardened steel ball with 100 kg total load is used for B-scale; while diamond cone (brale) with 150 kg total load is used for C-scale.

Rockwell hardness number on B-scale,

$$R_B = 130 - \frac{d}{0.002}$$

and on C-scale,

$$R_C = 100 - \frac{d}{0.002}$$

where d = depth of indentation

(c) **Vickers Hardness Test:** Vickers hardness test uses a surface based diamond pyramid with 136° apex angle between facets. It is forced into the surface with a suitable load ranging from 5 to 120 kg in steps of 5 kg. The test can be used for material with extreme hardness and case hardened parts. The Vickers hardness number (VHN) or diamond pyramid hardness (DPH) is defined as the load (P) per unit area of the impression and is given by the equation:

$$DPH = \frac{2P\sin\left(\frac{\alpha}{2}\right)}{d^2}$$

where P = load applied on the diamond, kgf

d = diamond of the impression, mm

$\propto$ = angle between the opposite faces of the diamond (136°)

Creep Test

The progressive deformation of material at constant stress is called *creep*. To determine the creep behaviour of the material, a constant load is applied to a specimen maintained at a constant temperature, and the strain (extension of specimen) is determined as a fraction of time. The creep test data is presented in the form of creep curve. The slope of the curve ($d\varepsilon/dt$) is referred to as the creep rate.

9.6.2 Dynamic or Impact Tests

Impact tests are designed to measure ability of the equipment or its parts to withstand shock loads. It involves a pendulum swinging down from a specified height, h_1 to strike the test piece and fracture it. The length, h_2 to which the pendulum rises after breaking the test piece is a measure of the energy used in breaking. After calculating the energy required to break the specimen, the impact strength can be found by the formula.

$$\text{Impact strength} = \frac{\text{Energy in fracture}}{\text{Corss section area in notch}}$$

Charpy Test: For Charpy test, a notched specimen is mounted as a simple supported beam, and heavy pendulum is allowed to strike the specimen from fixed height. The notch is on the face directly opposite to where the pendulum strikes the test piece.

Izod Test: In this case a notched specimen is held in a vice in the form of cantilever and a heavy pendulum is allowed to strike the specimen from a fixed height. The blow is struck on the same face as the notch at a fixed height above it.

9.6.3 Fatigue Test

Fatigue failure is caused by repeated stress-cycle. The most commonly used fatigue test is rotating beam test in which specimen is loaded and rotated. In this case, the specimen is subjected to alternating tension and compression stresses of equal magnitude.

BUSTLE - 9.2

Discuss a few destructive tests you know. Relate them to apply for plant maintenance of any industry you have chosen.

___

___

___

___

Test Your Understanding 9.2

1. Discuss various approaches to implement CBM.
2. How are destructive tests used for fault diagnosis under diagnostic maintenance? Briefly explain various destructive tests used for fault diagnosis.
3. Explain different methods of hardness testing. What are its applications in maintenance engineering?

9.7 NON-DESTRUCTIVE TESTS

Everyone is familiar with the word *Eureka* (means 'I found it'). The Archimedes' principle and the story behind it are also known to all. When his king asked him to find out without destroying, whether the crown is made up of pure gold or any impurity is present, he took the hardships, but finally made it possible to find the purity without destroying. Perhaps, that might be the first non-destructive test (NDT) in the modern science or it might be the root for development of NDT techniques if not the first. And then onwards, a variety of competitive NDT techniques for fault detection have long been established, and many more are being added to the list everyday. A few are explained in this section.

The NDT methods have gained popularity in the modern high-tech environment, as these are more reliable, safe, and economical. In some cases, no other way than the NDT is to be employed for example, the bomb testing, Brick strength-testing, etc. In the present context, NDT has two important roles to play in diagnostic maintenance. These are:

(a) From these investigations, the presence of defects is established. It is also useful in estimating the nature and the extent of defects and arriving at a conclusion to run or shut down the machine.

(b) NDT also helps in measuring the defect size and distribution, which plays a crucial role in taking the final decision to continue or replace the spares or even the machine also in few cases.

A number of such tests are available to detect the defects in metallic components. The most popularly used NDT techniques are

- Liquid Penetration Inspection
- Magnetic Particle Testing
- Eddy Current Inspection
- Ultrasonic Inspection
- Radiography (X-ray and Gamma Ray)

9.7.1 Liquid Penetration Inspection

This method is used to detect large cracks or openings, cold shuts, fatigue cracks, and pits. There are two types of liquid penetration inspections.

(a) Dye penetration inspection

(b) Fluorescent liquid penetration inspection

(a) **Dye Penetration Inspection:** The liquid dye penetrant is sprayed on to the clean surface of test piece. Excess amount of dye is removed; surface is washed with water and thoroughly dried.

Then a developer is sprayed on the surface, which brings out the colour in the dye penetrant that has penetrated into cracks or pinholes.

(b) **Fluorescent Liquid Penetration Inspection:** Fluorescent liquid is applied to the surface to be inspected. After removing the excess liquid, surface is washed and dried. Then the test surface is viewed under black light whose wavelength lies between the visible and ultra violet regions of the spectrum, which causes the penetrant to glow clearly in the dark. Some solids used in the cleaners and developers contain a high percentage of chlorine to make the liquid non-inflammable, but requires a great care in view of health hazards associated with chlorine. Penetrant testing has the advantage that it can be used for all types of materials.

9.7.2 Magnetic Particle Testing

There are several methods of magnetic particle testing to detect welding, casting, and forging defects in iron and steel parts. The part to be tested is magnetised by external magnetic yoke coil or passing an electrical current and the inspection medium is applied on it. In dry method of inspection, a special fine ferromagnetic powder is applied on the surface by means of hand shaker or vibrating screen so that the powder is uniformly distributed on the surface. In wet method, a liquid like kerosene containing suspended fine ferromagnetic particles is sprayed or brushed. Due to the formation of magnetic poles at the crack or flaw, the magnetic powder concentrates in that area.

9.7.3 Eddy Current Inspection

Eddy current testing is one of the methods of detecting discontinuities and flaws. The method is based on the principle that when the alternate current carrying conductor coil is brought near a metallic/conductive specimen, eddy currents are induced in the specimen due to electromagnetism. These eddy currents produce their own magnetic field that opposes the field of the current carrying coil, thereby increasing the impedance (resistance). Coil impedance can be measured whose variation indicates the crack or flaw.

9.7.4 Ultrasonic Inspection

Ultrasonic inspection is fast, reliable NDT method which employs electronically produced high frequency sound waves that will penetrate metals, liquids, and many other materials at speeds of several thousands feet per second. Piezo electric materials (e.g., quartz) are used to produce ultrasonic waves. (These materials undergo a change in physical dimension when subjected to electric field. This conversion of electrical energy to mechanical energy is known as Piezo electric effect). If an alternating electric field is applied to a piezoelectric crystal, the crystal will expand during the first half of cycle and contract when the electric field is reversed. By varying the frequency of the alternating electric field, we can vary the frequency of the mechanical vibration (sound wave) produced in the crystal. There are two common ultrasonic test methods, the through-transmission and the pulse-echo methods. Ultrasonic inspection is used to detect and locate defects such as shrinkage in cavities, internal bursts or cracks, porosity, and large non-metallic inclusions. Wall thickness can also be measured in closed vessels or in cases where such measurement cannot otherwise be made.

9.7.5 Radiography (X-ray and Gamma Ray)

Radiography is one of the most useful NDT methods, which can be applied for inspection of cracks, porosity, blowholes, etc., in welds of all types and thickness ranging from minute welds in electronic components to welds up to half metre thick employed in heavy fabrications. The principle of the technique is based on exposing the components to short wavelength radiation in the form of X-rays (wavelength less than 0.001 × 10^{-8} cm) or gamma rays (wavelength about 0.005 × 10^{-8} to 3 × 10^{-8} cm) from a suitable source such as an X-ray tube or cobalt-60. The position of the weld where defects are suspected is exposed to X-rays. During exposure, X-rays penetrate the welded object and thus affect the X-ray film kept behind. Since most defects (such as blowholes, porosity, cracks, etc.) possess lesser density than the parent metal, they transmit X-ray better than the surround parent metal does. Therefore, the film appears to be darker where defects are in line of the X-ray beam.

Gamma rays are used for detecting defects in welded plates thicker than those inspected by X-rays. Gamma rays can inspect a number of welded objects at a time. Radium and its salts (now-a-days the radio isotopes such as cobalt-60 are used for the purpose) decompose at a constant rate, emitting gamma rays which are of shorter wavelength and more penetrating than ordinary X-rays. Unlike X-rays, gamma rays from its source are emitted in all directions; therefore, a number of separate welded objects having cassette containing film, fastened to the back of each object, are disposed in a circle around the source placed in a central position.

A comparison chart of all the above techniques is given in Table 9.1 for quick reference with their advantages, limitations, when and where to apply. The other NDT techniques are not in the scope of the book and hence are not discussed here.

Table 9.1 Comparison of NDT techniques

S. No.	*Testing method*	*When to use*	*Where to use*	*Advantages*	*Limitations*
1	Liquid Penetrant	To locate surface cracks, porosity, laps, leaks in weld bonds, fatigue, grinding cracks.	All metals, glass, ceramics, castings, forging, machine parts and cutting tools; field inspections.	Simple to apply; portable; fast; low cost; results are easy to interpret; requires no elaborate set up.	Limited to surface defects.
2	Magnetic Particle	To detect surface or shallow substance flaws, cracks, porosity, non-metallic inclusions and weld defects.	Only for Ferro magnetic materials, parts of any size, shape, composition or heat treatment.	Economical; simple in principle; easy to perform; portable (for field-testing); fast for production testing.	Limited to magnetic materials, needs to de-magnetise after test; needs power source; needs to be cleaned before test.
3	Ultra sonic	To find internal defects, cracks, leaks in bonds, laminations, inclusions, porosity, grain structure and thickness.	All metals and hard non-metals; sheets, tubing, rods, forging, castings; field test; inservice part test; braze and adhesive bond joints.	Fast; dependable; easy to operate; tends itself to automation; test results are instant; portable; highly accurate; sensitive.	Requires contact or immersion of part; interpretation of results requires training.
4	Eddy Current	To find variation in coating thickness; longitudinal seam/ crack in tubing; defects due to HT and composition.	Tubing and bar stock, parts of uniform geometry, flat stock or sheets and wire.	High speed; non-contact; automatic.	False indications result from many variables; only good for conductive materials; limited depth of penetration.

5a.	Radio graphy: X-rays	To detect internal flaws and defects; to find weld flaws, cracks, seams, porosity, inclusions, holes; to measure thickness variation.	Assemblies of electronic parts, casting, welded vessels; field testing of welds; corrosion surveys, components of nonmetallic materials.	Provides permanent record on film; works on thin sections; high sensitivity; fluoroscopy techniques available; adjustable energy level.	High initial cost; power source required; radiation hazard, trained technicians needed.
5b.	Radio graphy: Gamma rays.	Detecting internal flaws, cracks, seams, holes, inclusion, weld defects; measuring thickness variations.	Forging, coatings, tubing, welded vessels; field testing welded pipes; corrosion surveys.	Detect variety of flaws; provide a permanent record; portable; low initial cost; source is small, good for inside shots; make panoramic exposure	One energy level per source; radiation hazard; needs trained personnel; source loses strength continuously.

9.8 LEO APPROACH

The LEO approach has been put forth by Briton as a simple formalised way of selecting the appropriate technique by assessing effectiveness. Three separate factors are considered and then combined to provide a measure of the size of the fault that must be detected for the technique to be effective.

L: **L**evel of critical parameter such as shocks/vibrations

E: **E**xperience factor

O: **O**peration efficiency factor

(a) The critical parameter size, L is the cause of the particular failure being considered. In the context of cracks this can be the crack depth. If the failure is seizure of a rolling bearing it is unlikely that the size of the damage in the races can be considered. More appropriately it can be correlation between vibration and shock pulse level and occurrence of seized bearings. Thus, L will be the vibration/shock level at which experience as shown bearing will seize.

(b) The experience factor, E to take account of:

- The amount of knowledge possessed of the cause and mechanism of the failure.
- The rate of progression of the failure in relation to monitoring frequency.
- Both these points of experience amount to a measure of extent of previous trouble encountered.

(c) The monitoring operational efficiency factor, O which takes account of:

- The sensitivity and reliability of the method and equipment in relation to the characteristics of the machine and its environment.
- The skill of the monitoring personnel.

'O' will vary from one where failures had been thoroughly monitored, to zero where it had never been possible to detect the fault in the past.

The product **LEO** provides a measure of the monitored parameter level or change which the monitored technique must be able to detect to be certain of providing sufficient warning of impending failure. Unfortunately, although data have been collected for NDT applications, the approach has not yet been taken up for general CM and no exact information is available on L, E, or O factors applicable to behaviour, wear debris, or vibration monitoring. However, the maintenance engineer will find it useful to develop his own factors for the technique he has yet had. In such case, the proposals for monitoring operational efficiency factors derived from Briton can be used as the guidelines.

BUSTLE - 9.3

List some occasions you used or experienced the NDT (to repair the machine parts or otherwise):

____

____

____

____

Test Your Understanding 9.3

1. Explain the following NDT techniques.
   (a) Liquid penetration
   (b) Radiography
2. How are the following useful to fault diagnosis? Explain the following:
   (a) Ultrasonics
   (b) Eddy current
3. What do you understand by LEO approach? Explain.
4. Explain any five NDT techniques to estimate the condition of equipment.
5. Prepare a comparative chart of any five NDTs with reference to their principles, application, merits, and demerits.
6. What is diagnostic maintenance? Explain its features.

9.9 CONCLUSION

Whenever a failure occurs or likely to occur, it is essential to think of its causes and the situations that leading to result in such state. Hence, a plant engineer (or maintenance personnel) should be capable of predicting, identifying, and analysing the causes for the failures, for which he/she needs a widespread knowledge on various tests of failure diagnosis.

9.10 RECAP

Various philosophies of maintenance are to be examined before adopting a particular philosophy. The philosophies such as failure-based, time-based, condition-based, reliability-centered, etc., are discussed. The condition-based maintenance techniques and approaches are discussed in this chapter.

The diagnostic maintenance, one of the approaches to CBM, uses the scientific methods to failure diagnosis.

9.11 KEY CONCEPTS

1. **Failure:** The termination of the ability of an item to perform its required function.
2. **Failure-based Maintenance:** It is to restore the condition when the machine fails either suddenly or gradually due to wear and tear.
3. **Time-based Maintenance:** The machine is overhauled at regular intervals and parts replaced if any, otherwise cleaned and reassembled.
4. **Reliability-centered Maintenance (RCM):** RCM uses the statistics of past failure data and the probability concepts so as to extend the availability of the machine with reliable performance.
5. **Condition-based Maintenance:** A preventive maintenance carried out at irregular intervals based on the actual condition of the machine.
6. **Total Productive Maintenance or Physical Assets Management or Tero-technology:** The maintenance from "design to discard" concept.
7. **Diagnostic Maintenance:** Using the scientific methods and analysis of failures to detect and rectify the faults.
8. **Destructive Test (DT):** Testing procedure used to test for the required quality by destroying the product, e.g., fatigue test, tensile test, etc.
9. **Non-destructive Test (NDT):** Testing the materials for the required properties by not destroying the products, e.g., X-ray test, ultrasonic test, etc.
10. **LEO Approach:** A measure of the monitored parameter level or change which the monitored technique must be able to detect to be certain of providing warning of impending failures. It is estimated by the product **LEO**, where, **L** stands for **L**evel of critical parameter such as shocks/vibrations; **E** stands for **E**xperience factor; **O** stands for **O**peration efficiency factor.

REVIEW QUESTIONS

1. Discuss the central theme and general procedure of the following philosophies of maintenance.
   (a) Time-based Maintenance
   (b) Failure-based Maintenance
   (c) Condition-based Maintenance
   (d) Reliability-based Maintenance
2. Illustrate the significance of condition-based maintenance with suitable example.
3. Describe any four destructive testing methods to evaluate the worn out part of a machine.
4. Explain any four non-destructive techniques to detect and diagnose the worn out part of a machine.
5. Explain the LEO approach used to assess the equipment effectiveness.

CHAPTER 10

Condition Monitoring

PREAMBLE

This chapter defines and explains various philosophies of maintenance. Various philosophies of the maintenance management, their central themes, principles, and their evolution are extensively discussed. The significance of condition monitoring, the principles, the method, and the techniques of CBM are detailed with appropriate supports. This chapter helps understand the diagnosing methods to find the fault in a machine. It also explains various destructive and non-destructive tests to estimate the machine condition. The advantages of CBM are also discussed here.

CHAPTER OUTLINE

10.1 INTRODUCTION

Machines are one of the major resources of production process and play a vital role in converting the inputs into desired outputs. Any failure of these machines will cause an unpredictable losses or damages. Therefore, it is worthwhile to maintain these machines so as not to fail. In this direction, there are several policies/methods developed and have been used from years, yet felt insufficient. Moreover, the condition-based maintenance (CBM) methods have gained importance in the recent past. In this chapter, the concepts and methods of condition-based maintenance are dealt with.

10.2 PRINCIPLES OF CONDITION-BASED MAINTENANCE

Condition-based maintenance is carried out by monitoring the condition of the machine regularly. Often condition monitoring is carried out in two fold principles namely—

- ***Condition*** Checking
- Trend ***Monitoring***

10.2.1 Condition Checking

Condition checking principle involves in checking the condition of the equipment at regular intervals. This is carried out by preparing a checklist of various points or parameters of the machine by

Table 10.1 Checklist and certificate of condition checking

<table>
<tr><td colspan="4">N V S & Co.
Department of Plant Engineering
Certificate of Condition Checking</td></tr>
<tr><td colspan="2">Machine code: L-29-12/01</td><td colspan="2">Name of the Machine: Lathe</td></tr>
<tr><td colspan="2">Machine Location: Tool Room</td><td colspan="2">Checked by: Prasanna</td></tr>
<tr><td>Sl. No.</td><td colspan="2">Parameter or Check Point</td><td>Condition</td></tr>
<tr><td>1.</td><td colspan="2">Gear oil</td><td>OK</td></tr>
<tr><td>2.</td><td colspan="2">Spindle run out</td><td>OK</td></tr>
<tr><td>3.</td><td colspan="2">Starting trouble</td><td>OK</td></tr>
<tr><td>4.</td><td colspan="2">Leakages (oil)</td><td>OK</td></tr>
<tr><td>5.</td><td colspan="2">Back lash error</td><td>OK</td></tr>
<tr><td>6.</td><td colspan="2">Speeds/feeds and mechanism</td><td>OK</td></tr>
<tr><td>7.</td><td colspan="2">Vibration</td><td>OK</td></tr>
<tr><td>8.</td><td colspan="2">Lubrication/grease points</td><td>OK</td></tr>
<tr><td>9.</td><td colspan="2">Chuck/head stock/tail stock/tool post</td><td>OK</td></tr>
<tr><td>10.</td><td colspan="2">Others (specify)</td><td>—</td></tr>
<tr><td colspan="2">Level of Condition:
Normal/<s>alarm/shutdown/breakdown</s></td><td colspan="2">Recommendation for Utilisation:
Recommended/ <s>Not Recommended</s></td></tr>
<tr><td colspan="2">Certified by: N.V.S. Raju.
Signature: Жξ</td><td colspan="2">Comments of certifier (if any):
—</td></tr>
<tr><td colspan="2">Checked Date and Time:
11.05.2009 – 2.00 pm (2nd Shift)</td><td colspan="2">Valid Up To:
11.08.2009 – 9.00 am (1st Shift)</td></tr>
</table>

which the machine condition can be defined. Usually, this activity is assigned to the maintenance department or the operator who maintains or repairs/restores the condition (in the event of failures) on the machine. The maintenance operator is supposed to check the condition at the very beginning of working on the machine or while handing over the equipment to the user department certifying about the condition. The condition checking includes the trial runs and test runs (called dynamic checking) also in addition to the regular checklist tests (known as static checking). An exemplary checklist is given in Table 10.1.

10.2.2 Trend Monitoring

Trend monitoring is continuous or regular measurement and recording of data, collected during operation, to indicate variations in the conditions of the machine or its components, in the interest of safe and economical operation. This needs thorough knowledge in the selection of suitable and measurable parameters to indicate the deterioration of machine or component. The trend in this measurement that can happen with running time when deterioration exceeds a critical stage and the way in which such trend monitoring can give a lead time before the deterioration reaches a shutdown level are interpreted in Figure 10.1.

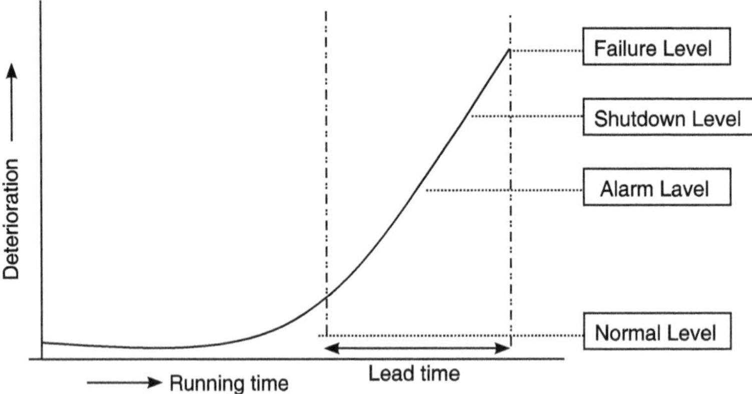

Figure 10.1 Trend monitoring

Various states the equipment is expected to undergo before breaking down (shown in Figure 10.1) are exlained here.

(a) **Normal Level:** It is the state of the equipment in which it is running satisfactorily and production is within the normal range.

(b) **Alarm Level:** It is the state developed in the equipment due to early stage of occurrence of the failure/fault and is noticed by some predetermined symptoms of the failure. For example, it is most likely to get sneeze or cough or head-ache before getting fever. Low production, lower quality of the output, heavy noise, vibrations, bad-smell, discolouration, etc., are some of the symptoms for noticing the alarm level of the failure or fault of the equipment.

(c) **Shutdown Level:** If neglected even after reaching the alarm level, the fault grows and reaches a state at which the equipment will not cooperate or a considerable loss of output is observed. This state is called shutdown level. This state is at the threshold of the breakdown level. The equipment should necessarily be closed or shutdown and should not be run any longer in the interest of preventing the failure.

(d) **Breakdown Level:** This is the ultimate and final failure state of the equipment at which stoppage of production or catastrophe even with a possible damage can occur.

(e) **Lead Time:** It is the duration between the points of start of failure to breakdown in process of deterioration of the equipment. In other words, it is time gap between last point of the normal level and breakdown level of the equipment. The lead time is one of the main merits of using trend monitoring rather than the simple alarms of automatic shutdown devices used in permanent monitoring.

In most cases, both condition checking and trend monitoring methods are used together. With the use of sensors, automation of monitoring is possible. A permanent monitoring by such sensors can prevent catastrophic failures by alarming or stopping the machine automatically when the parameter (selected to indicate the machine condition) exceeds a predetermined level. Of course, the pre-selected warning/alarm levels, based on the experience of the users show a way to optimise the life of wearing machine parts. Thus, in condition monitoring, the repairs are carried out only when the machine condition deteriorates to predetermined level. In other words, the repairs or replacement of parts take place only when it has definitely been proved that a fault exists and left as 'not-repaired' will result in unsatisfactory operation such as decrease of production or catastrophic breakdown with possible damage to other machine parts and disruption of production.

BUSTLE - 10.1

1. *Choose a machine undergoing deterioration and observe various states it experiences before it breaks down. Record the time/duration and explain the various states you observed.*
2. *Prepare checklists for various machines for obtaining certificate of condition checking for the equipment in an organisation of your choice.*

____

____

____

____

Test Your Understanding 10.1

2. What do know about condition checking and trend monitoring? Explain.
3. What is lead time? Give a brief account on lead time as applied in trend monitoring.
4. List the principles of CBM and explain.

10.3 FOUR STEPS TO DYNAMIC PREDICTIVE MAINTENANCE

The condition monitoring, also known as dynamic predictive maintenance is composed of several functional aspects on which the maintenance is carried out. However, there are four important schematic and functional steps observed in any implementation of condition monitoring activity. These functional aspects are often called four pillars of condition monitoring.

(a) Detection
(b) Diagnosis
(c) Prognosis (or Prediction)
(d) Programme (or Plan)

[A **D**ynamic **P**redictive Maintenance, the other name of CBM can be simplified by the two letters twice, viz. **DDPP**—**D**etection, **D**iagnosis, **P**rognosis, and **P**rogramme. Refer Figure 10.2.]

(a) **Detection:** This pillar of DPM details 'when' of the developing fault has just arisen. This focuses to answer the following 'when' questions.

- When might the fault have started?
- When (or how early) is the fault noticed?
- When does this fault result in a disaster if left unattended?

(b) **Diagnosis:** Diagnosis concentrates on knowing 'what' of the origin of the fault so that spare parts can be ordered.

- What is origin of the fault?
- At what stage is the fault now?
- What is the growth rate of the fault?
- What are the components or parts that have been suffering from this fault?

(c) **Prognosis:** This pinpoints what could happen if the fault is left unattended, i.e., the forecast of after effects. It is to estimate the severity of the fault, possibility, and/or frequency recurrence of the fault and operational problems. Subsequent measurement will then establish the trend to indicate what can happen. This function puts the effort to answer the following queries.

- What is the severity of the fault?
- What could be the results of the fault?
- What could be the production loss?
- What is cost of the fault in terms of hours of production and maintenance or repair?
- What are the side effects?
- What is the trend of the failure? (progressive/constant/retrogressive)
- What is the lead time?

(d) **Programme:** This function is concerned with preparation of *schedule* to enable the repair, schedule to be planned. This function attempts to reply the following type of questions.

- When should the equipment be taken for maintenance or repair?
- What maintenance policy should be adopted?
- When should the restoring action start and likely time when it can be finished?
- Who will maintain or repair so as to restore the condition?
- What is the action plan?

Permanent monitoring is generally applied where the most important economic factor is damage to the machine while CM is most suitable where the loss in production is the most important economic factor. Some situations can best be served by both methods used in parallel.

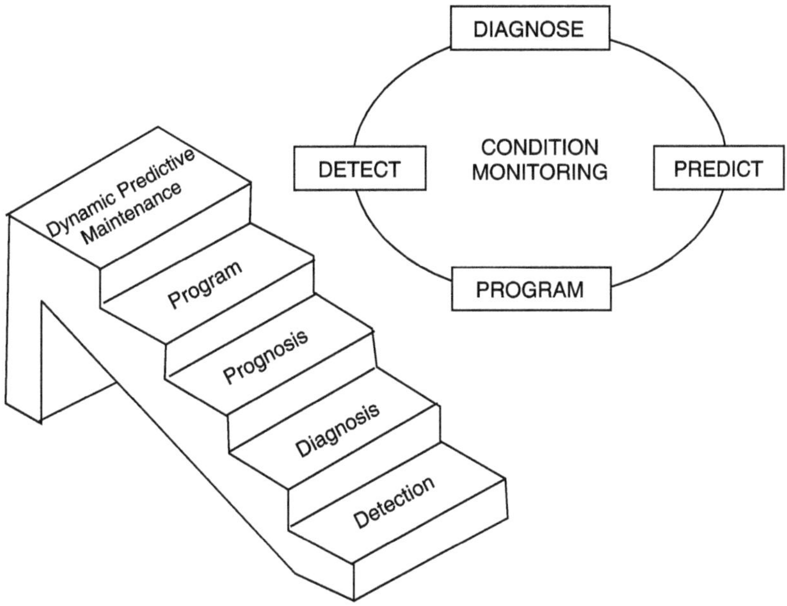

Figure 10.2 Four steps to reach dynamic predictive/condition-based maintenance

Condition checking is employed where a check measurement is taken with the machine running, using some suitable parameter. And these are then used as measures of the condition of machine at that particular instant. For effectiveness, the measurement should be qualitative, accurate, and quantifiable with known limiting values. Fixing these values requires a large amount of recorded past experience for the particular type of machine. This makes the method less flexible than the trend monitoring, particularly, if it is required to give lead time as well as machine knowledge. However, it is preferable in case, there are several similar machines operating together because comparative checking can be done between the machines under monitoring and other new/good machines. These two methods of condition monitoring are compared in Table 10.4. And the resulting advantages in terms of the provision of lead time and better machine knowledge are shown in Table 10.5.

Bustle - 10.2

Visit an organisation observing condition-based maintenance for their equipment. Write notes on how the following activities are performed.

(i) Trend Monitoring: ______

(ii) Permanent Monitoring: ______

(iii) Condition Checking: ______

(iv) Detection: ______

(v) Diagnosis: ______

(vi) Prognosis: ______

(vii) Program: ______

Test Your Understanding 10.2

1. Explain the four steps to reach condition-based/dynamic predictive maintenance.
2. Discuss the importance of diagnosis and prognosis in dynamic predictive maintenance.

10.4 SETTING UP A CONDITION MONITORING ACTIVITY

There are a number of aspects to be taken into account, if an industry is considering whether to apply CM to its plant, machinery, or equipment. The following suggested procedure can be helpful.

- The management should become familiar with the principles of CM and its benefits.
- Then they should consider various factors affecting the application of CM in relevance to their situation and also the actual merits/benefits which they intend to materialise (see Table 4.3 and 4.4 in Chapter 4).
- Then the possible financial savings are to be assessed based on the general financial statistics of the organisation.
- While making a general financial assessment they also should identify what general or organisational arrangements appear to be appropriate.
- After considering various aspects 1 to 4, the management should discuss with their senior engineers concerned with the operation and maintenance of their plant and equipment. The engineers should be asked to make a more precise estimate of the likely savings and to recommend an appropriate level of monitoring activity.
- If these more precise figures appear to be satisfactory, a decision should be made to proceed.
- A senior engineer should then be asked to survey the plant and equipment, to recommend which items should be monitored and by what methods and then prepare cost estimates.
- A final decision should then be made on the actual expenditure and approval is to be given to order the necessary equipment or services and to train the required staff. CM can then be brought in complete package.
- The management support and administrative work for setting up befitting procedures will be necessary during the first year of operation to ensure the smooth integration of CM in existing operation and the most effective utilisation of machine data that is collected.

10.5 IMPLEMENTATION OF CONDITION-BASED MAINTENANCE

The procedure for implementation of CBM is detailed in the following steps:

(a) Listing and numbering all machines, so as to have their identification and location.

(b) Selecting critical machines. A critical machine is that whose shutdown can cause production losses or dangerous to personnel. This distinction does not depend only on quantity but also on product quality. The degree of criticality of the machine is one of the factors influencing examination frequency.

(c) Establishing programmes and methods specifying the parts to be examined. Reliability of CBM depends not only on instrument but also mostly on the skill and sense of responsibility of the examiner who is entrusted with a group of machines.

(d) Establishing for each part of the machine the severity limits of the machine condition parameter (vibration, sound, temperature, contamination, etc.) to be measured.

(e) Selecting proper examination frequencies.

To arrive at the best frequency of examination, each machine must be considered.

- Its criticality in the process flow chart.
- The availability of the standby machines.
- The standardisation of items.
- The operating conditions.
- The failure statistics (MTBF-MTTR).
- The cost of examination.
- The overall cost of failure.
- The cost of maintenance.

The following formula has been suggested in order to optimise examination frequency.

$$e^{-\lambda t} - \lambda t = 1 + \lambda(C_1/C_2)$$

where λ = failure rate

t = examination interval

C_1 = examination cost

C_2 = unitary down time (maintenance cost + production losses)

This formula is valid only in the cases of constant failure rate (FR), while in the case of a variable FR the formula is more complicate, therefore, the empirical methods are applied.

(g) Recording data.

(h) Training examiners.

The examiners should have a high standard of experience, a thorough knowledge of machinery, and particularly endowed with analytical skill. However, it is necessary to train the personnel for this job by:

- Making clear the objectives of CBM.
- Illustrating the principle and working of the instruments.
- Going through essential mathematics and principles of physics to perform the job.
- Attending refresher courses on professional matters, like lubrication, ball bearing assembly, vibration analysis, etc.

BUSTLE - 10.3

Choose an organisation and prepare an implementation plan of CBM for the equipment.

____

____

____

____

Test Your Understanding 10.3

1. What are the different aspects to be considered to apply condition based maintenance?
2. What are the forces and factors to be considered when a CBM activity is set up in an industry? How do they influence the implementation plan?
3. What are the bottlenecks in setting up the CBM activity? How do you encounter them?
4. Explore the consequences you expect with implementation of CBM?

10.6 THE DESIGN OF MAINTENANCE INFORMATION SYSTEM (MIS)

The maintenance information system or MIS gains a significant role when CBM is introduced in an industry in two important aspects. The first aspect is collecting and processing of great quality of information that is never available before, regarding the condition of each part of a machine. This information is used to detect the faults or tendency of deterioration or probable time that a machine is expected to fail. Obviously, this simplifies the maintenance manager's job in optimising the frequency of examination or inspection on the machine. The second aspect of the information system stresses on the necessity to provide information so as to carry out maintenance works within lead time, i.e., in a period of time, which may be very short elapsing between the alarm and emergency shutdown. This becomes a tool for the maintenance manager to schedule or programme the maintenance works within the lead time effectively by reducing the waiting costs, shortage costs, logistic costs, etc. This is clearly interpreted in Figure 10.3.

Thus, a good maintenance information system puts the maintenance manager in a comfortable position to take the decision in the following two conditions.

- The condition of the machine is near to the lead time; procedure will be followed through planning office (work requisition, work order with specification programming, carried out work).
- The condition of machine is already within lead time (next to the shutdown); the information will be directly passed to the foreman for emergency maintenance.

To implement the CBM effectively, efficiently, and to operate correctly, the information should be very transparent. In other words, the information, which the maintenance manager gets from the inspection/examination team, should flow essentially from the maintenance manager also to the maintenance foremen as well as the shop floor foremen and everybody concerned in the following aspects.

- Condition of machine
- Probable part of defective machine
- Probable defect
- Time during which the fault has to be repaired

This can be better achieved by job report, which is another very important and fundamental feature in order to assure the success of the new philosophy. Perhaps, this can be made possible through the security and correction of diagnosis versus actual work.

CBM uses the lubrication team as the major source of information as this team systematically controls the machinery. Examination and lubrication, therefore, must be strictly connected as they

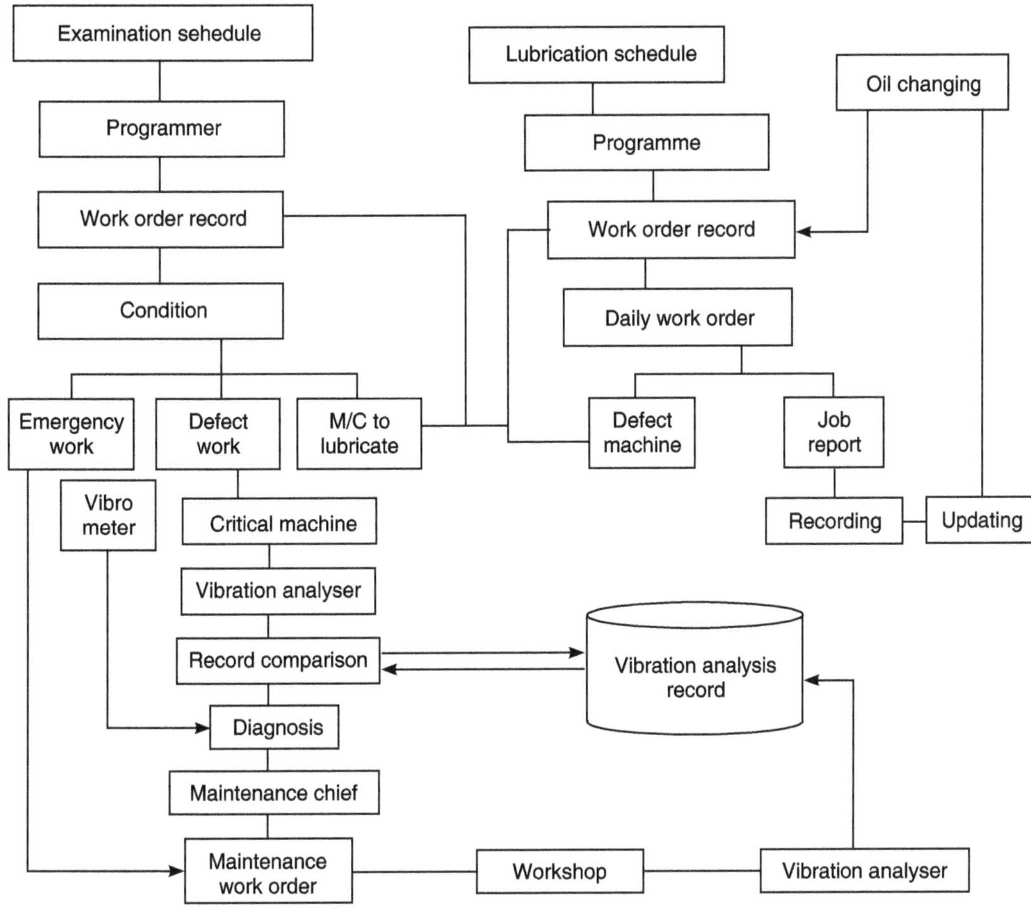

Figure 10.3 Maintenance information system for CBM

may transfer each other information on the conditions of the machine. Flow of information under CBM is illustrated in the upper side of Figure 10.5. Hence, the organisation of maintenance becomes more flexible as it is based on a daily input of data while the short-term programming becomes more important than the long-term one. As a result of this, the estimation of maintenance budget becomes somewhat difficult, since most of the work will be done only after an examination instead of basing it on a rigid PM schedule. Therefore, the budget can be based only on statistical data and probabilistic.

The role of the maintenance manager becomes very crucial from the moment CBM is adopted. Hitherto, his main duties have been to provide coordination and technical aid to his assistants in maintenance planning through a guesswork but now he becomes really the 'manager' of the service. However, three important functions required to be performed by the chief of the maintenance are listed below.

- To control the examiners' training or his eventual deviations from instructions.
- To improve the correlation between parameters chosen and real defects.
- To get severity curves specific for each machine.

Application of Computers in Maintenance Information System

The use of computer has become very essential today. The computer systems are used in plant engineering to provide the information listed below to the maintenance-in-charge and from plant engineering department to other departments.

- Maintenance master schedules
- Failure analysis report
- Cost analysis report
- Budget report
- Status of PM
- Materials stock report
- Materials (urgent) requirement report
- Work-order status report
- Pending maintenance work report
- Work-In-Progress (WIP) report
- Delay in execution (reasons) report
- Equipment-wise repair cost analysis report
- Equipment-wise down time or availability report
- Ratio of emergency work to planned work report
- Emergency work report
- Man-hours spent on maintenance work vs planned man-hours
- Details of information of experts in maintenance management
- Capacity planning (CRP)/manpower planning (HRP)/material planning (MRP) reports

BUSTLE - 10.4

1. *Prepare a flowchart of maintenance information system in an organisation with the prevailing systems and procedures.*
2. *Prepare an information system considering the production department, plant engineering department, planning department, materials department, purchase department, accounts and finance departments.*

____

____

____

____

Test Your Understanding 10.4

2. Discuss the role of maintenance information system with reference to the implementation of CBM?
3. What is maintenance information system? How will you design it?

10.7 TECHNIQUES OF CONDITION MONITORING

In fact, there are several techniques available to monitor the machine condition. But for the ease of understanding and implementing, all these can be studied under five main heads.

(a) **Visual Monitoring:** It is the monitoring technique, which involves inspection and recording of surface appearance. Normally, monitoring is done by naked eyes, but it can require the use of telescopes, microscopes, and various other optical equipments. Sometimes, it may even involve the use of borescopes to see inaccessible places, or the use of photography or surface imprinting for record purpose.

(b) **Vibration Monitoring:** The noise and vibration are the most important parameters to monitor a machine, particularly in the moving parts such as shafts, rotors, cutting tools, gears, etc. The vibration level is recorded by attaching a transducer like velocity probe, accelerometer, or proximity probe to the machine. Special equipment is also available for using the output from the sensor to indicate the nature of vibration problem and even its precise cause. In some cases, it may become necessary to use the principles of Sonics and acoustics.

(c) **Wear Debris Monitoring:** This works on the principle that the working surfaces of a machine are washed by the lubricating oil, and any damage to them should be detectable from particles of wear debris in the oil. If the debris consists of relatively large ferrous lumps such as those generated by the fatigue of rolling element bearings and gears or the pitting of cams and taproots, these can be picked up by removable magnetic plugs inserted in the oil return lines. For small debris particle, spectrographic analysis or microscopic examination of oil samples after magnetic separation are commonly used techniques. Another popular technique is SOAP analysis for debris monitoring.

(d) **Performance and Behaviour Monitoring:** Here we check the performance of a machine or component to see whether it is behaving correctly. This may, for example, involve monitoring the performance of a bearing by measuring its temperature to see whether it is carrying out its function of transmitting load between moving surfaces with minimum friction.

(e) **Corrosion Monitoring:** This is usually applied to fixed plant containing aggressive materials and is intended to monitor the rates of internal corrosion of the walls of the plant. This may be done by drilling sentinel holes part away, through the wall, which can be plugged when they leak or by inserting readily removable coupons of material of which the corrosion rate is assumed to relate to that of plant.

In the case of the first four techniques of monitoring which are applied primarily to various forms of rotating plant, the way in which they are used is that the existence of a problem is usually detected from the general level of the measurement, and its rate of change, while the nature of the problem can generally be determined from a more detailed analysis of the measurements obtained. This is outlined in Table 10.2.

Table 10.2 Application of monitoring methods
(For detection and analysis of problem)

Sl. No.	*Monitoring Methods*	*Direction of Problem by*	*Determination of Nature of Problem by Analysis of Measurement*
1.	Visual Monitoring	Overall Appearance	Colouring/Shape/Texture
2.	Performance Monitoring	Rate of Output	Uniform Quality Level/Rate of Output/Uniformity
3.	Vibration/Noise Level Monitoring	Overall Noise/ Vibration Level	Frequency of Noise Level/Signal Waveform/Signal Statistics
4.	Wear Monitoring	Amount of Debris/ Friction Colour	Shape/Size/Size Distribution of Debris/Chemical Composition
5.	Corrosion Monitoring	Colour/Chemical Analysis	Variation in Coating Thickness/Chemical Composition

The techniques/methods of monitoring described in Table 10.2 are in fact a mechanism of communication between a machine and a monitoring engineer. It will be observed that the visual method of monitoring in relation to the others requires negligible technological backup. This is because the necessary analytical facilities already exist with the human observer. The lack of these natural analytical facilities in the other three methods of monitoring is the main reason why more complex technological facilities and methods need to be applied. The different techniques within the main categories of monitoring methods are listed in Table 10.3 as a guide to their general selection for various parts of the machine parts.

Table 10.3 Selecting methods of monitoring

Component to be monitored		*Possible general monitoring method*			
		Visual	*Behaviour*	*Vibration*	*Wear debris*
STATIONARY	Castings	◆	◆		
	Mountings and foundations	◆		◆	
	Tanks and containers	◆	◆		
	Pressure vessels	◆	◆		
	Pipes	◆	◆		
	Heat exchanges	◆	◆		
	Screens and separators	◆	◆		
	Stator blades	◆			

ROTATING	Machine rotors			◆	◆
	Turbine blades	◆	◆	◆	
	Impellers and propellers	◆	◆	◆	
	Wheels	◆		◆	
	Gears	◆		◆	◆
	Chain drives			◆	◆
	Flexible couplings	◆		◆	◆
	Pulleys and belts	◆		◆	
	Governors		◆		◆
RECIPROCATING	Pistons				◆
	Linkage and levers	◆			◆
	Cams and tappets			◆	◆
	Valves		◆	◆	◆
	Cables and chains	◆			◆
	Bellows	◆	◆		
	Diaphragms	◆	◆		
	Springs	◆		◆	◆
	Guides and slides	◆		◆	◆
	Splines	◆			◆
FRICTION COMPONENTS	Brakes	◆	◆		
	Clutches	◆	◆		◆
	Vibration dampers	◆	◆	◆	◆
	Plain bearings		◆	◆	◆
	Rolling bearings			◆	◆
	Flexure	◆	◆		
	Seals				
	Lip		◆		
	Mechanical		◆		
	Packed glands		◆		
	Wingback		◆		◆
	Labyrinth		◆	◆	◆
	Piston rings		◆		◆
	Hard wear	◆			
	Elastic	◆			

TOOLS	Cutting tools	◆	◆		
	Metal working tools	◆	◆		
	Casting and moulding tools	◆	◆		
FLUIDS	Hydraulic	◆			
	Cooling and heat transfer	◆	◆		
	Lubricants	◆			

BUSTLE - 10.5

Select an organisation with at least 50 machines. What monitoring methods do you recommend to the plant items/machine parts of organisation if condition monitoring is going to be implemented? What instrumentation and tooling do you require for this? Design a system.

____

____

____

____

Test Your Understanding 10.5

1. Explain the following condition monitoring techniques. Discuss their merits and demerits.
   (a) Visual monitoring
   (b) Vibration monitoring
   (c) Corrosion monitoring
   (d) Performance monitoring
2. Discuss various techniques employed for monitoring the condition of the equipment.
3. Give brief notes on wear debris monitoring and corrosion monitoring.

10.8 COMPARISON BETWEEN CONDITION MONITORING (CM) AND FAULT DIAGNOSIS (FD)

Though these two have a close relation and look alike, some differences exist between condition monitoring and fault diagnosis. These differences are interpreted in Table 10.4:

Table 10.4 Comparison between of condition monitoring and fault diagnosis (diagnostic maintenance)

Measurement	*Trend Monitoring*	*Condition Checking*	*Diagnostic Maintenance*
Frequency	Readings are taken at regular time intervals while the machine is running.	Readings are taken at one time while the machine is running.	When the problem has become manifest or after failure has occurred.

Qualitative	Skilled operators can perform subjective trend monitoring if they are close enough to their machines.	Typical activity of an engineer when checking a machine during operation.	When machine is stopped, inspection of components can indicate the cause of the problem.
Quantitative	The taking of regular measurements and their recording and analysis gives a lead time on machine problem.	Numerical values allow comparison with established standards or other similar machines to give knowledge of machine condition.	Measurements may be analysed in considerable detail to provide guidance on possible causes of the problem.
	CONDITION MONITORING		

10.9 ADVANTAGES OF CONDITION MONITORING OVER OTHER PHILOSOPHIES

The advantages of condition monitoring are simplified and shown in a tabular form in Table 10.5.

Table 10.5 Advantages of condition-based monitoring

Advantages Obtained			*Methods by which CBM gives these advantages*	
			Lead Time	*Better M/C Knowledge*
OUTPUT/PRODUCTIVITY	*Increased machine availability*	*More running time*	Enables machine shutdown for maintenance and reduces various consequential losses by avoiding unexpected shutdown.	Allows time interval between planned machine overhauls to be maximised.
		Less mainten-ance time	Enables machine to be shut down without destruction or major damage. Enables maintenance crew to be ready with spare parts, to start work as soon as machine is shut down.	Reduces inspection time after shutdown and speeds up the start of correct remedial action.
	Increased rate of net output		As the shutdown time is warned well in advance, the plan can be revised or enables aggregate planning.	Allows some types of machine to be run at increased load and/or speed. Can detect loss in machine efficiency or increased energy consumption.
	Improved quality of product or service		Allows advanced planning to reduce the effect of impending breakdowns on the customer for the product or service, and thereby enhances company reputation.	Can be used to reduce the amount of product or service produced at sub-standard quality levels.
SAFETY	*Reduced injuries and fatal accidents to personnel caused by machinery*		Enables plant to be stopped safely when instant shutdown is not permissible.	Machine condition, as indicated by an alarm is adequate if instant shutdown is permitted.

BUSTLE - 10.6

Suppose, there are two similar machines in your industry—say P, the imported one and say Q, the in-house manufactured. In certain urgency you have used the spares of the imported machine P to the machine Q. Now, there is a breakdown of P and is for the want of the same spares. There is a chance of damage if the replaced part is removed and the in-house made spares do not suit the imported machine. What will you do in such a situation? Does the fault lie with design or manufacture or maintenance? Design an information system for resolving such situations considering design to discard stages.

Test Your Understanding 10.6

1. Explain the significance of condition-based maintenance. How is it more meritorious over other maintenance philosophies?
2. Distinguish between fault diagnosis and condition monitoring?
3. What are the factors that are to be considered to decide the optimum frequency of examination in condition monitoring? How do they influence the frequency?

10.10 CONCLUSION

The importance of condition based maintenance is greatly realised by the industries in India. Many organisations have already adopted and are successfully implementing while some others are in the direction of adopting it. Though all the techniques of monitoring such as visual, vibration, noise, performance, corrosion monitoring are not useful in all the organisations, the organisation can select a few of them and can mould and design them according its suitability and requirement.

10.11 RECAP

Various philosophies of the maintenance management, their central themes, principles, and their evolution are extensively discussed. The significance of condition monitoring is illustrated. The principles, the method, and the techniques of CBM are then detailed with appropriate supports. For the ease of maintenance managers who wish to adopt this policy, the procedure of adoption, and the implementation plans are laid down. A clear distinction is made between diagnostic maintenance and CBM to avoid confusion. A few destructive and non-destructive tests are also explained which are applications of those discussed in Chapter 9. The advantages of CBM are shown with a high clarity.

10.12 KEY CONCEPTS

1. **Dynamic Predictive Maintenance:** Prediction of possibility of occurrence of a failure by regular monitoring dynamically and attending immediately after the fault occurred and much before possibility of breakdown. This is also called condition based monitoring.
2. **Diagnostic Maintenance:** Maintenance by finding the root causes and analysing the faults occurred.
3. **Detection:** Finding the faults in the machines or any other parts of the machine.
4. **Diagnosis:** Analysing the faults before planning for maintenance of the machine.
5. **Prognosis:** Forecasting of the possible causes of faults and maintenance of the equipment before undertaking for maintenance.
6. **Programme:** Preparation of planning and scheduling for maintenance activities.
7. **Maintenance Information System (MIS):** The system of collecting the information about the various stages of maintenance activities in the plant.
8. **Lead Time:** The period of time elapsed between the alarm and emergency shutdown of the machine.
9. **Visual Monitoring:** The technique involved in inspecting and recording of surface appearance of the parts or equipment.
10. **Wear Debris Monitoring:** The technique involved in testing the weariness in machine parts by using the magnetic debris and lubricating oil or soap water.
11. **Performance and Behaviour Monitoring:** The method used to check the functional behaviour of the machine parts.

REVIEW QUESTIONS

1. Briefly explain the basic principles of condition monitoring.
2. Explain various states found during its breakdown while monitoring the trend of machine.
3. Discuss the terms detection, diagnosis, prognosis, and program in achieving condition monitoring?
4. Explain the steps to be followed to implement condition-based maintenance.
5. Explain the procedural steps in setting up a CBM activity.
6. Discuss the role of maintenance management (or maintenance manager) in an industry which wants to adopt CBM.
7. Discuss the role of maintenance mainagenment (or maintenance manager) in an industry which just implemented CBM.
8. Discuss the limitations and hurdles to implementation of condition monitoring. How do you overcome them?
9. What is maintenance information system? What is its significance in manintenance management? Explain how you design it.
10. Explain various monitoring techniques for evaluating the condition of a machine.
11. Distinguish between the condition checking and trend monitoring. Discuss their conceptual focus.
12. Distinguish between the condition monitoring and diagnostic maintenance?

13. What are the advantages of condition-based maintenance?
14. Dstinguish betwen
    (a) Prognosis and diagnosis
    (b) Condition checking and trend monitoring
    (c) Condition monitoring and diagnostic maintenance
15. Write short notes on
    (a) Vibration Monitoring
    (b) Debris Monitoring
    (c) Corrosion Monitoring

CASE STUDY

Performance Monitoring and Performance Evaluation of Group of Machines —A Case Study with Load Hauling Dump (LHD) Vehicles

INTRODUCTION

Production managers and plant engineers are often interested to know whether their equipment is running up to the set standards. At the same time, the top managers who have less time to involve deeply would like to focus their attention on the area working at lower efficiency for which they need some tools that can guide them to acting. Therefore, there is a necessity of establishing a method of evaluating the machine behaviour and monitoring the performance which can act like an acid test or litmus test. Further, this test should not consume much time but can be able to give reliable information.

We know that machines play a vital role in every industry and no machine is totally immune to breakdowns. In order to predict performance of machines and their behaviour in terms of breakdowns, many statistical tools are already available and are being employed. Out of those, graphical methods using conventional statistical techniques to plot the probability distribution of available and utilised hours are not only a fairly good representation of repairable machines but also provide information about their reliability of operation.

Moreover, the graphical method developed by Nelson using Mean Cumulative Repair Function (MCRF) from the repair data of machines based on the numbers of dumpers put into operation in a given period can be used to provide an insight into the system operation. This provides a method of representation of operational data and overall efficiency through graphical methods using MCRF. The major advantage of this method is that the estimation of system reliability does not need any depth of investigation of failure hours that are often cumbersome to calculate. This method is discussed here and applied on a group of vehicles operating in a large-scale coal manufacturing company to monitor their performance.

Background

Generally, in an industry, if there are similar or same types of machines, the maintenance department often chooses to follow the maintenance policy of interchanging by the Least Replacement Assembly (LRA) units. Suppose a machine is under breakdown due to engine failure and another due to tyre failure, then the maintenance department will interchange the tyres of the second machine with that of first one, if it is compatible, in the interest of maintaining the productivity. Thus, instead of keeping both machines under idle/breakdown condition, at least one of them can be kept under working condition. This production orientation (productive maintenance) can help in increasing the efficiency of production department and maintenance department, but induces the confusions in estimating the reliability of specific equipment, since the failure of one machine is transferred to another, thus, affecting the reliability of the equipment of interest. Therefore, the reliability studies should focus on the total group of machines of the same type as one unit. A maintenance department of a large-scale coal producing company is also adopting such LRA maintenance policies. This study is focused on this area and concentrates in this direction by analysing the group reliability using some NHPP graphical models such as Mean Cumulative Repair Function (MCRF).

Assumptions Made for the Analysis

The following assumptions are made for this analysis and calculation.

1. All machines are alike and of same running age.
2. The parts or components are interchangeable.
3. If the machine is not kept under use due to rain, want of spares, not required, unavailability of operator, unavailability of raw material or work, play day (off or holiday), National holiday or otherwise holiday, workers' strike or unrest, etc., the machine is treated as available, i.e., on role. Though such machine is not treated as breakdown, it does not add any production.
4. The road conditions have been considered to have no effect on the availability of the machine.
5. The workers behaviour to operating the vehicle is ignored.
6. Small breakdowns, i.e., those which take not more than one hour to repair are ignored. These are not considered as breakdowns.
7. In general, a day is counted from 6 a.m. onwards since shift timings are 6 am to 2 pm (first shift), 2 pm to 10 pm (second shift), and 10 pm to 6 am (third shift). If a particular organisation observes different timings, it can be considered accordingly for calculations.
8. The available (working) time of any machine is considered as 20 hours only per day and other 4 hours are attributed to and accounted for lunch/tea breaks, shift change over periods, and rest pauses.
9. If a machine is kept under breakdown for more than 10 hours in a day it is considered to be not available for the whole day, otherwise considered available.
10. This test is an acid test and cannot give detailed analysis or history of any individual machine.
11. This test cannot provide the information which machine is contributing to the loss of production or fall/raise of overall performance of the department.

Reliability Modelling of Performance Monitoring of Repairable Equipment

Reliability is associated with dependability at successful operation. While modelling a system to characterise by its reliability, one should first distinguish between repairable and non-repairable items.

For repairable items, the equipment is expected to be renewed to "as good as new" state after every repair, thus can assume independent and identical distribution (i.i.d) and so the system is modelled in homogeneous Poisson process (HPP) or renewable process (RP). However, the reliability of items is dependent on several other factors such as maintenance, environmental conditions, dependence on components and so forth. The repairable items cannot execute the original or new condition, therefore, it is appropriate to assume "as bad as old". Hence, the suitable reliability modelling is non-homogeneous Poisson process (NHPP).

Intensity Function of Non-homogeneous Poisson Process (NHPP)

If the data sets of repairable equipment indicate presence of trend, NHPP can be used to model failure data. One of the most commonly used NHPP models is Power Law Process (PLP) discussed by Rigdon and Basu.

Intensity of NHPP is the probability of failure in a short time interval, divided by length of interval. It is given by

$$U(t)\,dt = P\text{ (a failure in '}t\text{' to '}t + dt\text{')} \qquad \text{... (1)}$$

The failure intensity function depends upon cumulative time 't' (not the local time between failures). Therefore, the intensity function of Power Law Process (NHPP) model is given by

$$U(t) = (\beta/\alpha)\,(t/\alpha)^{\beta-1} \qquad t > 1 \qquad \text{... (2)}$$

Where α and β are scale and shape parameters and 't' is the global running time.

1. When $\beta > 1$, intensity function increases, which corresponds to a situation when time between failures (TBFs) become smaller, i.e., frequent occurrence of failures. In other words, it represents the case of deterioration of the machine (increasing failure rate).
2. When $\beta < 1$, intensity function decreases, which means the machine is improving (i.e., decreasing failure rate).
3. When $\beta = 1$, it corresponds to constant failure rate and it can be inferred that Power Law Process (PLP) becomes homogeneous Poisson process (HPP) with mean time between failures equal to α. It can be seen that for $\beta = 1$, NHPP also represents renewal process.

Graphical Analysis for Repairable System of Dumpers

A non-parametric graphical method based on Mean Cumulative Repair Function (MCRF) developed by Nelson is used to interpret interval data of inter arrival time for repair of repairable systems, i.e., dumpers. The information available in logbooks of maintenance department and various maintenance reports are used to establish the repair performance using graphical method. The merit of fitting non-homogeneous Poisson process (NHPP), the frequently used reliability model for repairable system is that the data collected for analysis need not be independently and identically distributed (i.i.d) random variables. In NHPP, the intensity is given by Power Law function. The derivative of MCRF has the same meaning of intensity function.

Therefore, $d/dt\,(\text{MCRF}) = (\beta/\alpha)(t/\alpha)^{\beta-1}$ ... (3)

Then from equation (3), MCRF is given by:

$$\text{MCRF} = (t/\alpha)^{\beta} \qquad \text{... (4)}$$

Hence, from equation (4), we get

$$\text{Log (MCRF)} = \beta\log(t) - \beta\log(\alpha) \qquad \text{... (5)}$$

The above equation (5) is in the form of a straight line $y = mx + c$ where the slope (m) directly gives the value of β. The estimate of α can hence be obtained by the value of 't'.

The estimate of reliability at age t when the random variable is, time to first repair, can be now obtained by

$$R(t) = \text{Exp}\{-(t/\alpha)\}^{\beta} \quad \text{... (6)}$$

Mean Cumulative Repair Function

Repair data of vehicles is taken in the form of number of vehicles put into operation in a period and the number of vehicles not in operation in the period and the repair function is the ratio of number of vehicle not in operation to the total number planned. Thus, the Cumulative Repair Function and the Mean Cumulative Repair Function (MCRF) can be estimated. This function provides useful information with engineering insight in to the system reliability.

This graphical analysis presents a novel method of analysis of repair data of repairable vehicles irrespective of 'system age'. The method based on MCRF developed by Nelson, is a function of time t. It is the average of cumulative repairs at time 't'.

In the MCRF tables (Refer Table 1 to 3):

1. Column 1 represents operating period (i).
2. Column 2 represents number on role (N_i).
3. Column 3 represents working.
4. Column 4 represents on repair (r_i), at operating period i.
5. Column 5 represents mean repairs, which is obtained by dividing column 4 by column 2.
6. Column 6 represents MCRF, which is cumulative of column 5.

Table 1 Mean cumulative repair function of dumpers (Daywise)

Day No.	*Total No.*	*No.of Working*	*No. of Not Working*	*Mean Repair Function*	*MCRF*	*Ln (day)*	*Ln (MCRF)*
(1)	(2)	(3)	(4)	(5) = (4)/(2)	(6) = Σ(5)	(7) = Ln(1)	(8) = Ln(5)
1	23	4	19	0.826086957	0.826086957	0	–0.191055
2	23	11	12	0.52173913	1.347826087	0.6931472	0.298493
3	23	5	18	0.782608696	2.130434783	1.0986123	0.7563261
4	23	8	15	0.652173913	2.782608696	1.3862944	1.0233889
5	23	10	13	0.565217391	3.347826087	1.6094379	1.2083112
6	23	7	16	0.695652174	4.043478261	1.7917595	1.3971053
7	23	6	17	0.739130435	4.782608696	1.9459101	1.5649861
8	23	10	13	0.565217391	5.347826087	2.0794415	1.6766901
9	23	7	16	0.695652174	6.043478261	2.1972246	1.7989797
10	23	10	13	0.565217391	6.608695652	2.3025851	1.8883863
15	23	6	17	0.739130435	10.04347826	2.7080502	2.3069235
20	23	10	13	0.565217391	13.47826087	2.9957323	2.6010781
25	23	8	15	0.652173913	17.13043478	3.2188758	2.8408567
30	23	3	20	0.869565217	20.65217391	3.4011974	3.0278206
35	23	6	17	0.739130435	23.95652174	3.5553481	3.1762406
40	23	11	12	0.52173913	27.13043478	3.6888795	3.3006562

Day No.	Total No.	No.of Working	No. of Not Working	Mean Repair Function	MCRF	Ln (day)	Ln (MCRF)
(1)	(2)	(3)	(4)	(5) = (4)/(2)	(6) = Σ(5)	(7) = Ln(1)	(8) = Ln(5)
45	23	11	12	0.52173913	30.43478261	3.8066625	3.4155861
50	23	7	16	0.695652174	33.73913043	3.912023	3.5186583
55	23	9	14	0.608695652	37.26086957	4.0073332	3.6179437
60	23	8	15	0.652173913	40.7826087	4.0943446	3.7082557
65	23	4	19	0.826086957	44.86956522	4.1743873	3.8037597
70	23	6	17	0.739130435	49.04347826	4.2484952	3.8927072
75	23	8	15	0.652173913	52.95652174	4.3174881	3.9694712
80	23	1	22	0.956521739	57.17391304	4.3820266	4.0460977
85	23	7	16	0.695652174	61.13043478	4.4426513	4.1130099
90	23	3	20	0.869565217	65	4.4998097	4.1743873
95	23	10	13	0.565217391	68.47826087	4.5538769	4.2265163
100	23	3	20	0.869565217	72.39130435	4.6051702	4.2820862
105	23	3	20	0.869565217	76.60869565	4.6539604	4.3387106
110	23	4	19	0.826086957	80.7826087	4.7004804	4.3917617
115	23	4	19	0.826086957	84.47826087	4.7449321	4.4364942
120	23	5	18	0.782608696	88.04347826	4.7874917	4.4778308
125	23	4	19	0.826086957	92.17391304	4.8283137	4.5236772
130	23	3	20	0.869565217	95.91304348	4.8675345	4.563442
135	23	7	16	0.695652174	100.3913043	4.9052748	4.6090756
140	23	0	23	1	104.1304348	4.9416424	4.6456443
145	23	3	20	0.869565217	108.3478261	4.9767337	4.6853467
150	23	2	21	0.913043478	112.7391304	5.0106353	4.7250766
155	23	5	18	0.782608696	116.7391304	5.0434251	4.7599418
160	23	5	18	0.782608696	120.2608696	5.0751738	4.7896633
165	23	8	15	0.652173913	123.9565217	5.1059455	4.8199309
170	23	4	19	0.826086957	127.9130435	5.1357984	4.8513507
175	23	3	20	0.869565217	132.0434783	5.164786	4.8831312
180	23	6	17	0.739130435	135.6956522	5.1929569	4.9104145
185	23	0	23	1	140.6086957	5.2203558	4.9459808
190	23	0	23	1	145.6086957	5.2470241	4.9809229
195	23	0	23	1	150.6086957	5.2729996	5.0146851
200	23	7	16	0.695652174	154.6086957	5.2983174	5.0408974
205	23	5	18	0.782608696	158.2608696	5.32301	5.0642447
210	23	5	18	0.782608696	161.8695652	5.3471075	5.0867909
220	23	5	18	0.782608696	168.9130435	5.3936275	5.129384
225	23	7	16	0.695652174	172.4782609	5.4161004	5.1502712
230	23	9	14	0.608695652	175.9130435	5.4380793	5.1699898

Day No.	*Total No.*	*No.of Working*	*No. of Not Working*	*Mean Repair Function*	*MCRF*	*Ln (day)*	*Ln (MCRF)*
(1)	(2)	(3)	(4)	(5) = (4)/(2)	(6) = Σ(5)	(7) = Ln(1)	(8) = Ln(5)
235	23	7	16	0.695652174	179.3043478	5.4595855	5.1890846
240	23	8	15	0.652173913	182.2173913	5.4806389	5.2052004
245	23	9	14	0.608695652	185.5652174	5.5012582	5.2234064
250	23	8	15	0.652173913	189.0869565	5.5214609	5.242207
255	23	9	14	0.608695652	192.2608696	5.5412635	5.2588531
260	23	11	12	0.52173913	195.2608696	5.5606816	5.2743365
265	23	7	16	0.695652174	198.9130435	5.5797298	5.2928678
270	23	7	16	0.695652174	202.2173913	5.598422	5.3093433
275	23	6	17	0.739130435	205.826087	5.6167711	5.3270316
280	23	10	13	0.565217391	209.0869565	5.6347896	5.3427502
285	23	6	17	0.739130435	212.3913043	5.6524892	5.3584303
290	23	7	16	0.695652174	215.2608696	5.6698809	5.3718506
295	23	11	12	0.52173913	218.1304348	5.6869754	5.3850932
300	23	6	17	0.739130435	221.5652174	5.7037825	5.400717
305	23	7	16	0.695652174	224.9130435	5.7203118	5.4157139
310	23	7	16	0.695652174	228.4347826	5.7365723	5.4312508
315	23	7	16	0.695652174	231.6521739	5.7525726	5.445237
320	23	7	16	0.695652174	235.3043478	5.768321	5.4608798
325	23	8	15	0.652173913	238.4782609	5.7838252	5.4742782
330	23	5	18	0.782608696	242.2608696	5.7990927	5.4900151
335	23	5	18	0.782608696	246.2173913	5.8141305	5.5062149
340	23	6	17	0.739130435	249.9565217	5.8289456	5.521287
345	23	1	22	0.956521739	253.9565217	5.8435444	5.5371631
350	23	4	19	0.826086957	258.2173913	5.8579332	5.5538018
355	23	4	19	0.826086957	262.3043478	5.8721178	5.5695055
360	23	5	18	0.782608696	266.2173913	5.886104	5.5843132
361	23	5	18	0.782608696	267	5.888878	5.5872487
362	23	4	19	0.826086957	267.826087	5.8916442	5.5903378
363	23	3	20	0.869565217	268.6956522	5.8944028	5.5935793
364	23	4	19	0.826086957	269.5217391	5.8971539	5.5966491
365	23	4	19	0.826086957	270.3478261	5.8998974	5.5997094

The term 'period' in the above points may mean any convenient duration such as a shift, day, week, or month. In the present study, all three varieties, i.e., operating day (Table 1 and Figure 1), operating week (Table 2 and Figure 2) and operating month (Table 3 and Figure 3) vs. MCRF are plotted for 23 dumpers of 85T.

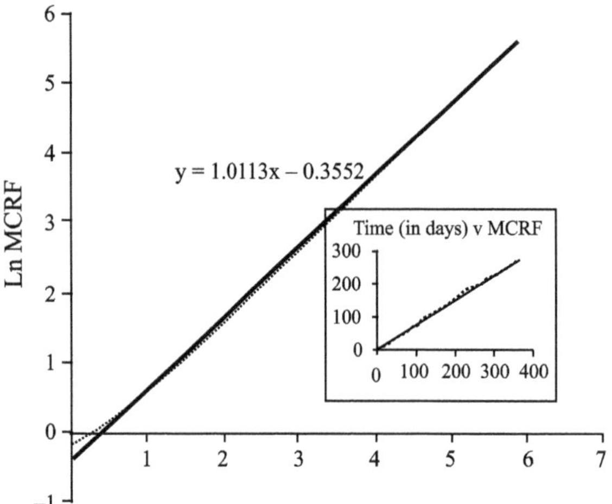

Figure 1 Logmean cumulative repair function vs logtime (day)

Figure 2 Log operating week vs log MCRF

Table 2 Mean cumulative repair function of dumpters (Week-wise)

Opp week	*No. on role*	*No. Working*	*No. of Rep*	*Mean Rep*	*MCRF*	*Log (op week)*	*Log (MCRF)*
1	21	12	9	0.428571	0.428571	0	-0.36798
2	21	13	8	0.380952	0.809523	0.301029996	-0.09177
3	21	11	10	0.47619	1.285714	0.477121255	0.109144
4	21	8	13	0.619048	1.904761	0.602059991	0.279841
5	21	12	9	0.428571	2.333333	0.698970004	0.367977
6	21	12	9	0.428571	2.761904	0.77815125	0.441209
7	21	13	8	0.380952	3.142857	0.84509804	0.497325
8	21	13	8	0.380952	3.523809	0.903089987	0.547012
9	21	11	10	0.47619	4	0.954242509	0.60206
10	21	6	15	0.714286	4.714285	1	0.673416
11	21	13	8	0.380952	5.095238	1.041392685	0.707164
12	21	13	8	0.380952	5.47619	1.079181246	0.738479
13	21	10	11	0.52381	6	1.113943352	0.778151
14	21	10	11	0.52381	6.523809	1.146128036	0.814501
15	21	10	11	0.52381	7.047619	1.176091259	0.848042
16	21	9	12	0.571429	7.619047	1.204119983	0.881901
17	21	12	9	0.428571	8.047619	1.230448921	0.905667

Opp week	*No. on role*	*No. Working*	*No. of Rep*	*Mean Rep*	*MCRF*	*Log (op week)*	*Log (MCRF)*
18	21	9	12	0.571429	8.619047	1.255272505	0.935459
19	21	9	12	0.571429	9.190476	1.278753601	0.963338
20	21	12	9	0.428571	9.619047	1.301029996	0.983132
21	21	11	10	0.47619	10.09524	1.322219295	1.004117
22	21	11	10	0.47619	10.57143	1.342422681	1.024134
23	21	9	12	0.571429	11.14286	1.361727836	1.046997
24	21	10	11	0.52381	11.66667	1.380211242	1.066947
25	21	10	11	0.52381	12.19048	1.397940009	1.086021
26	21	13	8	0.380952	12.57143	1.414973348	1.099385
27	21	13	8	0.380952	12.95238	1.431363764	1.11235
28	21	11	10	0.47619	13.42857	1.447158031	1.12803
29	21	11	10	0.47619	13.90476	1.462397998	1.143164
30	21	15	6	0.285714	14.19048	1.477121255	1.151997
31	21	10	11	0.52381	14.71429	1.491361694	1.167739
32	21	13	8	0.380952	15.09524	1.505149978	1.17884
33	21	13	8	0.380952	15.47619	1.51851394	1.189664
34	21	13	8	0.380952	15.85714	1.531478917	1.200225
35	21	12	9	0.428571	16.28571	1.544068044	1.211807
36	21	14	7	0.333333	16.61905	1.556302501	1.220606
37	21	15	6	0.285714	16.90476	1.568201724	1.228009
38	21	14	7	0.333333	17.23809	1.579783597	1.236489
39	21	14	7	0.333333	17.57143	1.591064607	1.244807
40	21	13	8	0.380952	17.95238	1.602059991	1.254122
41	21	12	9	0.428571	18.38095	1.612783857	1.264368
42	21	12	9	0.428571	18.80952	1.62324929	1.274378
43	21	13	8	0.380952	19.19048	1.633468456	1.283086
44	21	13	8	0.380952	19.57143	1.643452676	1.291623
45	21	12	9	0.428571	20	1.653212514	1.30103
46	21	15	6	0.285714	20.28571	1.662757832	1.30719
47	21	13	8	0.380952	20.66667	1.672097858	1.31527
48	21	12	9	0.428571	21.09524	1.681241237	1.324184

Table 3 Mean cumulative repair function of dumpers (Month-wise)

Month	*Month*	*Total Dumpers*	*Tot working*	*On repair*	*MRF*	*MCRF*	*ln(Month)*	*ln(MCRF)*
Apr-00	1	690	475	215	0.3115942	0.3115942	0	–1.166054
May-00	2	713	229	484	0.67882188	0.99041608	0.693147	–0.00963
Jun-00	3	690	135	555	0.80434783	1.79476391	1.098612	0.584873
Jul-00	4	713	161	552	0.77419355	2.56895746	1.386294	0.9435
Aug-00	5	713	128	585	0.82047686	3.38943432	1.609438	1.220663
Sep-00	6	690	153	537	0.77826087	4.16769518	1.791759	1.427363
Oct-00	7	713	109	604	0.84712482	5.01482001	1.94591	1.612398
Nov-00	8	690	228	462	0.66956522	5.68438523	2.079442	1.737723
Dec-00	9	713	233	480	0.67321178	6.35759701	2.197225	1.84965
Jan-01	10	713	257	456	0.63955119	6.9971482	2.302585	1.945503
Feb-01	11	644	189	455	0.70652174	7.70366994	2.397895	2.041697
Mar-01	12	713	140	573	0.80364656	8.5073165	2.484907	2.140927

Table 4 Mean cumulative cost function of dumpers

Month	*Month No*	*dcost*	*Total cost*	*MCF*	*MCCF*	*LN(Month)*	*LN(MCCF)*
Apr-00	1	2633906	8342543	0.316	0.31572	0	– 0.500698
May-00	2	2300430	14216979	0.162	0.47752864	0.30103	– 0.321001
Jun-00	3	2368111	9923199	0.239	0.71617255	0.47712125	–0.144982
Jul-00	4	2917536	7599156	0.384	1.1001015	0.60205999	0.041433
Aug-00	5	2807863	6695081	0.419	1.51949344	0.69897	0.181699
Sep-00	6	2442156	6129374	0.398	1.91792825	0.77815125	0.282832
Oct-00	7	2215344	19576044	0.113	2.03109432	0.84509804	0.30773
Nov-00	8	2322694	6150846	0.378	2.4087162	0.90308999	0.381786
Dec-00	9	2505204	6610449	0.379	2.78769257	0.95424251	0.445245
Jan-01	10	2590748	7106517	0.365	3.15225203	1	0.498621
Feb-01	11	2370507	6551774	0.362	3.51406351	1.04139269	0.54581
Mar-01	12	4793422	16634634	0.288	3.80222266	1.07918125	0.580038

Table 5 Hypothesis testing by Chi-square on the comuted MLEs β = 1, α = 1.5

Boundary value	*Observed Freq.(O)*	*Probability*	*Expected Freq. (E)*	*(O-E)*	*(O-E)2*	*(O-E)2/E*
2.6	281	0.824987	301.12017	–20.1	404.82136	1.3443847
5.2	63	0.145117	52.967814	10.03	100.64476	1.9001116
7.8	13	0.024836	9.0652093	3.935	15.482578	1.7079118
10.4	5	0.004209	1.5361255	3.464	11.998426	7.8108372
13	3	0.000709	0.2586497	2.741	7.5150017	29.054752
	Cal.Value	41.818			Crit.Value	43.8

The Hypotheses are: Null Hypothesis Ho: Failure times are Weibull with β = 1, α = 1.5

Alternate Hyothesis Ha: Failure times are not Weibull with β = 1, α = 1.5

Test statistic is Ei = 365 Pi = 365 {Exp [– (ai-1/α)β] – Exp[–(ai/α)β]}

for i = 1,2,3,4 and E5 = 365 [1-P1-P2-P3-P4]

Hence the Hypothesis is accepted. So β = 1 and MTBF is 1.5.

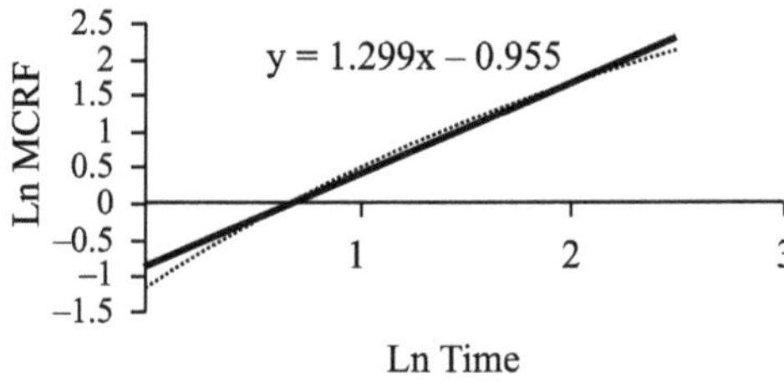

Figure 3 LN time vs LN SCRF

Figure 4 Log time vs Log MCCF

Fitting the NHPP Model

In repairable system reliability methodology, a NHPP model is used to model the repairs. The advantage of using NHPP model is that *time between repairs* are not necessarily independently and identically distributed random variables. This may be closer to real world situation. In NHPP, ROCOF is commonly modelled by a power law function, $\alpha\ \beta\ t^{\beta-1}$ where α & β are parameters of the NHPP model. The derivative of MCRF has the same meaning of ROCOF. Therefore, if we fit NHPP to our case, we have $d\,(\text{MCRF})/dt = \alpha\ \beta\ t^{\beta-1}$. MCRF is then given by

$$\text{MCRF} = \alpha\ T^{\beta} \qquad \text{... (7)}$$

hence

$$\text{Log (MCRF)} = \beta \log \alpha + \beta\ \log t \qquad \text{... (8)}$$

A straight line of the form $y = mx + c$ is fitted by plotting ln (MCRF) vs Ln (operating week). The estimates of α and β can be obtained from the graph. The slope of the straight line gives value of β, whereas the estimate of α is given by exponential of (intercept/slope of straight line).

$$\beta \text{ Log } \alpha = \text{intercept or } \alpha = \text{Exp (intercept/slope)} \qquad \text{... (9)}$$

The estimate of reliability at age 't' when the random variable is time to first repair can be obtained by

$$R(t) = \text{Exp}[-\ \alpha\ t^{\ \beta}] \qquad \text{... (10)}$$

Based on the above equation, the repair organisation reliability value for dumpers is calculated and the plot is given.

Given the system is of age 't', the reliability that the system will function additional time 'D' without a repair is usually of interest. The following equation can be used to obtain reliability after additional time 'D'.

$$R(t) = \exp\{-[\alpha\,(t + D)^{\beta} - \alpha\, t^{\beta}]\} \quad \text{... (11)}$$

Plotting MCCF Curves

In the similar fashion as that of MCRF, the repair costs of a population of systems are usefully expressed in terms of C (t), the population Mean Cumulative Cost Function (MCCF) versus system age 't'. The above function is obtained by the following data. At any particular age 't', each population system has accumulated a total repair cost. These cumulative costs usually differ because different systems undergo different numbers and types of repairs. The graphical plot between cumulative cost and age is generally viewed as a straight line for ease of viewing. In reality, these are staircase functions where the rise of each step is the additional repair cost at the time of repair. However, staircase functions are hard to view in such a plot. Consequently, there is a population distribution of cumulative costs at age 't' which is continuous. The distribution at age 't' has a mean C(t). The mean cumulative cost per system for large population, C(t) increases smoothly with age, which is generally plotted as heavy line. C(t) is regarded a continuous function assumed to have a derivative, i.e., the slope of the above plot and is given by

$$c(t) = dC(t)/dt \quad \text{... (12)}$$

Where c(t) is the mean rate at which repair costs accumulated at age 't'; it is expressed in rupees per unit time per system. Refer Table 4 and Figure 4.

Recommendations

1. If the data can be collected for larger period like two to three years, more authentic information about the reliability of equipment and performance in future can be extrapolated.
2. Similar data collection from different machines and analysis will give a reasonable picture of adequacy of repair facilities at different places. This can be related to the availability/ utilisation of equipment at different sets of similar machines.
3. Reliability of repairable equipment for future functioning can be predicted using the techniques given above. The graphs of MCRF and MCCF indicate the health of the system of the equipment as a whole.
4. The advantage of this technique is that it can be calculated at any month, week or day and does not require the failure history of each individual vehicle. Thus, it is comparatively easy to collect (taking the data that how many are under operation on the given day) the data and calculate. Further, it becomes a helpful tool in maintenance information system for maintenance managers in addition to act as the controlling aspect of the operations.

The applications may be further extended to evaluate any group of machines such as a fleet of buses (used for public transport) operated by a depot, trains or tram cars, a group of lorries or trucks operated in a cement industry of construction companies.

CHAPTER 11

Trend Analysis

PREAMBLE

This chapter focuses on the concept of trend and its computation. It analyses the failure patterns of equipment. Behaviour of machines, machine life cycle (MLC) and bathtub curve are used to explain different types of failure patterns. Various types of failures based on the volume, the mode of failure are discussed. The trends and patterns of failures with time of repairable systems are described using graphical methods such as cumulative plot test, eye ball analysis, and analytical trend tests such as Laplace test, Mil-hdbk-189 test are elucidated. Test for presence of correlation using serial correlation coefficient of correlation test can also help in tracking the trend.

CHAPTER OUTLINE

11.1 INTRODUCTION

In Chapter 10, it is learnt that one can predict or assess the occurrence of a failure on a machine by close monitoring of the trend. In previous chapters, 'trend monitoring' is interpreted as a part of *condition monitoring techniques*. However, this trend analysis of failures on a machine can also be used to predict the life cycle characteristics of the equipment, maintenance scheduling, and replacement analysis and design the reliability of the equipment.

In this chapter, we shall discuss various types of failures and various methods for the analyses of their trends. The use of this trend in estimation of replacement age, reliability, life cycle characteristics, and maintenance scheduling will be discussed in the chapters to follow.

11.2 FAILURE PATTERNS

No machine is immortal and completely immune to any failures, no matter how safely it is run, how closely the instructions of the manufacturer or supplier are followed, how best it is maintained up to its standards and specification. Perhaps, the occurrence of failure can only be prevented or prolonged if the probable reason for its occurrence is known. Even in such cases, sometimes one can temporarily stop the failure, but not permanently. Even this forestalling of failures is not possible unless there is complete knowledge of the failures that can occur on the equipment. Further, it is required to know their causes, effects or costs, and the remedial measures also. The awareness of the equipment failures often make the engineer so confident that after the rectification of the failures he/she will be able to assure the production manager about its running condition. Therefore, let us first understand the term failure and their types and then the patterns how they occur.

11.2.1 Definition of Failure

Failure is defined in many ways. Of them, the most popular and appropriate are given below:

Failure is defined as *inability of a machine or equipment to perform the intended or specific job under specified conditions.* A failure can be *an event that changes a product or machine or equipment from an operational condition to a non-operational condition.* Failure can be further defined as *"non-conformance" to some defined performance criterion.*

11.2.2 Failure Patterns in Machines Behaviour

Almost all machines (or equipment) irrespective of their nature such as electrical, mechanical, electronic, etc., are assumed to behave in almost the same manner. These machines are normally expected to have one of the three types of behaviours with references to the failures that occur. These three behaviours are as follows:

(a) The rate of failures is decreasing, i.e., **decreasing failure rate** or **DFR** (decreasing trend—machine condition is improving).

(b) The rate of failures is constant, i.e., **constant failure rate** or **CFR** (no trend—machine condition may or may not be consistent but machine is giving required average output).

(c) The rate of failures is increasing, i.e., **increasing failure rate** or **IFR** (increasing trend—machine condition is deteriorating).

Amazingly, every machine is found to exhibit all the three behaviours significantly and distinctly in certain period of its lifetime. Therefore, it has gained much significance in the studies of trend analysis and replacement analysis.

11.3 MACHINE LIFE CYCLE (MLC)—BATHTUB CURVE

As already discussed (in Chapter 7), machine life cycle is classified into in three phases, which is analogous to the three phases of human life cycle as shown below.

1. *Infancy phase:* Early failures or infant failures—DFR.
2. *Youth phase:* Random failures or rare event failures—CFR.
3. *Old age phase:* Wear out failures or old age failures—IFR .

11.3.1 Early Failures at Infant Stage of Machine

A baby, just born cries… why? It is because of the change in environment. The baby had a safe and secured environment when it was in the warm amniotic fluid of mother's womb, and now when it came in to the world, it experiences a different weather and feels some sort of discomfort. So, also the machine, immediately after brought from the original equipment manufacturer (OEM), may not work with full efficiency due to various reasons such as the initial friction between the moving or rotating parts, not adjusting to the environment, lack of skill and knowledge of the person working on it, and so forth. (For a better understanding of the concept, a comparison between the human life cycle and machine life cycle is given in Table 11.1).

Table 11.1 A comparison between human and machine life cycles

Life Stage	*Human Life Cycle*	*Machine Life Cycle*
Infancy	*Just born baby cries due to*	*Machine gives trouble due to*
	Change of environment	Environmental change
	Handling is improper	Improper care on operating
	Ignorant child care	No knowledge on machine care
	Food/water unhygienic	Defective raw material, fuel impurity
	Genetic problems	Design faults
Youth	Rarely becomes sick	Random (rare-event) failures occur
	Active to work	Machine can work at full efficiency
Old age	Age related problems such as diabetes, BP, etc.	Age related problems such as worn-out parts, weakened components
	Slow or less efficient	Slow or less productive
	Cannot work or tiresomeness	Cannot work or heats up due to friction between worn out parts

To control these failures one has to be aware of the machine thoroughly and adhere to the instructions strictly. In this consequence, the output may be slow or low or delayed. However, it is necessary in the interest of good running and long life of the equipment. For example, when you buy a personnel computer system, it may give certain problems in its performance initially due to incompatibility with parts, hardware, software, environment (like temperature, humidity, etc.), and so forth. This obviously restricts or slows down the job and does cause inconvenience to the user. Such problems generally referred to as infant or teething or start-up problems or early failures are usual to any new equipment. These failures will come down and may finally disappear over period of time. This period of specific machine behaviour, called infancy (or early failure period) is not same to all machines. However, this period is estimated by OEM out of experience or through scientific tests (accelerated run test, endurance test, etc.) and researches and usually is declared as the warrantee or guarantee period. The OEM believes that these failures will vanish within this period and hence bears the responsibility and accountability to the failures.

Since these failures are decreasing with time, the trend can be stated as decreasing failure rate (DFR) or machine is improving (see Figures 11.1, 11.4 and Table 11.2). During this failure period, failures of the machine are referred to OEM as the machine is covered under guarantee/warrantee or any contractual maintenance.

11.3.2 Random or Rare-event Failures

A young person hardly experiences any ailment. This may be because of his/her active age and growing biological features. Further, the working efficiency peaks up in this stage. A machine also seldom experiences any failures after passing through the infancy. Hence, the user derives maximum output in this period as the machine runs with its full efficiency. The length of this youth period depends on how the user utilises and maintains the machine. The effective usage and appropriate maintenance can enhance this portion of its life. User's care may come down in this stage due to the fact that the user might have got boredom or monotony or some sort of negligence as he observes the machine will be working at its best even though no care is taken. Of course, any ignorance or negligence may not result in immediate failure but its effect piles up and finally blows out in its next stage of life often which may foster reaching its old age phase earlier or even affect its life. However, the immediate failures that occur rarely (by chance) are known as rare-event or random or chance failures. Obviously, the reliability of machine will be very high in this stage. As discussed in Chapter 7, section 7.3.2, usually preventive maintenance (PM) and breakdown maintenance (BM) are adapted in this stage to low cost failures while RCM (reliability-centred maintenance) and CBM (condition-based maintenance) techniques are employed on the machines that are high cost or in case the failure may lead to high cost or damage.

It can be noticed over a period of time that almost constant failures (i.e., more or less same number of failures) occur on the machines during this stage and hence, the machine is said to be showing constant failure rate (CFR). Therefore, 'no trend' is detected in this stage. (See Figure 11.2 and Figure 11.4 and Table 11.2).

11.3.3 Old Age of Wear Out Failures

When a man reaches his old age, various ailments such as diabetes, hypertension, and rheumatic complaints open up and obstruct every part of his work. The seeds of these ailments in fact start sometime during the youth but are suppressed due to the growing health body and shoot up when body becomes weak. In the same way, in a machine also, the youth stage ends and the old age phase opens when recurrently fail due to worn out or weakened parts.

The maintenance department has to put their highest efforts at this stage as costs and time of failure go high if this stage is not detected. Generally, calendar age is considered to detect this

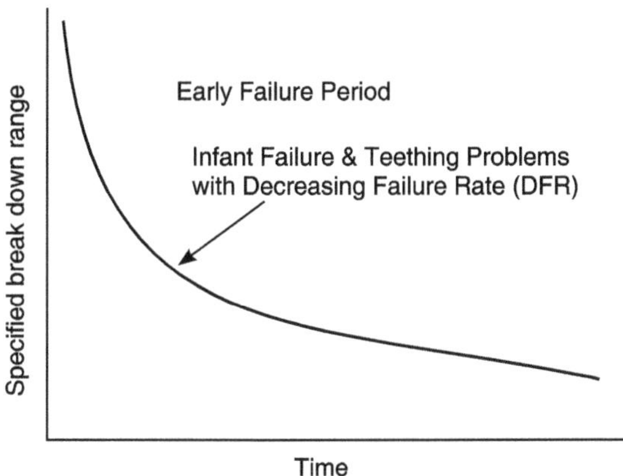

Figure 11.1 Failure pattern in early failure period of machine life cycle

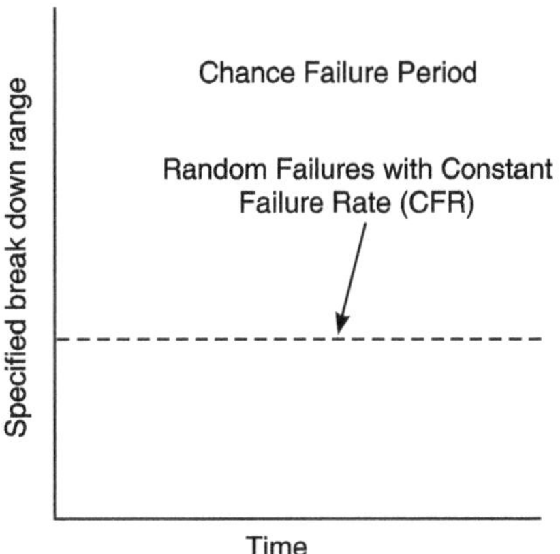

Figure 11.2 Failure pattern of chance failure period in machine life cycle

stage but this is not always correct. For instance, a doctor's car (used sparingly) and a travel agent's car (used continuously) would not reach the old age at the same time, though bought on the same day. Hence, one should notice that a machine reaching the old age depends on various factors. A few general aspects are listed below.

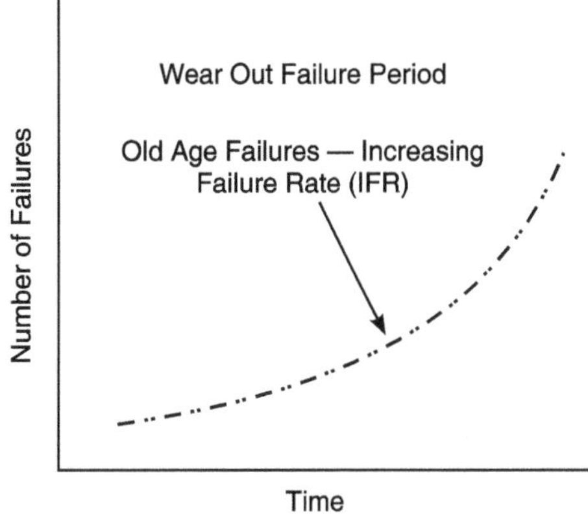

Figure 11.3 Failure pattern of old age failures in machine life cycle

- The running age
- The calendar age
- The operational period
- Maintenance strategies employed such as PM, BM, CBM, etc.
- Operating conditions
- Environmental effects
- Conditions maintained while usage
- Quality and strength parts and components
- Material defects
- Design faults
- Care and efforts put on it to increase its life during its infant and young stages, etc.

These failures are mostly because of the worn out parts of the machines and hence are called wear out failures or old age failures. As failures tend to increase in their severity or frequency or volume during this period, the increasing trend, or increasing failure rate (IFR) is shown by the machine and is said to be deteriorating. (See Figure 11.3, Figure 11.4, and Table 11.2.) As soon as this stage is observed, the maintenance engineer has to think of various alternatives (refer chapter 7, section 7.3.3) such as replacement, reconditioning, updating, design out maintenance, operate to failure, and corrective maintenance or scrapping.

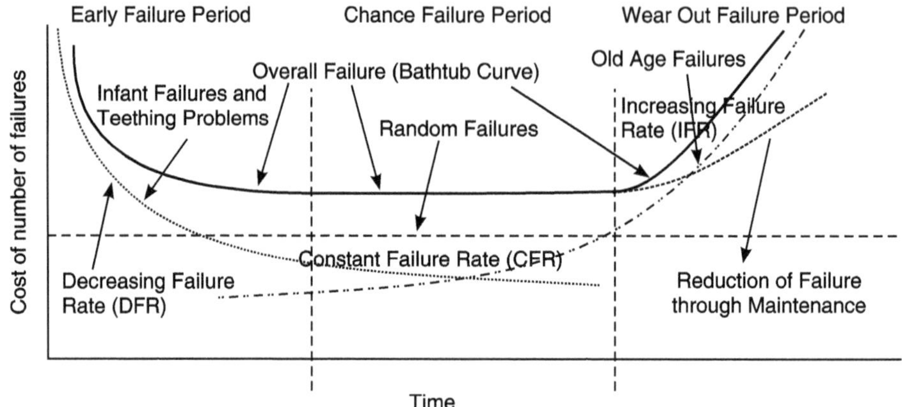

Figure 11.4 Failue patterns—machine life cycle—bathtub curve

Table 11.2 Failure analysis through machine behaviour and machine life cycle

Phase	*Type of Failure*	*Failure Rate (Failure Trend)*	*Probable Cause of Failure*	*Cost of Failures*	*Suitable Maintenance Policy*
Infant Phase	Early failures or infant failures	Decreasing failures rate (DFR)	Faulty design, erratic operation, environment problems, installation errors, heavy friction	Medium to high	(Under warrantee/ guarantee) Refer to OEM
					Contractual maintenance (no warrantee)
Youth Phase	Random or chance or rare event failures	Constant failure rate (CFR)	Operation errors, fatigue due to heavy workload, overrun	Low	Breakdown maintenance, preventive maintenance
				Medium to high	Reliability-centred maintenance, condition-based maintenance
Old Age Phase	Wear out or age failures	Increasing failure rate (IRF)	Wear, tear, creep, fatigue, weakened parts	Low	Operate fail and corrective maintenance
				High	Replacement, scrapping or reconditioning

BUSTLE - 11.1

Collect failure data for few machines of your choice in an industry with which you are familiar and classify them into three phases based on their frequency or cost of failures.

1. Infant stage:

____

2. Youth stage:

____

3. Old age phase:

____

Test Your Understanding 11.1

1. What do you understand by 'rare event failures'? Explain with an example.
2. What are 'infant failures'? What are the causes for the occurrence of infant failures?
3. List the reasons for old age failures in a machine life cycle. What remedial measures do you suggest in case of more old age failures?
4. Discuss the most suitable maintenance policies for a machine that experiences its youth phase.
5. Discuss the most suitable maintenance policies for a machine that experiences its old age phase.

11.4 TYPES OF FAILURES BASED ON VOLUME OF FAILURE

(a) **Small Failures:** These are the failures, which can be rectified in few minutes. These failures are very common in any machine. These will not have any considerable impact on the machine performance and on the operator.

(b) **Minor Failures:** These are the failures, which can be rectified in a few hours or with a little effort. This kind of failures will have a little considerable effect on the productive work and can hardly damage the machine or men or environment.

(c) **Major Failures:** These failures take a few days to rectify. They also may require large work force, knowledge, and skill to rectify. These failures can cause considerable damage to the machine, minor injuries to men, and affect the regular work.

(d) **Catastrophic Failures:** These are the costliest failures. The occurrence of such failures can cause damage to the machine, men, and sometimes the environment also. Rectification or recovery from the losses of this failure may take a few weeks or even months.

11.5 TYPES OF FAILURES BASED ON MODE OF FAILURE

(a) **Sudden Failures:** These types of failures occur in items after giving some period of desired service rather than deterioration while in service. This period of giving desired service is not constant but follows some frequency distribution, which can be progressive, retrogressive, or random in nature.

- *Progressive failure*: If the probability of failure in the beginning of an item is less and gradually increases in its life, then such failure is called *progressive failure*. For example, light bulbs and tubes fail progressively.
- *Retrogressive failure*: If probability of failure in the beginning of the life of an item is more but as time passes the chances of its failure become less, then such failure is said to be retrogressive.
- *Random failure*: In this type of failure, the constant probability of failure is associated with items that fail from random causes such as physical shocks, not related to age. For example, vacuum tubes in air burn equipment have been found to fail at a rate of the age of tube.

(b) **Gradual Failures:** A gradual failure is progressive in nature, i.e., as life increases, its operational efficiency also deteriorates resulting in increased running (maintenance and operational) costs. They also cause decrease in the resale or salvage value. Mechanical items like pistons, rings, bearings, etc., and automobile tyres fall under this category.

11.6 REPAIRABLE AND NON-REPAIRABLE SYSTEMS

While maintainability and reliability characteristics are designed for a system, one should first identify the repairable or replaceable features in it. Thus, the system is to be first defined whether it is a replaceable (non-reparable) system or a repairable system.

For a non-repairable system, the reliability is the survival probability over the item's expected life or for a period during its life. In non-repairable items, the first failure itself is the last failure, since there will be no more repairs on the item (sometimes it is not possible to repair and some other times not advisable to repair also). So, the term 'failure rate' seems to be somewhat inappropriate. Thus, when only one failure can occur during lifetime, the instantaneous probability of the first and only failure is called the 'hazard rate'. Non-repairable system can be individual parts (light bulb, transistor, etc.) or a machine (e.g., a fan or a lathe or milling) or systems comprised of many subsystems or parts (e.g., spacecraft). When even one component fails in a non-repairable system, the total system fails (usually) and system reliability is, therefore, a function of the time to the first (part) failure.

For the items which are repaired when they fail, reliability is the probability that failure will not occur in the period of interest. When more than one failure occurs, it can also be expressed as failure rate or rate of occurrence of failure (ROCOF). However, the failure rate expresses the instantaneous probability of failure per unit time when several failures can occur in a time continuum.

Repairable system reliability can also be characterised by mean time between failures (MTBF), but only under the particular condition of constant failure rate. The availability of repairable system should also be taken into consideration since repair takes time. Availability is affected by the rate of occurrence of failures and by maintenance time. Maintenance includes corrective (repair) or preventive (to reduce the likelihood of failures) measures.

Sometimes an item may be considered both repairable and non-repairable, for example, a missile is repairable system while it is in store and subjected to tests, but it becomes a non-repairable system when it is launched. Similarly, a vehicle is considered to be repairable as long as the cost of replacement is more than that of its repair, while it can be considered to be replaceable after it reaches a certain age. These separate states must be taken into account while carrying out trend analysis of such systems.

11.7 THE PATTERN OF FAILURES WITH TIME IN REPAIRABLE SYSTEMS

The failure rates or rate of occurrence of failures (ROCOF) of repairable systems can also vary with time and an important implication can be derived from these trends.

A constant failure rate (CFR) is indicative of externally induced failures, as in the constant hazard rate situation for non-repairable systems. A constant failure rate (CFR) is also typical of complex system if subjected to repair and overhaul where different parts exhibit different patterns of failures with time and parts have different ages since repair or replacement. Repairable system can show a decreasing failure rate (DFR) when reliability is improved by progressive repair as defective parts which fails relatively early are replaced by good parts. An increasing failure rate (IFR) occurs in repairable systems when wear out failure mode parts begin to predominate. The pattern of failures with repairable systems can also be illustrated by the use of the bathtub curve as shown in Figure 11.4, but with the failure rate (ROCOF) against age instead of hazard rate.

The impact of designed reliability is greatly influenced by variations in operating environment which can be controlled through monitoring the trend. Also, to estimate the reliability, it is essential to know the failure patterns and the behaviour of the machine to determine the distribution from which the sample has come. Experiences say that the simple way to predict most of the distributions is the trend analysis.

BUSTLE - 11.2

With the data you collected for Bustle 11.1, classify them into the following categories.

1. *Small Failures:*

____

2. *Minor Failures:*

____

3. *Major Failures:*

____

4. *Catastrophes:*

____

Test Your Understanding 11.2

1. Explain different types of failures based on the volume of failures.
2. With suitable examples distinguish between repairable and non-repairable systems.
3. Discuss various types of failures based on the mode of failures.
4. Describe the patterns of failures in repairable systems.
5. Distinguish between progressive and retrogressive failures with examples.

11.8 TREND

The main objective of trend analysis is directed to know whether the equipment is deteriorating or improving. This can be known by analysing the past failure data in terms of TBF (in case of repairable equipment) or TTF (in case of non-repairable equipment). From the analysis, one of the following three conclusions may be drawn.

11.8.1 Positive Trend

Positive trend implies that the machine is improving with time. This may be because the machine is new or in infant stage, i.e., first stage of bathtub curve (Figure 11.1 and Figure 11.4) or due to effective maintenance. It is observed that the machine performs well due to gradual decrease in failure times, i.e., decreasing failure rate (DFR). For example, a new machine gives trouble initially due to static friction, environmental change, or operational problems, etc. However, these will reduce over a period of time.

11.8.2 Negative Trend

If the machine is showing a negative trend, it implies that the machine is deteriorating with time. In other words, the equipment is subjected to frequent or long failures. It is indicated by gradual increase of failure times. It can be considered that machine showing negative trend is in the third stage of machine life cycle (bathtub curve) or old age (Figure 11.3 and Figure 11.4). It is due to worn out parts or inherent failures or inappropriate maintenance system. Whatever the reason may be, this situation disrupts the production and affects the productivity. Therefore, such equipment is of more concern to maintenance as well as production engineers. These failure patterns are categorised by increasing failure rate (IFR).

11.8.3 No Trend

This is one of critical situations faced by maintenance engineers. If failure times are neither increasing nor decreasing, it can lead to conclusion that the equipment is experiencing constant failure rate (CFR). But this is true in some cases only and the 'no trend' situation need not mean youth of machine (i.e., random failures period with constant failure rate). However, the no trend situation implies that the failure behaviour is independent of time. It may be in random failure stage or typical increasing failure rate due to independent and identical distribution (i.i.d.) of failure times. Therefore, this rate is to be further analysed to know this fact.

11.9 METHODS OF TREND ANALYSIS

Trend analysis is carried out basically in two ways, graphical and analytical. The following methods are most popularly employed in graphical and analytical methods for monitoring monotonic trend.

1. Graphical Methods For Trend Testing
   (i) Cumulative Plot Test
   (ii) Eye Ball Analysis
2. Analytical Trend Tests
   (i) Laplace Test
   (ii) MIL-HDBK-189 Test

11.9.1 Cumulative Plot Test

This is the most powerful test and gives easy understanding because of its pictorial nature. To perform this test, first of all, the time between failures (TBFs) of the equipment is collected in chronological order. The cumulative time between failures (CTBF) are then calculated and plotted against the cumulative number of failures. By the presence of trend in TBFs, it means that whether the equipment or the item is deteriorating or improving with age.

By simply plotting cumulative TBFs against the cumulative number of failure, it can be tested whether the machine or item under consideration is improving or deteriorating. If a curve concave upward is derived *(note that this is the opposite characteristic in TTT plot discussed in Chapter 12)*, this means that the TBFs are becoming shorter and shorter, that is to say, the machine is deteriorating. On the other hand, if a curve concave downwards is obtained, this means that the machine is improving.

The data which exhibits linearity can be considered to have no trend. Such trend plot is known to exhibit independently and identically distributed (i.i.d.) data in statistics and has to be further analysed by statistical distributions.

11.9.2 Eye Ball Analysis

This is a simple analysis of testing presence of trend. In this we pass our eye through chronological TBFs and search for increase or decrease of the failure rate, i.e., if the TBFs are showing increasing failure magnitude towards the end, it shows decreasing failure rate. If the magnitude of TBFs decreases towards end, then it indicates increasing failure rate. If the magnitude of TBFs is approximately constant throughout the period, then it shows constant rate of failure.

A slight modification can be done to this test. First, tabulate the cumulative frequency and divide the total period into equal number of parts usually from 5–10. (The class interval can be judged by using Sturge's formula.) Then find the number of failures in each period of operations. If there is increasing number of failures for each period, then it indicates increasing failure rate and so on. Further, this can be easily judged by fitting a linear trend line.

11.9.3 Laplace Test

This test is highly useful for distinguishing between homogenous Poisson process (HPP) and a monotonic trend. Its high degree of accuracy keeps in the top position of all trend tests. However, it is tedious and cumbersome to calculate, particularly when the data is large. Further, it requires calculation from the beginning at every time while the graphical tests can provide instant answers at any point of the data.

The specific situation is discussed where one system is run until a pre-specified number of failures, m has occurred. Under HPP assumption, the first m–1 arrival times, T_1, T_2, ...T_{m-1} are the order statistics from a uniform distribution on (0, t_m).

Hence the test statistic

$$U_L = \frac{\left[(1/k)\cdot\sum_{i=1}^{k} T_i\right] - (T^*/2)}{T^*\cdot\sqrt{(1/12k)}} \qquad \text{... (1)}$$

approximates a standardised normal variate, the approximation being adequate at 5%, level of significance for $m \geq 4$.

[In equation (1), $k = m-1$ and $T^* = t_m$ if observation stops at the last failure, $k = m$ and T^* = the time at which observation stops, otherwise. At a significance level α there is evidence of trend if: $U_L > Z_{\alpha/2}$ (reliability deterioration) $U_L < -Z_{\alpha/2}$ (reliability improvement)]

Compare the calculated '*U*' value with normal variate at 10% level of significance. If calculated value is within –1.645 to +1.645, accept the hypothesis, i.e., no trend, else trend exists. [At 5% level of significance the critical values are -1.96 to +1.96 and 1% LOS the critical values of normal variate are -2.58 to +2.58].

11.9.4 MIL-HDBK-189 Test

The test is based on test statistic

$$U = 2\sum_{i=1}^{k} \ln(T_n / T_i) \qquad \text{... (2)}$$

Under the null hypothesis of an HPP, ν is distributed as χ^2, with $2(m-1)$ degrees of freedom.

If calculated χ^2 value is less than Chi-squared table value at $2(m-1)$ degrees of freedom 5% LOS, trend is present, else no trend.

For degrees of freedom (ν) greater than 30, the quantity $(\sqrt{2\chi^2} - \sqrt{2\nu - 1})$ may be used as a normal variate with unit variance.

$$Z = (\sqrt{2\chi^2} - \sqrt{2\nu - 1}) \qquad \text{... (3)}$$

If calculated Z value is in the following range of table (critical) values, then trend is present, else no trend.

Critical Values Normal Distribution Tables

–1.645	to	+1.645	(at 10% level of significance)
–1.96	to	+1.96	(at 5% level of significance)
–2.58	to	+2.58	(at 1% level of significance)

BUSTLE - 11.3

With the data you collected for Bustle 11.1, conduct trend analysis.

____

____

____

____

Test Your Understanding 11.3

1. What is meant by trend? Explain different methods of trend analysis.
2. Explain the three types of failure behaviours of a machine with reference to the trend.
3. What are positive trend, negative trend, and no trend behaviours of machine failures? Explain.
4. What do you understand by 'no trend'? Explain with an example.
5. Describe the procedure to conduct trend analysis by cumulative plotting.
6. Explain the method of eye ball analysis for identifying the trend in machine failure data.
7. Write short notes on the following trend analysis techniques:
   (i) Laplace Test (ii) Mil-HDBK-189 Test

11.10 TEST FOR PRESENCE OF CORRELATION

It is important to test the successive inter-arrival times for independence by testing them for serial correlation. To test the data for serial correlation, (i-1)th TBFs are to be plotted against ith TBFs of the data. If the plotted points are randomly scattered without any pattern, it can be interpreted that the TBFs are free from serial correlation.

11.10.1 Testing for the Presence of Serial Correlation

Before modelling the reliability data, it should be tested for the mutual independence by testing it for the presence of serial correlation. The presence of serial correlation can be tested by plotting the ith TBF say, x_i against the (i–1)th TBF, x_{i-1}. If the plotted points are randomly scattered without any pattern, it can be interpreted that the TBFs are free from serial correlation. In case the plot reveals serial correlation, then the TBFs should be plotted at greater lags such as x_i against x_{i-2}, x_{i-3}, etc., to search for serial correlation over greater lags (Bendell and Wall 1985).

11.10.2 Analysis with Coefficient of Correlation Test

The data can also be analysed with the Karl Pearson's Coefficient of Correlation Test between ith v (i–1)th TBFs and ith v (i–2)th and so on. The degree of association can be concluded with the value of the coefficient of correlation that ranges between –1 to 1 via zero. The value close to –1 can be understood as the negative close association, which means that the increase in the first quantity induces the decrease in the second quantity. If the values is close to +1, it means that the there is proportional increasing correlation between the quantities. If the value is close to zero, it indicates poor correlation, which means that the quantities are independent (i.i.d. assumption is not contradicted).

11.10.3 Analysis of Data Free from Trends and Correlation

When the data are free from the presence of a trend and serial correlation, the next step is to choose a best-fit probability distribution model using "total time on test" (TTT) plots or "goodness of fit" tests to study its statistical characteristics. Both graphical (TTT) and analytical (goodness of fit) methods used for this purpose are explained in Chapter 12.

ILLUSTRATION 11.1:

The failure data in terms of time between failures (TBF) in hours, obtained for the equipment in a process industry is as follows:

14, 12, 9, 13, 7, 24, 11, 21, 6, 15, 19, 31, 36, 27, 57, 42, 69, 48, 81, 98

Conduct the trend analysis using

a. Cumulative Plot Test
b. Eye Ball Analysis
c. Serial Correlation
d. Coefficient of Correlation

Solution.

The data for cumulative plot test and eye ball analysis is shown in the following table. Relevant calculations are shown in the table itself. (TBF: time between failures; CTBF: cumulative time between failures; OTBF: orderled time between failures)

Table 11.3 Trend analysis

S.no	*TBF*	*CTBF*	***OTBF***
0	0	0	0
1	14	14	6
2	12	26	7
3	9	35	9
4	13	48	11
5	7	55	12
6	24	79	13
7	11	90	14
8	21	111	15
9	6	117	18
10	15	132	21
11	19	151	24
12	31	182	27
13	36	218	31
14	27	245	36
15	57	302	42
16	42	344	48
17	69	413	57
18	48	461	69
19	81	542	81
20	98	640	98
0		0	
20		640	

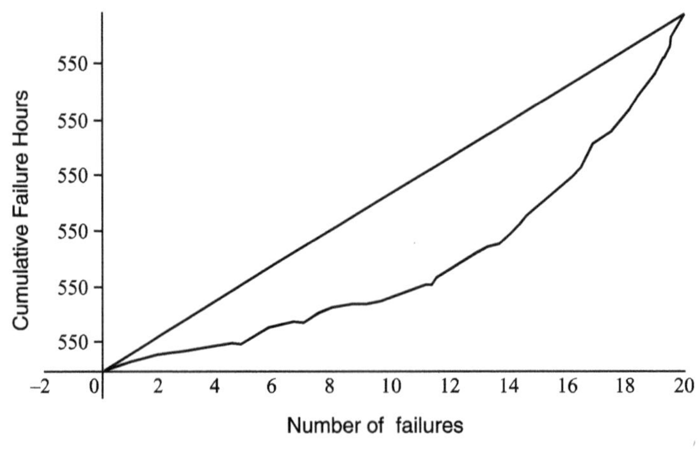

Figure 11.5 Trend analysis—cummulative plot test

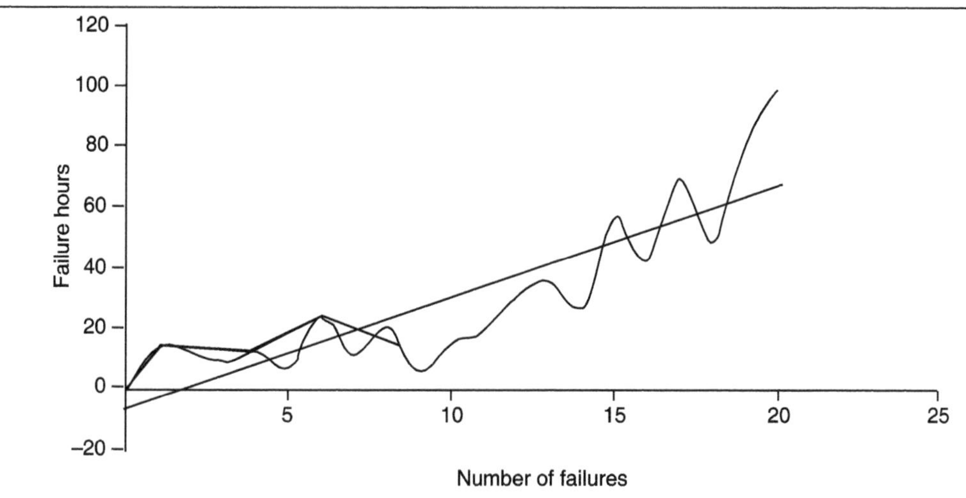

Figure 11.6 Eye ball analysis

The data for serial correlation test is given Table 11.4.

Table 11.4 Karl Pearson's coefficient of correlation and serial correlation

S.no	*TBF (i)*	*x = i–32*	*(i–1)*	*y = (i–1)–27.1*	*x**2*	*y**2*	*x*y*
0	0						
1	14	–18	0	–27.1	324	734.41	487.8
2	12	–20	14	–13.1	400	171.61	262
3	9	–23	12	–15.1	529	228.01	347.3
4	13	–19	9	–18.1	361	327.61	343.9
5	7	–25	13	–14.1	625	198.81	352.5
6	24	–8	7	–20.1	64	404.01	160.8
7	11	–21	24	–3.1	441	9.61	65.1
8	21	–11	11	–16.1	121	259.21	177.1
9	6	–26	21	–6.1	676	37.21	158.6
10	15	–17	6	–21.1	289	445.21	358.7
11	19	–13	15	–12.1	169	146.41	157.3
12	31	–1	19	–8.1	1	65.61	8.1
13	36	4	31	3.9	16	15.21	15.6
14	27	–5	36	8.9	25	79.21	-44.5
15	57	25	27	–0.1	625	0.01	-2.5
16	42	10	57	29.9	100	894.01	299
17	69	37	42	14.9	1369	222.01	551.3
18	48	16	69	41.9	256	1755.61	670.4
19	81	49	48	20.9	2401	436.81	1024.1
20	98	66	81	53.9	4356	2905.21	3557.4
	32		27.1		13148	9335.8	8950
Coefficient of correlation = sum x*y/sqrt{(sum x**2)*(sum y**2)} r = 0.8078252							

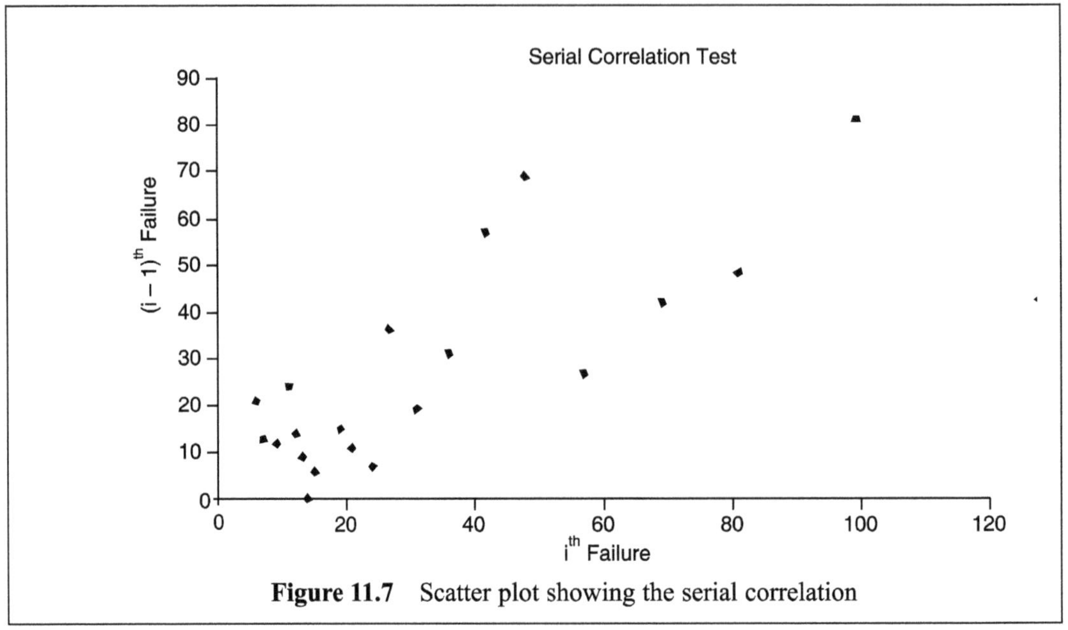

Figure 11.7 Scatter plot showing the serial correlation

BUSTLE - 11.4

With the data you collected for Bustle 11.1, conduct correlation test.

Test Your Understanding 11.4

1. What is meant by correlation? Explain the procedure to finding coefficient of correlation.
2. What is serial correlation? How do you find it?
3. What is serial correlation? Discuss its application in analysis of machine down time?
4. Explain the significance of Karl Pearson's coefficient of correlation. Explain the step by step procedure to conduct the test.
5. Explain the application of coefficient of correlation in machine down time analysis.
6. The time between failures for a machine are noticed as follows:

   12, 10, 7, 11, 5, 22, 9, 19, 4, 13, 17, 29, 34, 25, 55, 40, 67, 46, 79, 96

   Conduct the trend analysis by using cumulative plot test and confirm by eye ball analysis. Also check serial correlation and verify by Karl Pearson's coefficient of correlation.
7. The time between failures for a machine are noticed as follows:

   10, 18, 07, 2, 15, 62, 9, 39, 14, 19, 11, 20, 34, 25, 59, 80, 27, 65, 19, 06

   Conduct the trend analysis by using

(a) Cumulative plot test
(b) Eye ball analysis

8. The time between failures for a machine are noticed as follows:

10, 18, 07, 2, 15, 62, 9, 39, 14, 19, 11, 20, 34, 25, 59, 80, 27, 65, 19, 06

Conduct
(a) Serial correlation test
(b) Karl Pearson's coefficient of correlation

11.11 CONCLUSION

Maintenance department is often concerned with the critical decisions for which the plant engineer is required to be aware of and keep on tracking machine health. The trend analysis and cost analysis of failures can help the engineer in taking right decisions easily and timely. The trend analysis can help the engineer in taking the decisions of maintenance scheduling and planning for the preventive actions. Further, the trend analysis can help in estimating the reliability characteristics of the machine and life cycle characteristics. These are discussed in the chapter to follow.

11.12 RECAP

This chapter focuses on the concept of trend and its computation. It goes in the direction of analysing the length and frequency of failures. Behaviour of machines, machine life cycle (MLC) and bath tub curve are used to explain different types of failure patterns. Various types of failures based on the volume, the mode of failure are discussed. The trends and patterns of failures with time of repairable systems are described using graphical methods such as cumulative plot test, eye ball analysis and analytical trend tests such as Laplace test, Mil-hdbk-189 test are elucidated. Test for presence of correlation using serial correlation coefficient of correlation test can also help in tracking the trend. These are also given an appropriate place.

11.13 KEY CONCEPTS

1. **Failure:** Inability of a machine or equipment to perform the intended or specific job under specified conditions.
2. **Infant or Early Failures:** The failures that occur due to design deficiencies, environmental problems, and operational (or operators') faults.
3. **Chance or Random Failures:** The failures that occur randomly and rarely due to shocks or unpredicted reasons but usually not related to age.
4. **Wear Out or Old Age Failures:** The failure which occur due to wear out and aging.
5. **Reliability:** Reliability is the probability that failure will not occur in the period of interest.
6. **Failure Rate:** When more than one failure occurs, it can also be expressed as failure rate or rate of occurrence of failure (ROCOF).
7. **MTBF (Mean Time Between Failures):** It is the average duration between one failure to the next.

8. **Trend:** It is the manner in which the equipment is deteriorating or improving.
9. **Progressive Trend:** If the probability of failure gradually increases in its life, then such failure is called progressive trend.
10. **Retrogressive Trend:** If the probability of failure gradually decreases in its life, then such failure is called retrogressive trend.
11. **Positive Trend:** Failure times of machine decreasing with time, the machine is said to be improving.
12. **Negative Trend:** Failure times of the machine increasing with time or machine deteriorating with time.
13. **No Trend:** The failure times do not show any trend, i.e., it is neither increasing nor decreasing. The cumulative plot graph is (almost) linear.

REVIEW QUESTIONS

1. Explain the three stage of machine life cycle with the help of a hypothetical graph.
2. Illustrate various failure trends that a machine can exhibit in its life span with some examples.
3. What do you understand by the term 'trend'? Discuss different types of trends a machine can exhibit. How is trend useful in estimating the maintainability of machine?
4. Explain step by step procedure to estimate failure trend of a machine.
5. Explain various graphical methods to evaluate failure trend of a machine.
6. Explain various analytical methods to evaluate failure trend of a machine.

CHAPTER 12

Reliability Oriented Maintenance Models

PREAMBLE

This chapter focuses on identifying, collecting relevant data, and analysing a problem related to reliability of the equipment. A clear cut path is established to predict the repairable equipment behaviour in this chapter. It conducts trend analysis and examines various reliability-centred maintenance models for fitting the data. A goodness of fit test is performed which confirms the distribution of failure patterns. A suitable maintenance policy/strategy and schedules can be suggested based on the analysis. This chapter will help design and assign the reliability characteristics.

CHAPTER OUTLINE

12.1 INTRODUCTION

So far, we have discussed various maintenance models in which the failure intensity, failure frequency, failure duration, etc., are used to study the effect, the cost or damages due to failures. Further, we also have discussed various time-based, failure-based, and condition-based models in view of forestalling such failures. Of course, all these models speak about how to stop further deterioration of machine, but no thought is given towards a point that how long and at what level of confidence the repair or maintenance work carried out will work to keep the equipment without failing. That is to say how much reliable is the maintenance. Some of these reliability-oriented maintenance models are discussed in this chapter.

The failure intensity or the frequency can be reduced by employing suitable and timely actions. This is possible if the area that is causing more loss or damage or disturbance to the production is identified. At the same time, one should remember that it should be cost effective, therefore, it must first be decided whether it is worthwhile to study. After identifying the area of study, it is important to lay down a path and method of analysis.

The next objective is to fit a theoretical distribution to a random sample of failure or repair data, i.e., to perform a statistical test in order to accept or reject the hypothesis that the observed times come from a specified distribution. Now, the procedures for identifying and specifying candidate distributions for fitting the data are discussed. This chapter outlines the procedure of reliability models and a few goodness of fit tests available for conducting the testing of hypothesis. The particular probability distributions addressed include the exponential, Weibull, normal, and log-normal.

12.2 IDENTIFICATION OF THE PROBLEM AND DATA COLLECTION

Appropriate data collection after identifying the problem is one of the most important steps in reliability analysis. A data collection must be designed, developed, and maintained to facilitate correct and effective reliability analysis. The uses of the collected data may be one or more of the following:

- Detection of in-service problems leading to production losses.
- Evaluation of reliability and maintainability characteristics.
- Studying the economics of system breakdowns.
- Feedback to manufacturer for design modification if needed.

Data on reliability characteristics can be obtained by two methods, either directly from the field or from the sample testing in laboratories. For an effective reliability programme, the collection and analysis of the failure statistics such as failure modes, their frequencies, and possible causes of failures for each component and equipment type from a representative population operating under identical condition is needed.

Data collection in field is an expensive and time consuming work. Nevertheless, data acquired from field is the best representative of the system reliability and maintainability characteristics and all efforts should be made to collect data from the field. Some of the points which are of practical importance for development of field data collection systems are given here (Mahon, 1972; Ireson, 1988):

- The depth of reporting and data collection should vary according to the importance of the equipment or its components.
- Reporting should not be confined to facts. Opinions in the *"engineering judgment"* category should also be included.
- Attempt should be made to collect only those data that are likely to be useful.

During the last few years, many companies with high degree of mechanisation have realised the potential benefit of collection of machine performance statistics.

12.3 PROCEDURE FOR RELIABILITY–CENTRED MAINTENANCE MODELLING AND ANALYSIS

After distinguishing between the repairable and non-repairable items, the reliability study can be established as it suits to the requirements*. The most useful and universal procedure of reliability -centred maintenance modelling for repairable equipment is narrated here. The field data is collected for the equipment categorised under repaired items in the form of time between failures (TBFs) and time to repair (TTRs). The data inconsistencies and errors are removed and the refined data is analysed by both types of models viz. graphical and analytical models. However, more importance is given to graphical method since it provides better simple understanding and can be easily reproduced. The graphical tests such as eye ball analysis, cumulative plot test, and serial correlation determine the presence of trend. The machines, which exhibit presence of strong trend, are further analysed and fitted into non-homogeneous Poisson process (NHPP) model. Power law process (PLP) model, one of the most popular and commonly used NHPP models can be used for such study. If there is no trend, it confirms the independently and identically distributed (i.i.d.) assumption. The values obtained can be checked with some analytical tests such as Laplace test. The machines free from trend are further graphically analysed through total time on test (TTT) plots. The exponential fit that confirms homogeneous Poisson process (HPP) models can be known from this analysis, otherwise can be categorised as renewable process (RP) model which is in accordance with Weibull pattern. The goodness of fit (analytical) tests can confirm which pattern is assumed to be in accordance; otherwise can identify the actual distribution from which the data has come. Thus, reliability characteristics and maintenance schedules can be estimated. Further, their reliability growth plots can also be drawn to estimate the improvements. Algorithmically, it is explained as given below and followed by a flow chart in Figure 12.1.

Algorithm

Step 1: Identification and Definition of Problem or Setting the Hypothesis

Usually, the reliability studies consume more cost and time. Further, it requires higher level of engineering knowledge to understand the complex problems and models. Therefore, it has to be first decided whether the study is worthwhile. In view of this, it is often suggested to select the area where

---

*The procedure may vary slightly or completely depending on the situations and input/output patterns. The researchers in this field may make a note of this and make changes accordingly as per their requirements.

the cost of breakdown is very high or any failure is hazardous or leads to catastrophe that may claim human lives, assets, and/or environmental disorders. Once the decision is taken, problem should be defined clearly and the bounds of the study are also to be demarcated, otherwise it becomes an endless project. However, improvements can always be incorporated at the later stages based on the feedback reports. A suitable hypothesis is stated for the equipment upon which the reliability studies are to be conducted.

Step 2: Collection of Relevant Data

Relevant data is collected often in the form of time between failures, in short called TBF and time to repair, concisely called TTR (in the case of repairable items) or time to fail, in short called TTF (in case of non-repairable items). See Figure 2.2. While collecting such field data, there may be some practical problems. Suitable assumptions as applicable can be made to overcome such problems. The data so collected is to be tabulated.

Step 3: Removal of Inconsistencies and Errors in Data

To err is human. There is every possibility of occurrence of errors in the collected data. Such suspicious entries can be eluded. Further, in some situations there can also be some inconsistencies in the data of TBF values. For instance, a particular machine (say an automobile) is under breakdown/ non-available for the want of some spare parts (say a valve for a tyre). If there is any delay in supply of such spare part, the machine can be restored to up condition and the downtime will increase. Suppose, this continues for a very long period, the machine should be deleted from the list completely. Similar inconsistencies, if found, should be precluded for the analysis.

Step 4: Trend Analysis and Correlation Tests

These are already discussed in the previous chapter (Chapter 11). The trend analysis gives us the idea about whether the machine is deteriorating or improving, by which we can also understand which phase of life cycle (bathtub curve), the equipment is experiencing (infant/youth/old age). Various graphical (eye ball analysis, cumulative plot test, etc.) and analytical methods (Laplace test, MIL-HDBK-189 test) are available for testing the trend. The correlation and regression analyses also can aid the study in confirming the time independence of the distribution. If the trend is positive, assumption of independently and identically distribution (i.i.d.) is said to be not contradicted. Hence, the data can be fitted in non-homogeneous Poisson process (NHPP) models such as power law process (PLP). On the contrary, if the data does not exhibit any trend (or very weak trend which is not considerable), then further analysis should be carried out by total time on test (TTT) plotting or other suitable reliability models.

Step 5: TTT Plots to Examine Exponential Fit

Total time on test (TTT) graphs can be plotted in several ways. The most popular and convenient methods are maximum likelihood estimation (MLE), piecewise exponential estimation (PEXE), and product limit estimation (PLE) or Kaplan-Meyer estimation (KME). With the help of TTT plots, it enables to know whether the data has any exponential fit. If there is exponential fit, we can model the data in homogeneous Poisson process (HPP), otherwise it can be fitted in renewable process (RP) such as Weibull model.

Step 6: Fitting the Suitable Model

Using relevant statistical tools, the data is fit in suitable model. Its characteristics can be hence studied to design the reliability of the system.

Step 7: Confirmation Tests and Goodness of Fit

Further, the fit of the model can be confirmed by various statistical methods (hypothesis tests). The general goodness of fit tests such as Chi-square tests are universally applicable while some specific tests are available to check particular distributions. The specific goodness of fit tests such as Bartlett's test (exponential distribution), Mann's test (for Weibull distribution), and Kolmogorov-Smirnov test (for normal/log-normal distribution) are the few found among most useful tools in this study. The goodness of test not only confirms the distribution but also can authenticate the methodology of reliability analysis.

METHODOLOGY OF RELIABILITY ANALYSIS

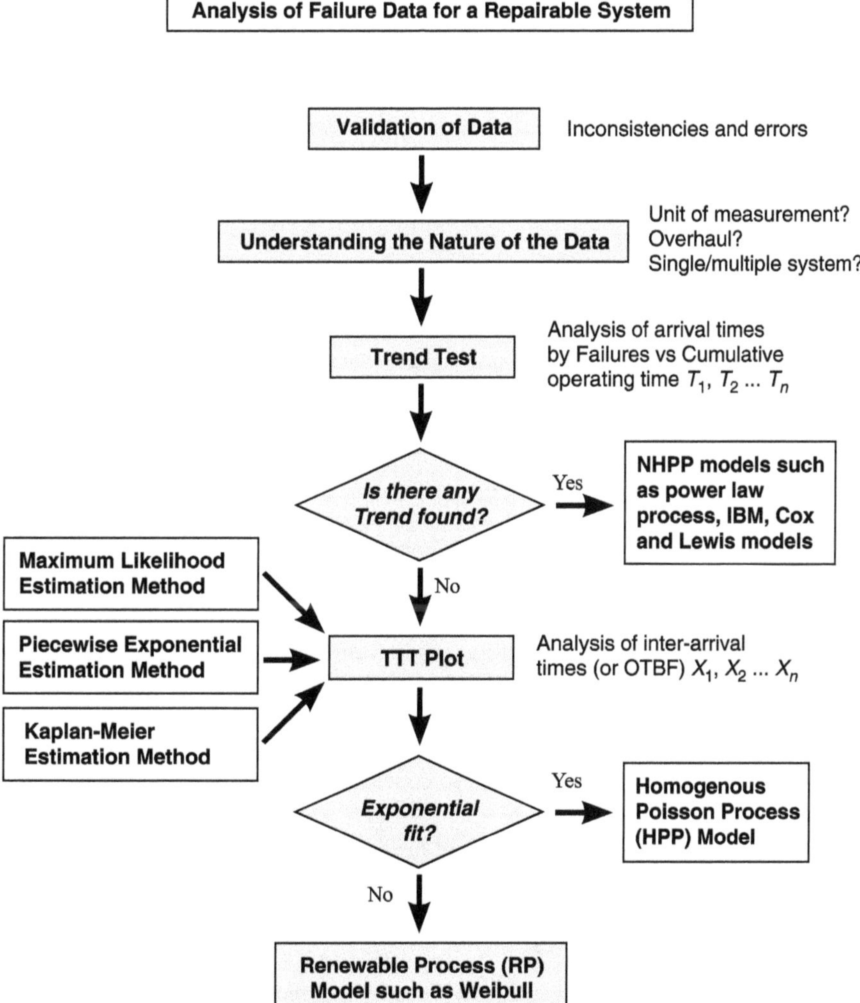

Figure 12.1 General procedure for analysis failure data of repairable system

Step 8: Reliability-centred Maintenance Planning and Scheduling

The reliability characteristics are then estimated. Various characteristics such as $T_{optimal}$, T_{median}, T_{mode}, B1 life, B.1 life, etc., can be estimated and hence maintenance planning and scheduling can be done.

Step 9: Reliability Growth Testing

Further the reliability growth can be tested using suitable growth models such as Duane's growth model, AMSAA models in test-fix-test-fix cycle.

This procedure is shown through the flow chart in Figure 12.1.

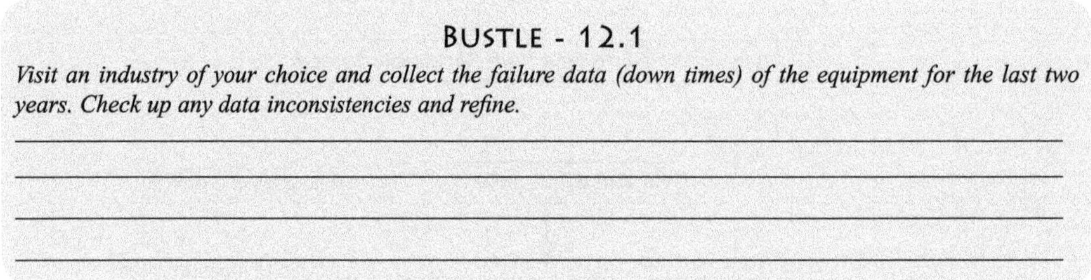

Test Your Understanding 12.1

1. What is the importance of data collection in reliability studies of reparable equipment? How can you make use of failure data of equipment for reliability studies?
2. Give the general procedure of reliability analysis of equipment failure data.
3. What do you understand by data inconsistencies? How can you eliminate them? Give some examples.

12.4 STATISTICAL PRELIMINARIES IN RELIABILITY EVALUATION

The probabilistic nature of system reliability arises from the fact that the performance parameter (which is a random variable) of the system follows a pattern, which can be expressed in probabilistic terms. The function, which describes this pattern or distribution of the performance parameter of a system, is known as the *distribution function*. If the particular set of performance (or failure) are well represented by a particular distribution it can be concluded that the failure characteristic follow that particular distribution.

A brief description of the basic fundamental theory is presented in the sections to follow.

12.5 PROBABILITY FUNCTIONS

The probability and availability studies involve use of mathematical and statistical models, which are probabilistic in nature. The probability functions important for development of reliability and availability theory are

- Probability Density Function (PDF).

- Cumulative Distribution Function (CDF).
- Reliability Function (R).
- Hazard Function (H).

12.5.1 Probability Density Function (PDF)

The probability density function (PDF) is the basis for predicting the behaviour of any probabilistic situation such as reliability or availability. Given the PDF, one can compute reliability function and the instantaneous density function f(t), usually referred to as failure density function.

12.5.2 Cumulative Distribution Function (CDF)

This function defines the probability that a random variable 't' lies between some lower limit (often zero) and upper limit 't'. For a continuous variable such as time to failure, the cumulative probability F(t) is given by the integration of the density function between the limits required.

$$F(t) = \int f(x)\, dx \qquad \text{... (1)}$$

In reliability studies, this function is also known as unreliability function. The cumulative distribution function increases from zero to unity as 't' increases from its smallest to largest value.

12.5.3 Reliability Function

The reliability function R(t), which is defined as the probability of a system or an item to perform or operate its required function without failure under given condition for an intended period of operation and it is mathematically given as

$$R(t) = 1 - F(t) = 1 - \int f(x)\, dx = \int R\, dx \qquad \text{... (2)}$$

12.5.4 Hazard Rate and Hazard Function

The hazard function (t) or instantaneous failure rate function is a conditional expression that an item in service for time "t" will be in the next instant of time "dt" given that it has not previously failed, survived up to time 't'.

i.e.,
$$P\,[\text{the unit will fail in } (t, t + dt)] \cong \lambda(t)\, dt \qquad \text{... (3)}$$

Where $\lambda(t)$ is the hazard rate at time 't' and 'P' is the probability of occurrence. The mathematical relationship between failure density function f(t) and hazard rate $\lambda(t)$ is

$$\lambda(t) = f(t)/[1 - F(t)] \qquad \text{... (4)}$$

The general equation showing the relationship between the reliability function R(t) and hazard rate

$$R(t) = \text{Exp}[- \int \lambda(x)\, dx] \qquad \text{... (5)}$$

When λ is independent of time as in the case of constant failure rate, expression reduces to

$$R(t) = e^{-\lambda(t)} \qquad \text{... (6)}$$

12.6 STATISTICAL DISTRIBUTIONS IN RELIABILITY STUDIES

Standard statistical distributions available in the literature are in use to model various reliability parameters. Though variety of distributions can be employed to model, it has been found that a relatively small number of these distributions satisfy the needs and relatively work most. Some of the most common distributions used in reliability-centred maintenance modelling are described below (also refer Table 12.1).

12.6.1 EXPONENTIAL DISTRIBUTION

The exponential distribution is the most widely used statistical distribution, which explains the general failure distribution of a system during its normal operating life period, when the failure occurs at random. Exponentially distributed items to failure are characterised by constant failure rate. The PDF for the exponential distribution is mathematically expressed as

$$F(t) = \lambda.\ e^{-\lambda t},\ t \geq 0 \qquad \ldots (7)$$

Where λ is the constant failure rate and 't' is the time.

The exponential cumulative distribution function is

$$F(t) = 1 - e^{-\lambda t} \qquad \ldots (8)$$

The exponential reliability function is

$$R(t) = e^{-\lambda t} \qquad \ldots (9)$$

12.6.2 Log-normal Distribution

The log-normal distribution can be a good fit to the distribution of component repair times and consequently is becoming an important distribution in the analysis of the repairable systems. The PDF for the distribution is

$$F(t) = 0.4343/t\sigma\sqrt{2\pi}\ {*}e^{-(x-\xi)/2.\sigma 2} \qquad \ldots (11)$$

Where σ = log standard deviation of t.

$X = \log_{10} t$; ξ = log mean value of x; $0.4343 = 1/\log_{10}$

Log-normal cumulative distribution function is

$$F(t) = \phi(\log_t - \xi/\sigma\ ) \qquad \ldots (12)$$

Where "ϕ" is the standard normal cumulative distribution function and can be given by

$$\phi(z) = \int (e^{-(u2)/2} / \sqrt{2\pi})\ du. \qquad -\alpha < Z < \propto$$

The log-normal reliability function is given by

$$R(t) = 1 - \phi\ (\log_t - \xi/\sigma); \qquad t > 0 \qquad \ldots (13)$$

12.6.3 Weibull Distribution (2-Parameter)

Weibull distribution fits in many cases where the hazard rate either increases or decreases, because this distribution has no characteristic shape. The probability density function for this distribution is expressed as

$$F(t) = (\ \beta/\alpha^{\beta}\ ).t^{(\beta-1)}.\exp[-(t/\alpha)^{\beta}] \qquad \ldots (14)$$

Where β = Shape parameter; α = Scale parameter; Assuming $\gamma = 0$ (location parameter).

The Weibull cumulative distribution function is

$$F(t) = 1 - \exp[-(t/\alpha)^{\beta}] \qquad \ldots (15)$$

$$T \geq 0 \text{ and } a, b \geq 0$$

Reliability function is given by

$$R(t) = 1 - F(t) = \exp[-(t/\alpha)^{\beta}\ ] \qquad \ldots (16)$$

Therefore, $1/[1 - F(t)] = \exp[(t/\alpha)^{\beta}\ ]$

Taking double logarithms

$$\text{Ln}\ [\ln\ \{1/1 - F(t)\}] = \beta \ln(t) - \beta \ln \alpha \quad \ldots (17)$$

This is a straight line of the form y = mx + c

Thus, the Weibull paper is constructed by having a log-reciprocal ordinate (y) scale representing cumulative probability of failure (or failure percentage) and the abscissa (x) scale representing the life value. The slope of the straight line plotted on this paper will be "β" the shape parameter.

The value of shape parameter defines the different phases of an item or system life characteristic as shown:

- $\beta < 1$ indicates decreasing failure rate or the phase of early failure.
- $\beta = 1$ indicates the constant failure rate or the normal useful operating life.
- $\beta > 1$ indicates the increasing failure rate or wear out life of component.

Many standard statistical distributions are used to model the reliability parameters. But it has been found that a relatively small number of these distributions satisfy the needs and work most relatively. Some of the distributions, which are most relevant, are reviewed in this chapter and their PDF, CDF, and RF are tabulated in Table 12.1.

Table 12.1 Functions of Probability Distributions

Distribution	*PDF*	*CDF*	*RF*
Exponential Distribution	$F(t) = \lambda.\ e^{-\lambda t}, t \geq 0$ Where λ is the cfr and 't' is time.	$F(t) = 1 - e^{-\lambda t}$	$R(t) = e^{-\lambda t}$
Normal Distribution	$\phi(z) = \int e^{-(u2/2)} / \sqrt{2\pi}\ du; \quad -\alpha < Z < \propto$		
Log-normal Distribution	$F(t) = 0.4343/t\sigma\sqrt{2\pi} * e^{-(x-\xi)/2.\sigma 2}$ Where σ = log std dev of t. $X = \log_{10} t$; ξ = log mean value of x; $0.4343 = 1/\log_{10}$	$F(t) = \phi(\log^t - \xi/\sigma)$	$R(t) = 1 - \phi(\log^t - \xi/\sigma)$; $t > 0$
Weibull Distribution (2-Parameter)	$F(t) = (\beta/\alpha).t^{(\beta-1)}.e^{-(t/\alpha)\beta}$ Where β = Shape parameter, α = Scale parameter, Assume γ = 0 (location parameter).	$F(t) = 1 - e^{-[(t)/\alpha]\beta}$ T ≥ 0 and a, b ≥ 0	$R(t) = 1 - F(t) = e^{-(t/\alpha)\beta}$

BUSTLE - 12.2

For the failure data (down times) of the equipment for the last two years you collected as asked under Bustle – 12.1, check if they fit in the exponential distribution. Fit a suitable straight line or curve.

____

____

____

____

Test Your Understanding 12.2

1. Explain the characteristics of exponential distribution with suitable examples.
2. Explain the characteristics of normal distribution. Give examples.
3. Write short notes on the following distributions with reference to reliability studies of equipment failures:
   (a) Exponential Distribution (b) Normal Distribution
   (c) Log-normal Distribution (d) Weibull Distribution
4. Define and explain the following terms with reference to reliability studies of equipment failures:
   (a) Probability Density Function (PDF). (b) Cumulative Distribution Function (CDF).
   (c) Reliability Function (R). (d) Hazard Function (H).

12.7 TREND TEST

These are discussed at length in Chapter 11. Any test discussed in Chapter 11 can be used to check the presence of trend. By checking the trend, it will enable us to understand if the machine is improving or deteriorating, i.e., DFR or IFR.

The TBFs of the equipment in chronological order can be analysed for the presence of trend to know whether the equipment or the item is deteriorating or improving with age. By simply plotting cumulative TBFs against the cumulative number of failures, it can be tested whether the machine or item under consideration is improving or deteriorating. If a curve concave upward is obtained, this means that the TBFs are becoming shorter and shorter that is to say, the machine is deteriorating. On the contrary, if a curve concave downwards is obtained, this means that the machine is improving.

If the data exhibits linearity, the plot is said to have no trend and the failure data is known as independently and identically distributed (i.i.d.) and can be further analysed by statistical distributions. Various trend tests have been illustrated in Chapter 11.

Eye Ball Analysis

In this we pass our eye through chronological TBFs and search for increase or decrease of the failure rate, i.e., if the TBFs are showing decrease in magnitude towards the end, it shows decreasing failure rate. If the magnitude of TBFs is approximately constant throughout the period, then it shows constant rate of failure. If there is increasing number of failures for each period, then it indicates increasing failure rate and so on.

12.8 GRAPHICAL EVALUATION OF NON-HOMOGENEOUS POISSON PROCESS (NHPP) PARAMETERS

In the NHPP power law process (PLP), the scale and shape parameters can be estimated by plotting logarithm of cumulative number of failures against logarithm of running time on a simple square graph paper or simply plotting the cumulative number of failures against the running time on a logarithm paper (log-log graph). The slope of the best-fitted line gives the value of the "β" and once "β" is known, the value of "α" can be estimated. The approach is based on PLP model where failure intensity (t) is as given in equation (18).

The mean value of the intensity function is

$$E[N(t)] = (t/\alpha)^{\beta} \qquad \text{... (18)}$$

Where N(t) is cumulative number of failures at time 't' (instantaneous). {Now, generalising by introducing n(t) as cumulative number of failure}, E[n(t)] is estimated by the observed number of failures at time 't' and thus, we have

$$\text{Ln}\,[n(t)] = \beta \ln t - \beta \ln \alpha \qquad \text{... (19)}$$

Above equation is of the form Y = m x + c where Slope = m = "β"

An estimate of "α" can be obtained by using the fact that if 't' takes a value such that ln [n(t)] is equal to zero (i.e., ln[N(t)] = 0) equation (19) takes the form

$$\beta \ln t_o - \beta \ln \alpha = 0$$

$$\beta \ln t_o = \beta \ln \alpha$$

and thus t_o is an estimate of "α".

12.9 METHOD OF TOTAL TIME ON TEST (TTT) PLOTTING

The technique of TTT plotting originally suggested by Barlow and Campo (1975) is very simple to use for machine failure data analysis. Like other probability methods, this method is also based on assumption that the time between failures (TBF) are independently and identically distributed (i.i.d.) so that the actual chronological ordering of the TBFs can be ignored. Thus, it is not useful in analysing a data set which *have structures and are positive to serial correlation tests.*

The theory of TTT plotting:

Suppose that $0 = t_0 < t_1 < t_2$--------$< t_n$ is ordered sample of times to failure of a unit, then the TTT is given by

$$S_j = \sum_{i=1}^{j}(n-i+1)(t_i - t_{i-1}) \qquad \text{... (20)}$$

Thus,

$$S_0 = 0;\ S_1 = nt_1;\ S_2 = n.t_1 + (n-1)(t_2 - t_1) = S_1 + (n-1)(t_2 - t_1)$$

$$S_3 = S_2 + (n-2)(t_3 - t_2) \text{ and } S_i = n.t_1 + (n-1)(t_2 - t_1)$$

$$+\text{----------}(n - i + 1)(t_i - t_{i-1})$$

Denote the sum or total time on test (TTT) at T_i, i = 1, 2, 3, 4…n

The TTT plot is obtained by plotting (i/n, u_i) for i = 0, 1, 2, 3…n, where $u_i = S_i/S_n$ on a unit square and connecting the plotted points by the line segments. If the plot is concave downwards, the machine is deteriorating and if the plot is concave upwards then the machine is improving (note that it is opposite characteristic in trend plot), and if the plot wriggles around the diagonal many times, the machine is experiencing constant failure rate.

The Maximum Likelihood Estimation (MLE)

This is the most popularly used term in reliability studies in which the probability of the sample is obtained by multiplying the density function evaluated at each data point. The product containing the data points and the unknown parameter is called the *likelihood function.* By finding the value of parameter that maximises this expression, the set of data observed "more likely" is made. In other words, we choose parameter values that are more consistent with our data by maximising the likelihood of the sample. Thus, the MLE method is equivalent to maximising the equation of several variables. If the assumption of independently and identically distributed (i.i.d.) is not contradicted for time between failures (TBFs), the MLE method provides fairly good estimate of the parameter of the theoretical distribution. This method requires computer programs to do the calculations efficiently and rapidly. Further, there are some more methods available for plotting TTT transforms such as Kaplan-Meier (KME) or product limit estimates, piecewise exponential estimates (PEXE).

12.10 TTT TRANSFORMS

The concept of total time on test (TTT) plots has proved to be a very useful tool in many reliability applications such as model identification, aging properties, age replacement policies, etc. Plotting data is the first step to understand the information contained in the data. In case of failure data of repairable equipment, these plots indicate increasing or decreasing or constant failure rates. The health monitoring of any equipment can be found by using these plots, whether the failure data satisfies i.i.d assumption or not. If the data confirms i.i.d assumption and increasing failure rate, then optimal age replacement interval is obtained by plotting the scaled TTT with cumulative distribution function (CDF). If the data shows any trend in the form of deterioration, optimum replacement interval is obtained from the TTT plot by considering the age of the equipment as discussed by Barlow. The ratio of cost of failure maintenance to cost of repair maintenance (cost ratio) is the deciding factor.

TTT plots can be constructed using different methods like Kaplan-Meier estimate, rank adjustment method, Nelson, piecewise exponential estimate, and maximum likelihood estimate methods. While the first four methods are non-parametric, the last method is parametric. It is observed that whether the data is censored or not and follows i.i.d assumption, TTT plots are more suitable when maintenance policies are evaluated. Sometimes, it is necessary to clean the data when it contains inappropriate and misleading events before synthesising it into a model to be used for future decisions and policy.

The applications of TTT transforms were advocated by Kelfsjo (1982, 1986) Kelfsjo and Westberg (1990). From the shape of the plot, one can easily know about failure rate of the equipment. If the plot shows convex upwards, then the machine is deteriorating (increase in failure rate) and if it is concave downwards, the machine is improving (decrease in failure rate) with time. If the plot crosses the diagonal several times, then the machine experiences a constant failure rate. TTT plots can be drawn using several methods namely maximum likelihood estimates (MLE), piecewise exponential estimates (PEXE), and Kaplan-Meier estimates (KME). The expressions are as follows.

Consider an equipment with a CDF of F(t). Let $0 = t_0 \leq t_1 \leq t_2 \leq \ldots\ldots\ldots\ldots t_n$ are ordered sample of times of failure events of size n from a cumulative life distribution F(t) with F(0) = 0 and survival function R(t) = 1 – F(t) then finite mean is given by,

The ith TTT transform is defined as,

$$TTT_i = \int_0^{ti} R(t_i)dt \qquad \mu_0 = \int_0^{\infty} R(t)dt \qquad \ldots (21)$$

with minimum value of $\mu_0(0) = 0$ and maximum value $\mu_0(\infty) = \mu$. The scaled TTT transform is defined as,

$$\varphi(t_i) = \frac{\int_0^{ti} R(t_i)dt}{\mu_0} \qquad \ldots (22)$$

The curve between φ(t) and F(t) from (0,0) to (1,1) is the TTT transform and it is the graphical representation of the analytical solution for parametric models. This transform is independent of scale.

Thus, total time on test S_i can be defined as,

$$S_i = n\int_0^{ti} R(t_i)dt \qquad \text{... (23)}$$

Where $S_o = 0$, R (0) = 1 and i = 1,2,3…n expression for S_n is given by,

$$S_n = n\int_0^{tn} R(t_i)dt \qquad \text{... (24)}$$

The scaled TTT, φ_i is equal to S_i/S_n [refer equation (18)] and CDF, F_i is equal to $1 - R_i$. TTT plot is obtained by plotting φ_i with F_i.

12.11 TTT PLOTS

It is a plot of line segments joined between two successive discrete points. These points of scaled TTT and unreliability (CDF) are obtained from the failure data. It gives a picture of failure data in a plot of unit square. If the plot is concave above the diagonal line, increasing the failure rate (convex below the diagonal line decreasing failure rate) can be inferred. When curve embraces diagonal line a constant failure rate can be assumed. Let t_1, t_2, …t_n be the n failure times of an equipment with suspensions between two few successive failure times. If all the units are put into test at the same time, then the total test time TTT_i for all units at time t_i is defined as,

$$TTT_i = T_i = \sum_{j=1}^{i}[n-(j-1)]\,(t_j - t_{j-1}) \qquad \text{... (25)}$$

which is equivalent to the cumulative time up to the end of t_i for, i = 1,2,…n where $t_0 = 0$. This is shown pictorially in Figure 12.2. The plot of CDF (which is equal to i/n) versus normalised TTT value (TTT_i/TTT_n) is the TTT plot.

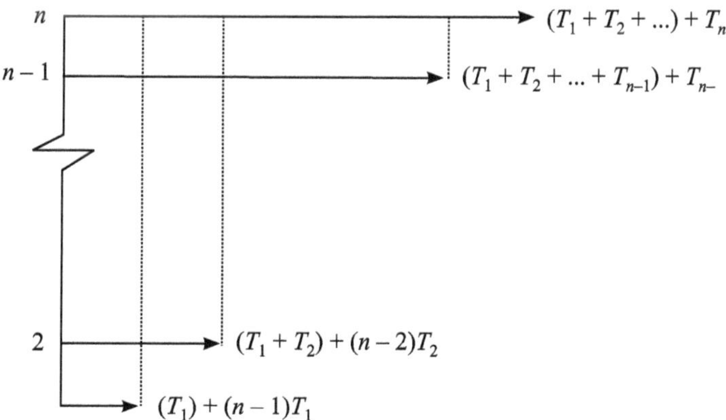

Figure 12.2 Total time on test calculation

When the sample size 'n' increases to infinity, the TTT plot converges to a life distribution F(t) from which the sample has come and it is the TTT transform. If t_n is not the last observed failure, then a pseudo time or last censored time after last failure, t_{n+1} is defined as,

$$TTT_{n+1} = TTT_n + \int_n^{\infty} R(t)dt \qquad \text{... (26)}$$

and the scaled TTT is, $\varphi_i = TTT_i/TTT_{n+1}$

The following sections show different methods of constructing TTT plots of censored or uncensored data.

12.11.1 Product Limit Estimator (PLE) or Kaplan-Meyer Estimation (KME) Method

Many a times, it is not possible to observe failure times on test. The times for those items, failures are not observed, are said to be censored, but the amount of time it is observed functioning is noted. In this method, the reliability $R(t_i)$ at failure time t_i is given by,

$$R(t_i) = \prod_1^i (1 - d_i / n_i) \qquad \text{... (27)}$$

where, n_i is the number at risk at time t_i and d_i is the number of ties. It is common to regard t_n as an observed failure time even if that is not the actual situation. The total time on test is given by equation (28). The scaled TTT is evaluated and plot is drawn for observed failure times only.

Thus, the survival function is $R(t_i) = (n - 1)/n$ and TTT is referred to as

$$S_i = \sum_{j=1}^{i} (n - j + 1)(tj - tj - 1) \qquad \text{... (28)}$$

Thus, the scaled TTT, φ_i is equal to S_i/S_n and F_i is equal to i/n.

12.11.2 Piecewise Exponential Estimation (PEXE) Method

The distribution function called *piecewise exponential estimator* was proposed by Kitchin (1980). In a data of size n, if $t_1, t_2, \ldots t_i$ are observed failure times and $t_{i1}, t_{i2}, \ldots t_{im}$ are suspended data between two successive failure times, then the failure rate estimation is given by,

$$h_i = \frac{1}{\sum_{j=1}^{m} (t_{ij} - t_i - 1) + \left(n + 1 - i - \sum_{j=1}^{m} m_j \right) \cdot (t_i - t_i - 1)} \qquad \text{... (29)}$$

The survival function is given by,

$$R(t) = exp(-h_1 t) \qquad 0 \le t \le t_1$$

$$\exp\left\{ -\left[\sum_{j=1}^{i} h_i \cdot (t_i - t_i - 1) + h_{i+1} \cdot (t - t_i) \right] \right\} \qquad t_i < t \le t_{i+1} \qquad \text{... (30)}$$

In this method, constant failure rate is fitted to each interval. The exponential survival functions are pieced together to form piecewise exponential estimator. This is a continuous function. In order to construct the TTT plot from PEXE an estimate of survival function after time t_n must be made. If there are suspensions after last failure, then final TTT can be obtained using equation (26). If there is evidence that failure distribution is increasing failure rate (IFR) class, then Weibull distribution can be assumed and the final term of TTT_{n+1} can be evaluated using the expression.

$$TTT_{n+1} = TTT_n + \int_{in}^{\infty} \exp[-(t/\alpha)^{\beta} \cdot dt] \qquad \text{... (31)}$$

where α (scale) and β (shape) are parameters of Weibull distribution. The method of evaluation of second term in equation (31) is described in the next section. [The PEXE can be seen as a generalisation of the Nelson estimator in its original form, i.e., without any suspensions, the PEXE is equal to the Nelson estimator].

Precisely, survival function is

$$R_i = \exp\left[-\sum_{j=0}^{i} 1/(n-j)\right] \qquad \text{... (32)}$$

and S_i is

$$S_i = n\sum_{k=1}^{i}(n-k+1)(t_k - t_{k-1})(R_{k-1} - R_k) \qquad \text{... (33)}$$

The scaled TTT, φ_i is equal to S_i/S_{n+1}, where S_{n+1} is given by

$$S_{n+1} = S_n + (\alpha/\beta)\cdot(t_n/\alpha)\cdot \text{Exp}\,[-(t_n/\alpha)^{\beta}]$$

And α and β are ML estimates of Weibull distribution.

12.11.3 Maximum Likelihood Estimation (MLE) METHOD

If the failure data is assumed to follow Weibull distribution and the ML estimates are known, then the failure rate is given by,

$$h(t) = (\beta/\alpha) \cdot (t/\alpha)^{\beta-1} \qquad \text{... (34)}$$

and the CDF is given by,

$$F(t) = 1 - \exp(t/\alpha)^{\beta} \qquad \text{... (35)}$$

The TTT statistic can be evaluated as,

$$TTT_i = \int_0^{ti} \exp[-(t/\alpha^{\beta})] \cdot dt \qquad \text{... (36)}$$

and

$$\int_{tn}^{\infty} R(t)\cdot dt = \int_{tn}^{\infty} \exp[-(t/\alpha)^{\beta}]\cdot dt \qquad \text{... (37)}$$

Transforming, $y = e(-t/\alpha)^{\beta}$, the equation becomes,

$$\int_{tn}^{\infty} R(t)dt = \alpha\Gamma\left(1+\frac{1}{\beta}\right)\left\{1-\frac{\Gamma\left[\frac{1}{\beta},\left(\frac{t}{\alpha}\right)^{\beta}\right]}{\Gamma\left(\frac{1}{\beta}\right)}\right\} \qquad \text{... (38)}$$

The above term is to be added to get TTT_{n+1} [Equation (26)]. This equation is an incomplete gamma function and can be evaluated using library function available in MATLAB or Microsoft Excel.

Summarily, the survival function is $R(t_i) = \exp[-(t_i/\alpha)^{\beta}]$

All other functions and parameters are same as defined in the PEXE method.

Consider Illustration 1 given in Chapter 11. Let us now extend the analysis using TTT plotting using maximum likelihood estimation (MLE).

ILLUSTRATION 12.1:

The failure data in terms of time between failures (TBF) in hours, is obtained for the equipment in a process industry is as follows:

14, 12, 9, 13, 7, 24, 11, 21, 6, 15, 19, 31, 36, 27, 57, 42, 69, 48, 81, 98

Conduct the trend analysis and identify the appropriate failure distribution model using TTT plots.

Solution. The required calculations are shown in the following table. (TBF: time between failures; CTBF: cumulative time between failures; OTBF: ordered time between failures)

Table 12.2 TTT transforms of failures by maximum likelihood estimation

S.no	*TBF (i)*	*CTBF*	*OTBF*		*s(i)*	*i/n*	*phi*	*SLOPE*
0	0	0	0		0	0	0	0
1	14	14	6	120	120	0.05	0.18779343	0.34144259
2	12	26	7	19	139	0.1	0.21752739	0.36254564
3	9	35	9	36	175	0.15	0.27386541	0.42133141
4	13	48	11	34	209	0.2	0.32707355	0.46724793
5	7	55	12	16	225	0.25	0.35211268	0.46948357
6	24	79	13	15	240	0.3	0.37558685	0.46948357
7	11	90	14	14	254	0.35	0.39749609	0.46764246
8	21	111	15	13	267	0.4	0.41784038	0.46426708
9	6	117	18	36	303	0.45	0.4741784	0.49913516
10	15	132	21	33	336	0.5	0.5258216	0.5258216
11	19	151	24	30	366	0.55	0.57276995	0.54549519
12	31	182	27	27	393	0.6	0.61502347	0.55911225
13	36	218	31	32	425	0.65	0.66510172	0.57834932
14	27	245	36	35	460	0.7	0.7198748	0.59989567
15	57	302	42	36	496	0.75	0.77621283	0.62097027
16	42	344	48	30	526	0.8	0.82316119	0.63320091
17	69	413	57	36	562	0.85	0.87949922	0.6514809
18	48	461	69	36	598	0.9	0.93583725	0.66845518
19	81	542	81	24	622	0.95	0.97339593	0.67130754
20	98	640	98	17	639	1	1	0.66666667
						0	0	
0		0		0		1	1	0.67130754
20		640		0				0.34144259
						-0.5	0	
						0.95	0.97339593	

Table 12.3 Karl Pearson's Coefficient of Correlation

S.no	*TBF (i)*	*x = i-32*	*(i–1)*	*y = (i –1) – 27.1*	*x**2*	*y**2*	*x*y*
0	0						
1	14	–18	0	–27.1	324	734.41	487.8
2	12	–20	14	–13.1	400	171.61	262
3	9	–23	12	–15.1	529	228.01	347.3
4	13	–19	9	–18.1	361	327.61	343.9
5	7	–25	13	–14.1	625	198.81	352.5
6	24	–8	7	–20.1	64	404.01	160.8
7	11	–21	24	–3.1	441	9.61	65.1
8	21	–11	11	–16.1	121	259.21	177.1
9	6	–26	21	–6.1	676	37.21	158.6
10	15	–17	6	–21.1	289	445.21	358.7
11	19	–13	15	–12.1	169	146.41	157.3
12	31	–1	19	–8.1	1	65.61	8.1
13	36	4	31	3.9	16	15.21	15.6
14	27	–5	36	8.9	25	79.21	–44.5
15	57	25	27	–0.1	625	0.01	–2.5
16	42	10	57	29.9	100	894.01	299
17	69	37	42	14.9	1369	222.01	551.3
18	48	16	69	41.9	256	1755.61	670.4
19	81	49	48	20.9	2401	436.81	1024.1
20	98	66	81	53.9	4356	2905.21	3557.4
	32		27.1		13148	9335.8	8950
Coefficient of Correlation = Sum x*y/sqrt((sum x**2)*(sum y**2)) r = 0.80782524							

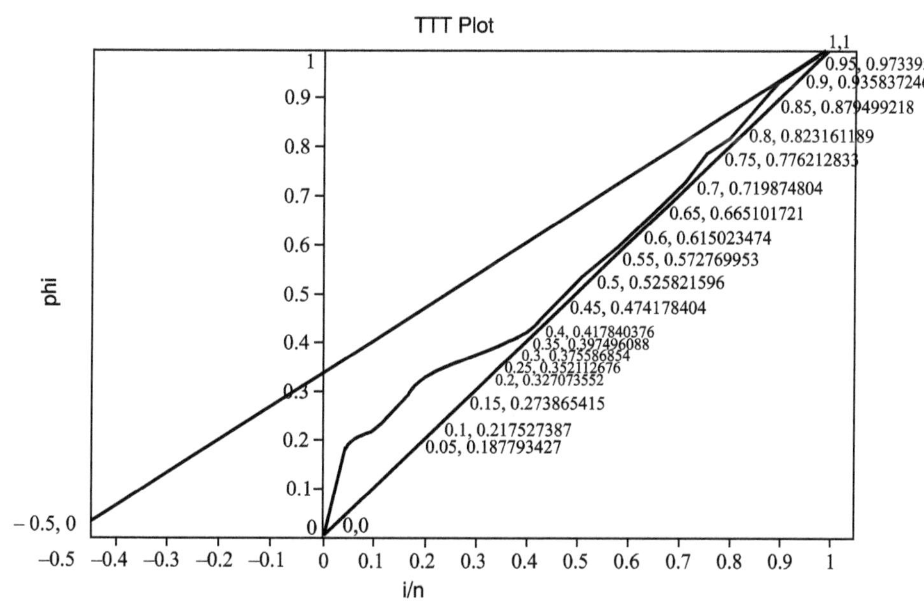

Conclusion: From trend analysis the presence of (positive) trend is evident. Further, from the above TTT plot, it can be concluded that the failure rate is increasing in the equipment. Also it can be predicted or expected that the data is fit in NHPP models such as power law process (PLP).

BUSTLE - 12.3

Choose an equipment of your choice and collect suitable data. Conduct the trend analysis and construct a TTT plot for the data.

____

____

____

____

Test Your Understanding 12.3

1. Explain the importance of trend analysis in repairable equipment reliability studies. Brief out any two methods of trend analysis.
2. What is TTT plotting? Discuss any two methods of TTT plotting.
3. The time between failures for a machine are noticed as follows:

   12, 10, 7, 11, 5, 22, 9, 19, 4, 13, 17, 29, 34, 25, 55, 40, 67, 46, 79, 96

   Draw TTT plot and comment.

   Answer: The failure rate is increasing (IFR) and may fit in NHPP/PLP.

12.12 FAILURE DISTRIBUTIONS AND GOODNESS OF FIT

The methods for deriving reliability distributions directly from failure data through graphical models such as TTT plots have already been established. Now, let us identify an appropriate theoretical distribution, perform a goodness of fit test and estimate the parameter(s) through analytical models. In general, fitting a theoretical distribution is preferred to empirically developing a model. The reasons for such preference on fitting distribution and testing the goodness of fit can be listed as given below.

- Empirical models do not provide information beyond the range of the sample data. In reliability engineering the tails of the distribution are of most interest to extrapolate, to find trend, etc.
- A sample is only a small random subset of the population of failure times, and it is the distribution from which the sample came, but not the sample itself that is to be established. Thus, it is of interest in determining the probabilistic nature of the underlying failure process.
- Generally, the failure process is a result of some physical phenomena that can be associated with a particular distribution. For example, the central limit theorem provides justification for using the normal distribution when additive effects are present or the log–normal

distribution when multiplicative effects are present. Chance or random process may generate an exponential time to failure.

- Small sample sizes provide very little information concerning the failure process. However, if the sample is consistent with a theoretical distribution, then much stronger results based on the properties of the theoretical distribution are possible.
- The theoretical reliability model can help in performing more complex analysis of the failure process. For example, the effect of changes on the MTBF can be easily accomplished.

12.13 IDENTIFYING CANDIDATE DISTRIBUTIONS

Identifying candidate distributions is both an art and a science. An understanding of the failure process, knowledge of the characteristics of the theoretical distributions, and a statistical analysis of the data will assist in selecting a failure or repair distribution. One suggested approach is given below:

- Construct a histogram of the failure or repair times.
- Compute descriptive statistics.
- Analyse the empirical failure rate.
- Use prior knowledge of the failure process.
- Use properties of the theoretical distribution.
- Construct a probability plot.
- Examine the goodness of fit.

12.14 GOODNESS OF FIT TESTS

Except for the final step in the above approach, all other steps are familiar to us. Hence, the final step in the selection of a theoretical distribution, i.e., to perform a statistical test for goodness of fit is explained here. Such a test compares a null hypothesis (H_0) with an alternative hypothesis (H_1 or H_a) having the following form:

H_0: The failure times came from the specified distribution.

H_1 or H_a: The failure times did not come from the specified distribution.

Table 12.4 Decision table for hypothesis tests

Occurrence	*Accept H_o*	*Accept H_a*
H_o true	Correct decision	Type I error (α)
H_a true	Type II error (β)	Correct decision

The test consists of computing a statistic over a sample of failure times. This statistic is then compared with a critical value obtained from relevant statistical tables. Generally, if the test statistic is less than the critical value (table value), the null hypothesis H_0 is accepted; otherwise, the alternative hypothesis H_1 is accepted.

The critical value depends on the level of significance of the test and the sample size. The level of significance is the probability of erroneously rejecting the null hypothesis in favour of the alternative hypothesis. Table 12.4 shows the four possible cases that may occur. Because of the randomness inherent in the sampling process, the test statistic has a probability of exceeding the critical value even though H_0 is true. This results in a Type I error having a probability of occurring equal to the level of significance (α). It is also possible for the test statistic to be less than the critical value even though H_1 is true. This results in a Type II error. It occurs with a probability that is usually controlled indirectly by the specification of the level of significance and the sample size.

There are two types of goodness of fit tests available:

(a) General tests

(b) Specific tests

A general test is applicable to fitting more than one theoretical distribution and a specific test is tailored to a single distribution. If available, specific tests will be more powerful (have a higher probability of correctly accepting/rejecting a distribution) than general tests. What follows is a discussion of a general test, the chi-square goodness of fit test, and specific tests for the exponential, Weibull, normal and log-normal.

12.14.1 Chi-square Goodness of Fit Test

This test is applicable to both continuous and discrete distributions and can be used when the parameters of the distribution are estimated from the maximum likelihood estimators (MLEs). The test is valid for large sample sizes only. The data must be grouped into classes.

The test statistic is

$$\text{Chi-square } (\chi^2) = \sum_{i=1}^{k} (O_i - E_i)^2 / E_i \qquad \dots (39)$$

k = Number of classes

O_i = Observed number of failures (or repairs) in the ith class

$E_i = n\,p_i$ = Expected number of failures (or repairs) in the ith class

n = Total number at risk (sample size)

p_i = Probability of a failure occurring in the ith class if H_0 is true

The probabilities are based on the distribution stated in the null hypothesis. To ensure an approximate chi-square distribution, nP_i should be at least 5 for all i or use Sturge's rule. The test statistic, χ^2 has a chi-square distribution with degrees of freedom equal to k-1 number of estimated parameters. Critical values are then obtained from statistical tables. On comparison, if calculated value is less than critical value, then null hypothesis is accepted, otherwise rejected.

12.14.2 Bartlett's Test for Exponential Distribution

A specific test for fitting exponential distribution is Bartlett's test. The hypotheses are

H_o: Failure times are exponential. *H_1: Failure times are not exponential.*

The test statistic is

$$B = 2r\left[\ln\left((l/r)\sum_{i=1}^{r} t_i\right) - (l/r)\sum_{i=1}^{r}\ln t_i\right] \Big/ [l + (r+l)/(6r)] \qquad \dots (40)$$

where t_i = time of failure of the ith unit

r = number of failures

The test statistic, B under the null hypothesis, has a chi-square distribution with r – 1 degrees of freedom. In this test,

$$\Pr\{\chi^2 < \chi^2_{1-\alpha/2,r-1}\} = \Pr\{\chi^2 > \chi^2_{\alpha/2,r-1}\} = \alpha/2 \quad ...(41)$$

If $\chi^2_{1-\alpha/2,r-1} < B < \chi^2_{\alpha/2,r-1}$ then the null hypothesis is accepted; otherwise the alternative hypothesis accepted. Observe that the failure rate of the exponential need not be estimated to apply this test. A sample size of 20 or more failures is recommended in using this test in order to achieve an adequate power for the test to reject the exponential when appropriate. However, if chi-square test rejects, the Bartlett's test need not be performed.

ILLUSTRATION 12.2:

It is observed till 20 failures occur in the thirty machines placed on test. The following are the failure times. Test whether the data can be fit in exponential distribution. (Given that for $\alpha = 0.05$, $r = 20$; the critical values $\chi^2_{1-\alpha/2,r-1} = 10.1$; $\chi^2_{\alpha/2,r-1} = 30.1$)

10.7	88.6	84.6	11.5	1.8	2.5	4.6	14.2	87.7	30.4
32.0	6.2	84.9	42.6	99.1	36.3	96.5	31.5	20.9	50.1

Solution.

Hypothesis: H_0: *Failure times are exponential.* H_1: *Failure times are not exponential*

Level of Significance: $\alpha = 0.05$,

and number of failure times (r) = 20

Estimator: The Bartlett's test is suitable to apply

The test statistic is given by

$$B = 2r\left[\ln\left((l/r)\sum_{i=1}^{r} t_i\right) - (l/r)\sum_{i=1}^{r}\ln t_i\right] \Big/ [l + (r+l)/(6r)]$$

where t_i = time of failure of the ith unit and r = number of failures.

Here, we can observe that $\sum_{i=1}^{r} t_i = 836.3$ and $\sum_{i=1}^{r} \ln t_i = 63.9385$

Then,

$$B = 2(20)[\ln(836.3/20) - 63.9385/20]/\{1 + (20+1)/[6(20)]\}$$

$$= 18.258$$

$$\chi^2_{1-\alpha/2,r-1} = 10.1 < B = 18.258 < \chi^2_{\alpha/2,r-1} = 30.1$$

Hence, the null hypothesis is accepted.

The failure data can be assumed to exhibit exponential distribution.

12.14.3 Mann's Test for Weibull Distribution

A specific test for the Weibull (2-parameter) failure distribution is a test developed by Mann, Schafer, and Singapurwalla [1974]. The hypotheses are

H_0: *The failure times are Weibull.* H_1: *The failure times are not Weibull.*

The test statistic is

$$M = k_1 \sum_{i=k_1+1}^{r-1} [(\ln t_{i+1} - \ln t_i)/M_i]/k_2 \sum_{i=1}^{k_1} [(\ln t_{i+1} - \ln t_i)/M_i] \quad \text{... (42)}$$

where $k_1 = [r/2]$ $k_2 = [(r - 1)/2]$

$M_i = z_{i+1} - z_i$ $Z_i = \ln [-\ln (1 - i - 0.5/n + 0.25)]$

and [x] is the integer portion of the number x, is an approximation. If $M < H$, H_0 is accepted and $M > F$ then H_1 is accepted. Values for F may be obtained from tables of the F-distribution, $2k_2$ degrees of freedom for the numerator and $2k_1$ degrees of freedom for the denominator. This test is for the two-parameter Weibull distribution. So if the alternative hypothesis is accepted, the three-parameter Weibull as well as other distribution should be considered. The data must be rank-ordered for the test statistic to be computed.

ILLUSTRATION 12.3:

Among the 50 placed for testing, the following 35 failure times were observed and the testing terminated at 35th failures. The data is believed to follow Weibull distribution. Test the claim at 0.05 LOS.

132.1	19.4	29.7	36.8	51.4	65.8	111.4
7.3	19.7	61.4	60	37	119.4	118.2
31.9	22.3	7.8	41.7	61.3	78.4	72.6
13.3	100.4	22.8	32.2	46.7	30.2	1.3
139.7	110.6	65.6	50.4	33	26.7	13.9

Solution.

H_0: *The failure times are Weibull.* H_1: *The failure times are not Weibull.*

Level of Significance (LOS) = 0.05

S.no.	*Ordered Failure Times* (t_i)	*Ln* t_i	M_i	$Lnt_{i-1} - Ln\ t_i$	$(Lnt_{i-1} - Lnt_i)/M_i$
1.	1.3	0.2624	1.1087	1.7255	1.5563
2.	7.3	1.9879	0.5211	0.0662	0.1271
3.	7.8	2.0541	0.3469	0.5337	1.5384
4.	13.3	2.5878	0.262	0.0441	0.1683
5.	13.9	2.6319	0.2115	0.3334	1.5762
6.	19.4	2.9653	0.1781	0.0153	0.0860
7.	19.7	2.9806	0.1544	0.124	0.803
8.	22.3	3.1046	0.1367	0.0222	0.1621
9.	22.8	3.1268	0.1229	0.1579	1.2845
10.	26.7	3.2847	0.112	0.1064	0.9504
11.	29.7	3.3911	0.1032	0.0167	0.1622
12.	30.2	3.4078	0.0959	0.0548	0.5715
13.	31.9	3.4626	0.0897	0.0094	0.1044

14.	32.2	3.472	0.0845	0.0245	0.29
15.	33	3.4965	0.08	0.1090	1.3625
16.	36.8	3.6055	0.0762	0.0054	0.0711
17.	37	3.6109	0.0728	0.1196	1.6429
18.	41.7	3.7305	0.0699	0.1132	1.6201
19.	46.7	3.8437	0.0674	0.0763	1.1319
20.	50.4	3.92	0.0651	0.0196	0.3017
21.	51.4	3.9396	0.0632	0.1547	2.4485
22.	60	4.0943	0.0615	0.0215	0.3493
23.	61.3	4.1158	0.06	0.0016	0.0268
24.	61.4	4.1174	0.0587	0.0662	1.1274
25.	65.6	4.1836	0.0576	0.003	0.0524
26.	65.8	4.1866	0.0566	0.0984	1.738
27.	72.6	4.2850	0.0558	0.0768	1.3768
28.	78.4	4.3618	0.0552	0.2474	4.4812
29.	100.4	4.6092	0.0547	0.0967	1.7632
30.	110.6	4.7059	0.0543	0.0072	0.1361
31.	111.4	4.7131	0.0541	0.0593	1.0987
32.	118.2	4.7724	0.054	0.0101	0.1863
33.	119.4	4.7825	0.0542	0.1011	1.8646
34.	132.1	4.8836	0.0545	0.0559	1.0256
35.	139.7	4.9395			

As per the given data $n = 50; r = 35; k_1 = k_2 = 17$

$$M = k_1 \sum_{i=k_1+1}^{r-1} [(\ln t_{i+1} - \ln t_i)/M_i] / k_2 \sum_{i=1}^{k_1} [(\ln t_{i+1} - \ln t_i)/M_i]$$

where $k_1 = [r/2]$ $\quad k_2 = [(r-1)/2]$

$M_i = z_{i+1} - z_i$ $\quad Z_i = \ln[-\ln(1 - i - 0.5/n + 0.25)]$

$$\text{Numerator} = k_1 \sum_{i=k_1+1}^{r-1} [(\ln t_{i+1} - \ln t_i)/M_i] = 352.368$$

$$\text{Denominator} = k_2 \sum_{i=1}^{k_1} [(\ln t_{i+1} - \ln t_i)/M_i] = 211.725$$

Therefore, $\quad M = 1.664$ with 34 dof

Since $\quad M = 1.664 < F_{crit\ 0.05,34,34}$, null hypothesis ($H_0$) is accepted.

12.14.4 Kolmogorov-Smirnov (K-S) Test for Normal and Log-normal Distributions

A goodness-of-fit test for use with the normal distribution when the parameters are estimated is a version of the Kolmogorov-Smirnov test developed by H.W. Lilliefors [1967]. It compares the empirical cumulative distribution function with the normal cumulative distribution function. The hypotheses are

H_0: The failure times are normal. *H_1: The failure times are not normal.*

The test statistic is

$$D_n = \max\{D_1, D_2\}$$

where

$$D_1 = \max \sum_{1 \le i \le n}^{n} \{\Phi(t_i - t)/s - i - 1/n\} \qquad \text{... (43)}$$

$$D_2 = \max \sum_{1 \le i \le n}^{n} \{i/n - \Phi(t_i - t)/s\} \qquad \text{... (44)}$$

$$\bar{t} = \sum_{i=1}^{n} t_{i/n} \qquad s^2 = \sum_{i=1}^{n} (t_i - \bar{t})^2/n \qquad \text{... (45a \& 45b)}$$

if $D_n < D_{crit}$, then accept H_0; if $D_n \ge D_{crit}$, then accept H_1. The values for D_{crit} may be found in statistical tables. This test is appropriate for complete samples only.

ILLUSTRATION 12.4:

The time between failures for machine are found to be 65.5, 63.4, 83.5, 75.3, 83.2, 73.0, 84.3, 77.1, 78.4, 70.0, 65.1, 72.5, 72.7, 72.3 and 61.6 hours. Test if the machine failure times are behaving normal.

Solution.

H_0: The failure times are normal. *H_1: The failure times are not normal.*

Given failure times (TBF) are

65.5, 63.4, 83.5, 75.3, 83.2, 73.0, 84.3, 77.1, 78.4, 70.0, 65.1, 72.5, 72.7, 72.3 and 61.6 hours.

Rank ordering the above TBFs, i.e., ordered time between failures (OTBF) are

61.6	63.4	65.1	65.5	70.0
72.3	72.5	72.7	73.0	75.3
77.1	78.4	83.2	83.5	84.3

We can calculate the mean (μ) = 73.20 and standard deviation (σ) = 7.04 for the above sample data.

The test statistic is $D_n = \max\{D_1, D_2\}$ where

$$D_1 = \max \sum_{1 \le i \le n}^{n} \{\Phi(t_i - \bar{t})/s - i - 1/n\} \qquad D_2 = \max \sum_{1 \le i \le n} \{i/n - \Phi(t_i - \bar{t})/s$$

$$\bar{t} = \sum_{i=1}^{n} t_{i/n} \qquad s^2 = \sum_{i=1}^{n} (t_i - \bar{t})^2/n$$

S.no. (i)	(i-1)/n	i/n	TBF (t_i)	Z = $(t_i - \mu)/\sigma$	Cum. Prob. of failure P(Z)	$D_1(i)$	$D_2(i)$
(1)	**(2) = [(1) – 1]/n**	**(3) = (1)/n**	**(4) given**	**(5) = [(4) –73.2]/7.04**	**(6)=from Normal Distribution table for (5)**	**(7) = (6) – (1)**	**(8) = (2) – (6)**
1.	0	0.0667	61.6	1.65	0.0495	0.495	0.0172
2.	0.0667	0.1333	63.4	1.39	0.0823	0.0156	0.0511
3.	0.1333	0.2	65.1	1.15	0.1251	–0.0083	0.0749
4.	0.2	0.2667	65.5	1.09	0.1389	–0.0621	0.1288
5.	0.2667	0.3333	70.0	0.45	0.3264	0.0597	0.0070
6.	0.3333	0.4	72.3	0.13	0.4483	0.1149	-0.0483
7.	0.4	0.4667	72.5	0.1	0.4602	0.0602	0.0065
8.	0.4667	0.5333	72.7	0.07	0.4721	0.0054	0.0612
9.	0.5333	0.6	73.0	0.03	0.4880	–0.0453	0.1120
10.	0.6	0.6667	75.3	0.43	0.6179	0.0179	0.0488
11.	0.6667	0.7333	77.1	0.55	0.7088	0.0422	0.0245
12.	0.7333	0.8	78.4	0.07	0.7704	0.0370	0.0297
13.	0.8	0.8667	83.2	1.42	0.9222	0.1222	–0.0555
14.	0.8667	0.9333	83.5	1.39	0.9279	0.0612	0.0055
15	0.9333	1.000	84.3	1.58	0.9430	0.0096	0.0570
N = 15			M = 73.2	σ = 7.04			

From the above table, we have

$$\text{Max } D_1 = 0.1222 \text{ and}$$

$$\text{Max } D_2 = 0.1288$$

So, K-S Test Statistic $D_n = 0.1288$

and $D_{crit} = 0.201$ for sample size = 15

Since $D_n < D_{crit}$,

We accept H_0, i.e., the sample is fit in normal distribution.

12.14.5 Cramer-von Mises Test for Power Law Process Model

Both the minimal repair process and the AMSAA (reliability growth) model assume a non-homogeneous Poisson process based on the power law formula. This model allows for the application of statistical techniques such as the calculation of the maximum likelihood estimators (MLE), confidence intervals, and the Cramer-von-Mises test statistic for goodness of fit. To determine whether the non-homogeneous Poisson process is a more appropriate model than the constant failure rate model (homogeneous Poisson process), a trend test on the failure times is performed. For the intensity function, $\rho(t) = abt^{b-1}$ and the hypotheses tested are

H_0: The intensity function is constant (b = 1). *H_1: The intensity function is not constant (b ≠ 1).*

The test statistic is computed from the expression

$$\chi^2 = 2n/\hat{b} \qquad \text{... (46)}$$

Where n = number of failures and *b* = MLE for the growth of deterioration rate.

12.15 WEIBULL PLOTS

From Weibull cumulative density function (CDF),
we have

$$F(t) = 1 - \exp[-(t/\alpha)^{\beta}] \text{ and } R(t) = \exp[-(t/\alpha)^{\beta}] \quad \text{... (47)}$$

So, $$\ln [1/\{1 - F(t)] = (t/\alpha)^{\beta} \quad \text{... (48)}$$

Taking logarithm again,

$$\text{Ln} \ln [1/\{1 - F(t)] = \beta \ln t - \beta \ln \alpha \quad \text{... (49)}$$

The above equation (49) is of the form y = mx + c where ordinate is $\text{Lnln}[1/\{1 - F(t_i)]$ and abscissa is $\ln(t_i)$. Thus, a plot between $\ln(t_i)$ and $\text{Ln} \ln [1/\{1 - F(t_i)]$ is representative to Weibull probability paper plot of $[t_i, F(t_i)]$.

Generally, a least-squares fit to the data is recommended over a manual plot of the data on probability paper. It is more accurate and less subjective than fitting a straight line to the data by eye. In addition, measures of how well the curve fits the data, such as the index of fit, are available. With the use of the personal computer and one of the many statistical applications available (MS Excel) with graphics capability, plotting data by hand is no longer necessary. Weibull graph paper is constructed so that data generated from a Weibull distribution will graph as a straight line. The abscissa is a logarithmic scale, and the ordinate, while labelled in terms of the cumulative percentage of failures, F(t), is scaled on the basis of $\text{Lnln}[1/\{1 - F(t)\}]$, which can be easily calculated using the Microsoft Excel.

Alpha (α) can be estimated from the point on the line that corresponds to 63.2 percent of the failures, since $F(\alpha) = 0.632$. The 63.2 percent line is often identified on the graph paper. It can be seen from equation (49) that β can be estimated from the slope of the plotted line. Once β is calculated, the value of α can be estimated using the intercept of the fitted straight line.

The index of fit (r) can be calculated from the following equation (12.50):

$$r^2 = \frac{\sum_{i=1}^{n} (\bar{y}_i - \hat{a} - \hat{b}\bar{x})^2}{\sum_{i=1}^{n} (y_i - \bar{y}_i^2)} \quad \text{... (50)}$$

12.15.1 Meaning and Importance of the Two Parameters (α and β) of Weibull Distribution

Alpha (α) is called *scale parameter* of Weibull distribution that influences both the mean and the spread or dispersion of the distribution. The effect of α can be straight away translated in terms of reliability. As α increases, the reliability increases at given point in time.

Each of the CDF and reliability curves passes through the same point where t = α, irrespective of the value of β, since from equation (47), we have $R(t) = \exp[-(t/\alpha)^{\beta}]$ and at t = α, we get $R(\alpha) = \exp[-(\alpha/\alpha)^{\beta}] = \exp(-1) = 0.368$. Therefore, 63.2 percent of all Weibull failures will occur by time t = α regardless of the value of shape parameter. Hence, this parameter is also called the *characteristic life* and has the units identical to those of failure time, T.

Beta (β) is referred to as *shape parameter* of Weibull distribution. For β < 1, the PDF is similar to the exponential and for large values of β (e.g., β ≥ 3), the PDF is somewhat symmetrical, like normal distribution. For 1 < β < 3, the density function is skewed. When β = 1, λ(t) is constant and the distribution is identical to exponential with $\lambda = 1/\alpha$ where α is the scale parameter of Weibull distribution. Thus, the hazard rate function can be increasing or decreasing depending on the value of β. This is summarised in the following Table 12.5.

Table 12.5 Weibull shape parameter (β) characteristics

S.No.	*Value of β*	*Property (Failure Rate)*	*Distribution*
1.	$0 < \beta < 1$	Decreasing Failure Rate (DFR)	Exponential
2.	$\beta = 1$	Constant Failure Rate (CFR)	Exponential
3.	$1 < \beta < 2$	Increasing Failure Rate (IFR)	Concave
4.	$\beta = 2$	Increasing Failure Rate (IFR)	Rayleigh
5.	$\beta > 2$	Increasing Failure Rate (IFR)	Convex
6.	$3 \leq \beta \leq 4$	Increasing Failure Rate (IFR)	Approximately Normal (Symmetrical)

ILLUSTRATION 12.5:

The ordered time between failures of given equipment were found to be as shown below.

4, 5, 6, 7, 8, 9, 9, 10, 10, 10, 11, 11, 11, 12, 12, 13, 13, 14, 14, 14, 14, 15, 15, 18, 28, 30, 33, 34, 35, 48

Draw a Weibull plot to fit a straight line and hence determine the two Weibull parameters.

Solution.

Table 12.6 Weibull plotting for dumper no. 201 for yr 2000–01

S.no(i)	OTBF(t)	f(t)	1/1–f(t)	(X) = ln(t)	(Y) = lnln (1/1–f(t))		
0	0			(x)	(y)		
1	4	0.02302632	1.02356902	1.38629436	–3.75949238	5.26842216	10.2809946
2	5	0.05592105	1.05923345	1.60943791	–2.85517959	3.02360384	5.29960092
3	6	0.08881579	1.09747292	1.79175947	–2.37504617	1.64915945	3.31951196
4	7	0.12171053	1.13857678	1.94591015	–2.04192194	0.80962291	2.2166105
5	8	0.15460526	1.18287938	2.07944154	–1.78407934	0.32126592	1.51532643
6	9	0.1875	1.23076923	2.19722458	–1.57195253	0.07457689	1.03807396
7	9	0.22039474	1.28270042	2.19722458	–1.39043267	0.07457689	0.70113709
8	10	0.25328947	1.33920705	2.30258509	–1.23073547	0.00010713	0.45919881
9	10	0.28618421	1.40092166	2.30258509	–1.08728563	0.00010713	0.28536137
10	10	0.31907895	1.46859903	2.30258509	–0.9563086	0.00010713	0.16258252
11	11	0.35197368	1.54314721	2.39789527	–0.83511642	0.05167657	0.079537
12	11	0.38486842	1.62566845	2.39789527	–0.72171316	0.05167657	0.02843263

13	11	0.41776316	1.71751412	2.39789527	–0.61456159	0.05167657	0.00377834
14	12	0.45065789	1.82035928	2.48490665	–0.51243711	0.19740701	0.00165293
15	12	0.48355263	1.93630573	2.48490665	–0.41433145	0.19740701	0.01925487
16	13	0.51644737	2.06802721	2.56494936	–0.31938588	0.41461699	0.05461919
17	13	0.54934211	2.2189781	2.56494936	–0.22684191	0.41461699	0.10644002
18	14	0.58223684	2.39370079	2.63905733	–0.13600231	0.68676133	0.17396496
19	14	0.61513158	2.5982906	2.63905733	–0.04619707	0.68676133	0.25694386
20	14	0.64802632	2.8411215	2.63905733	0.04324996	0.68676133	0.35562537
21	14	0.68092105	3.13402062	2.63905733	0.13305841	0.68676133	0.47080428
22	15	0.71381579	3.49425287	2.7080502	0.22403882	1.00151691	0.60393445
23	15	0.74671053	3.94805195	2.7080502	0.31716001	1.00151691	0.75734096
24	18	0.77960526	4.53731343	2.89037176	0.41365487	2.11822826	0.93460217
25	28	0.8125	5.33333333	3.33220451	0.51520189	6.53933201	1.14125479
26	30	0.84539474	6.46808511	3.40119738	0.62426864	7.44885627	1.38618133
27	33	0.87828947	8.21621622	3.49650756	0.74484254	8.80269905	1.68463766
28	34	0.91118421	11.2592593	3.52636052	0.8842595	9.2499839	2.0659833
29	35	0.94407895	17.8823529	3.55534806	1.05911385	9.69490873	2.59921214
30	48	0.97697368	43.4285714	3.87120101	1.32737139	15.220199	3.53614775
					–0.55309338	76.4249135	41.5387461
					b = beta =	1.8043	
					intercept =	–6.3054	
					alpha =	32.9388122	

Weibull plot for failure data of the given equipment

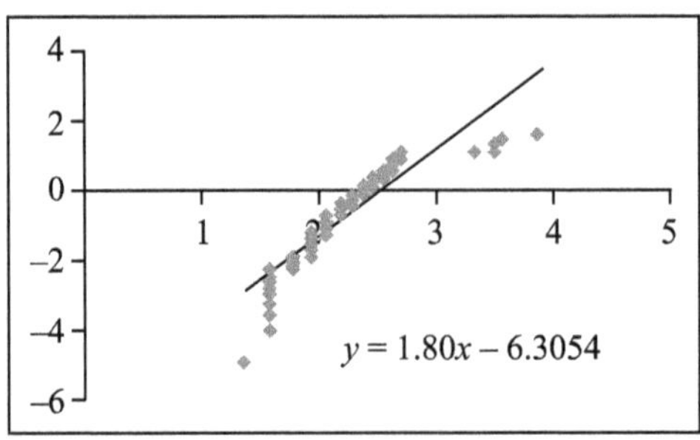

BUSTLE - 12.4

1. *Thirty bearings were placed on test for wear-out failures by an accelerated testing. The observed (complete) failure times are recorded as follows:*

278.2	*320.0*	*361.8*	*331.7*	*295.3*	*355.4*	*333.7*	*346.2*	*376.4*	*446.7*
340.3	*273.3*	*361.6*	*361.5*	*389.2*	*391.2*	*372.8*	*336.8*	*342.6*	*305.7*
339.3	*397.0*	*301.8*	*282.5*	*447.5*	*443.1*	*350.6*	*300.5*	*313.5*	*287.1*

   Test if the data has come from the exponential distribution.

   Answer: *Null hypothesis is Rejected*

2. *In a machine shop the TBF in hours of 25 machines were found as follows. Test whether they can be fit in log-normal distribution. Given $D_{critical}$ for a sample size of 25 is 0.165*

818.1	*1407.5*	*464.9*	*4991.0*	*452.0*
1773.0	*326.9*	*964.8*	*1677.8*	*282.3*
1725.4	*652.3*	*639.2*	*338.8*	*1847.8*
734.9	*220.2*	*1078.1*	*1077.3*	*629.4*
240.5	*511.8*	*1083.4*	*821.3*	*670.8*

   Answer: Max of $D_1 = 0.0736$ and $D_2 = 0.0756$, i.e., $D_{25} = 0.0756 < D_{crit} = 0.165$; Hence accept H_0.

Test Your Understanding 12.4

1. Fitting of theoretical distribution is generally preferred to empirical methods in reliability studies. Why?
2. Describe the method of identifying the candidate distribution. Also explain the decision table in identification of candidate distribution.
3. What is goodness of fit? Discuss any three methods of goodness of fit tests generally used in reliability-oriented maintenance planning.
4. Explain the goodness of fit test for the data assumed to have come from the following distributions:
   (a) Exponential distribution
   (b) Weibull distribution
5. Explain the meaning and significance of the two (scale and shape) parameters of the Weibull distribution? Discuss the characteristics of shape parameter.
6. Write a note on Weibull plotting.

12.16 DESIGN OF RELIABILITY

To a large degree, reliability is an inherent attribute of a system, component or product. As such, it is an important consideration in the engineering design process. When the life-cycle costs of a system are being analysed, reliability plays an important role as a major driver of these costs and has considerable influence on system performance. As it is seen, decisions made during conceptual and preliminary design will affect total life-cycle costs. The objective of this chapter is to describe a reliability design process that will establish and then achieve realistic reliability goals.

12.16.1 Reliability Design

Reliability design is an iterative process that begins with the specification of reliability goals consistent with cost and performance objectives. This requires consideration of the life-cycle costs of the system and the effect that reliability has on overall costs and system effectiveness.

Once the reliability goals have been established, these goals must be translated into individual component, subcomponent, and part specifications. This is not necessarily an easy task, and it generally requires a reliability block analysis. After individual component and part requirements have been determined, various design methods can be applied in order to meet the goals. Following completion of preliminary and detailed design and along with initial development and prototyping a failure analysis can be performed to determine whether specifications are being met and to provide a systematic approach for identifying, ranking, and eliminating failure modes. This requires the use of reliability testing, including, perhaps, a formalised reliability growth-testing program.

Once reliability goals have been achieved, verification that safety margins are also being met must be made. If either the reliability or safety goals are not met, the design process must continue. This may require reallocating reliability goals among the components if it is not possible to achieve desired component reliability. More often the entire system must be redesigned. The effect in design changes should then be verified through continued use of failure analysis and reliability testing.

12.16.2 Planning for Reliable Operation

Once the testing has been completed and the reliability goals have been met, the specifications arrived at, must be maintained during production. This requires conformance to design and manufacturing specifications through a good quality control programme.

Once the product has entered into operations, an active preventive maintenance programme, continuous engineering improvements and field modifications, and routine replacement of worn parts can significantly improve reliability.

12.16.3 Characteristics of Reliability Designing and Planning

The following characteristics are often chosen for reliability designing and planning for maintenance of repairable equipment as these are found more suitable and meaningful.

(a) **T (mode):** It is the most frequent probable time between failure in a sample space of time domain.

$$T\text{ (mode)} = \alpha * (1 - 1/\beta)^{1/\beta}$$

(b) **T (median)**: It is the median time between failure of an equipment among the sample space of time domain about a Weibull distribution at reliability (R) = 0.5.

$$T\text{ (median)} = \alpha * (-\ln 0.5)^{1/\beta}$$

(c) **T (Optimal)**: It is optimal preventive maintenance interval for a machine to be observed at a certain reliability values.

$$T\text{ (optimal)} = \alpha*(1/(\beta - 1))^{1/\beta}$$

(d) **Characteristic life:** It is the life (period) of equipment in which 63.2 % of Weibull failures will occur. It is independent of shape parameter and reliability.

$$T\text{ (Char)} = \alpha \quad \text{[Its value is equal to alpha } (\alpha)\text{]}.$$

(e) **B1 life:** It is the time at which 1 percent of the population will have failed at a reliability of 0.99.

$$B1\text{ life} = \alpha * (-\ln 0.99)^{1/\beta}$$

(f) **B.1 life:** It is the time at which 0.1 percent of the population will have failed at a reliability of 0.999.

$$\text{B.1 life} = \alpha * (-\ln 0.999)^{1/\beta}$$

ILLUSTRATION 12.6:

In a reliability analysis, it is found that the data is fit in 2-parameter Weibull distribution with the shape parameters as 1.2 and scale parameter as 3.5. Design the reliability in terms of characteristic life, optimal life, B1 life, B.1life, T (median) and T (mode).

Solution.

Given that

Shape parameter (α) = 1.2 time units

and Scale parameter (β) = 3.5 time units

The reliability design characteristics can be calculated as follows:

(a) **T (mode)**

$$T\text{ (mode)} = \alpha * (1 - 1/\beta)^{1/\beta} = 1.2 * (1 - 1/3.5)^{1/3.5}$$
$$= 1.09000974 = 1.09 \text{ time units}$$

(b) **T (median)**

$$T\text{ (median)} = \alpha * (-\ln 0.5)^{1/\beta} = 1.2 * (-\ln 0.5)^{1/3.5}$$
$$= 1.080694164 = 1.0807 \text{ time units}$$

(c) **T (Optimal)**

$$T\text{ (optimal)} = \alpha*(1/(\beta - 1))^{1/\beta} = 1.2 *(1/(3.5 - 1))^{1/3.5}$$
$$= 0.9236003753 = 0.924 \text{ time units}$$

(d) **Characteristic life**

$$T\text{ (Char)} = \alpha$$
$$= 1.2 \text{ time units}$$

(e) **B1 life**

$$\text{B1 life} = \alpha * (-\ln 0.99)^{1/\beta} = 1.2 * (-\ln 0.99)^{1/3.5}$$
$$= 0.3223856453 = 0.3224 \text{ time units}$$

(f) **B.1life**

$$\text{B.1 life} = \alpha * (-\ln 0.999)^{1/\beta} = 1.2 * (-\ln 0.999)^{1/3.5}$$
$$= 0.1667632909 = 0.1668 \text{ time units}$$

BUSTLE - 12.5

In a reliability analysis, it is found that the data is fit in 2-parameter Weibull distribution with the shape parameters as 1.0 and scale parameter as 2.0. Design the reliability in terms of characteristic life, optimal life, B1 life, B.1life, T (median) and T (mode).

____

____

____

____

Test Your Understanding 12.5

1. Enumerate the reliability characteristics used in designing of reliability-centred maintenance planning and scheduling. Briefly explain each of them.
2. Describe the following terms:
   (a) B1 Life
   (b) B.1 life
   (c) $T_{optimal}$
   (d) Characteristic life
   (e) T_{median}
3. Write a short note on reliability design of repairable equipment.
4. Evaluate the reliability characteristics for a failure data fitted in a 2-parameter Weibull distribution whose characteristic life is 1.2 units and scale parameter is 3.0 units.
5. In a reliability analysis, it is found that the data is fit in 2-parameter Weibull distribution with the shape parameters as 1.4 days and scale parameter as 2.5 days. Estimate and design the reliability in terms of characteristic life, optimal life, B1 life, B.1 life, T (median) and T (mode).
6. Design the reliability using the reliability characteristics in terms of optimal life, B1 life, B.1 life, modal and median life in the following cases and comment on the results and predict the shape of the curve.
   (a) Alpha = 10; Beta = 1.0
   (b) Alpha = 5; Beta = 1.5
   (c) Alpha = 1.5; Beta = 2.5
   (d) Alpha = 2.5; Beta = 2.0

12.17 CONCLUSION

In most industries, the failure data is not maintained properly for various reasons. Even if it is maintained, it has lot of inconsistencies. This creates a wrong lead to the conclusion and thus the maintenance scheduling goes wrong. It can cause damage and cost hugely. Thus, the failure data plays a vital role in predicting the behaviour of equipment. Therefore, it is very important to maintain the failure data. However, the failure data of one machine may not be identical with the other. And therefore, testing for each and every machine in large industry may not be economical. Conclusively, it depends on the engineer's wisdom to choose the maintenance policy on planning and scheduling.

12.18 RECAP

A clear cut path is established to predict the repairable equipment behaviour in this chapter. The methodology suggests that after removing the data inconsistencies, the trend analysis and correlation test can be performed by which it will enable us to check if the data is contradicting i.i.d. that leads to fit in NHPP. If i.i.d. assumption is not contradicted, then TTT transforms may be conducted to check if there is any exponential fit to fit HPP by which exponential, normal and log-normal can be thought of, otherwise RP (Weibull) can be fit. This can be crosschecked and can prove the reliability of the method by using various goodness of fit tests. Later the reliability design can be done with some pre-determined characteristics.

12.19 KEY CONCEPTS

1. **TBF:** Time Between Failures
2. **TTR:** Time To Repair
3. **TTF:** Time To Fail
4. **NHPP:** Non-homogeneous Poisson Process
5. **HPP:** Homogeneous Poisson Process
6. **PLP:** Power Law Process
7. **i.i.d.:** Independently and Identically Distributed
8. **TTT:** Total Time on Test
9. **RP:** Renewable Process
10. **MLE:** Maximum Likelihood Estimation
11. **PEXE:** Piecewise Exponential Estimation
12. **PLE:** Product Limit Estimation
13. **KME:** Kaplan-Meier Estimation
14. **PDF:** Probability Density Function: f(t)
15. **CDF:** Cumulative Distribution Function: $F(t) = \int f(x)\,dx$
16. **R:** Reliability Function: $R(t)=1 - F(t)=1- \int f(x)\,dx = \int R\,dx$
17. **H:** Hazard Function
18. **λ(t):** Hazard Rate $\lambda(t)= f(t)/[1-F(t)]$
19. **H_0:** Null Hypothesis
20. **H_1 or H_a:** Alternative Hypothesis
21. α: Scale Parameter
22. β: Shape Parameter
23. γ: Location Parameter.
24. **LOS:** Level of Significance, denoted by α
25. χ^2 **:** Chi-Square
26. **K-S Test:** Kolmogorov-Smirnov Test
27. **T_{mode}:** Most frequent probable TBF in a sample space of time domain. $T_{mode} = \alpha * (1 - 1/\beta)^{1/\beta}$
28. **T_{median}**: The median TBF of an equipment among the sample space of time domain about a Weibull distribution at reliability (R) = 0.5; $T_{median} = \alpha * (-\ln 0.5)^{1/\beta}$
29. **$T_{Optimal}$:** Optimal PM interval for a machine at a certain reliability values. $T_{optimal} = \alpha*(1/(\beta - 1))^{1/\beta}$
30. **Characteristic life:** The life (period) of equipment in which 63.2 % of Weibull failures will occur. It is independent of shape parameter and reliability and is equal to alpha (α). $T_{Char} = \alpha$
31. **B1 life:** The time at which 1 percent of the population will have failed at a reliability of 0.99. B1 life = $\alpha * (-\ln 0.99)^{1/\beta}$

32. **B.1 life:** The time at which 0.1 percent of the population will have failed at a reliability of 0.999. B.1 life = α * $(-\ln 0.999)^{1/\beta}$

REVIEW QUESTIONS

1. With the help of flow chart describe the procedure of reliability analysis of failure data.
2. How do you identify and confirm if a given failure data is distributed in Weibull pattern?
3. Explain the characteristics of normal and log-normal distributed failure data.
4. Distinguish between
   (a) Shape and scale parameters of Weibull distribution
   (b) B1 and B.1 Life
   (c) PDF and CDF
   (d) Type-I and Type-II Errors
5. What do you understand by 'goodness of fit'? Discuss any two 'goodness of fit' tests used in reliability analysis.
6. Illustrate various decision alternatives with suitable examples while identifying a candidate distribution of a failure data through hypothesis tests.

CHAPTER 13

Reliability Oriented Maintenance Systems and Evaluation

PREAMBLE

This chapter focuses on reliability, availability, and maintainability requirement as design goals of specifications. Various combinations of the components such as series system, parallel system, and combined (series and parallel) system are discussed that will help enhance or design the overall reliability of the systems as needed. The advantages of modular design or modularisation are explored and then quality and reliability are distinguished. Different types of availability such as intrinsic availability (A_i), achieved availability (A_a), operational availability (A_o), and the factors affecting availability and hence the reliability are discussed. The effect of redundancy and group redundancy on reliability is narrated through the numerical problems. The evaluation of overall equipment effectiveness (OEE), maintenance performance, and efficiency are also depicted through numerical problems.

CHAPTER OUTLINE

13.1 INTRODUCTION

The terms reliability, availability, and maintainability (R-A-M) are relatively new in the field of engineering. In present days, the reliability (dependability), availability (operability in need), and maintainability (reparability or ease of repair) have become very important considerations in purchase decisions of not only costly or heavy equipment but also in consumer products.

Reliability, availability, and maintainability are not only the important integral aspects of the engineering process but also necessary scales for formulating or evaluating features and characteristics of the equipment in terms of reliability improvement, life cycle costing, cost benefit analysis, operating and capability studies such as maintenance performance indices and even influence the incentive decisions too.

13.2 RELIABILITY IMPROVEMENT

It is known to all that the failure rate reflects unreliability. It is already explained about the shape (in bathtub curve) and failure characteristics for a machine at each stage of its life. The information (failure data) obtained from test results and field use experience exhibits one of the following three different types:

(a) The highest rate of failure occurs during initial use (called infant failures or teething troubles). It may be due to environmental change, previously undetected faulty sub-components or damage in transportation, etc., and slowly these problems vanish registering or exhibiting a decreasing failure rate (reliability improvement).

(b) Random failures can be seen now and then only indicating that the machine is undergoing the useful life that registers the constant failure rate. New product design is especially concerned with the failure rate during the useful performance life (the shallow portion of the bathtub curve).

(c) The phase that follows the useful performance life is wear out phase. This phase experiences the increasing failure rate and reaches highest failure rate finally becoming most unreliable.

Whatsoever may be the phase which the machine is undergoing, it does not perform properly (as desired) unless all of its sub-components function correctly. In such cases, the reliabilities of individual sub-components must be greater than the reliability desired for the final product. This situation exists whenever the chances of failure for each sub-component are independent of one another.

For example, consider the probability of starting a bike. It may not start due to various reasons, such as slip in kick rod, the failure of spark plug, the failure of magnetic system, generation current for the spark plug, or the carburettor not supplying fuel to the engine, etc. Thus, the probability of starting or not starting of the bike depends on the probability of the functioning or not functioning of its components.

13.3 RELIABILITY SYSTEMS

The functioning of a system (or machine) does not just depend on the reliability of the sub-systems (or components) but also depends on the fashion or orientation how they (the sub-systems or components) are connected.

Going more into the details, it is fundamentally three fashions in which different sub-systems or components of a system or machine may be connected:

(a) Series system
(b) Parallel system
(c) Series and parallel system or combined system

Now, let us take up each of these systems separately and learn how their reliability can be calculated.

13.3.1 Series System

Let's take a system composed of two sub-systems say, A and B connected in series as shown in Figure 13.1. Let the reliabilities of the two components be R_A and R_B. Now let us compute the reliability say, R of the overall system.

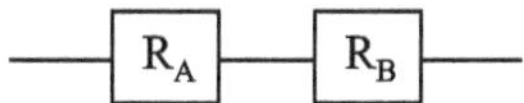

Figure 13.1 Series system of two components

Now, if the probability of failing the sub-systems A and B are denoted by F(A) and F(B), then we have

$$R_A = 1 - F(A) \quad \text{or} \quad F(A) = 1 - R_A$$

and

$$R_B = 1 - F(B) \quad \text{or} \quad F(B) = 1 - R_B$$

We know that there are three possibilities for the system not functioning, and there is only one possibility for the system to function, i.e., both the sub-systems function well, as given below:

(i) Sub-system 'A' fails : System fails
(ii) Sub-system 'B' fails : System fails
(iii) Both sub-systems 'A' and 'B' fail : System fails
(iv) Both sub-systems 'A' and 'B' function well : System functions

Thus, the probability of A not failing, i.e., $1 - F(A)$

And the probability of B not failing, i.e., $1 - F(B)$

The probability of both not failing is $[1 - F(A)]\cdot[1 - F(B)]$

So, the overall reliability of the system $R = [1 - F(A)]\cdot[1 - F(B)] = R_A \cdot R_B$

Now let us generalise the above concept for N components,

Let N = Number of sub-systems or components of a system connected in series

R_K = The reliability or the probability of a component 'K' in the system functioning at a given time, t

R_i = Reliability of the ith system where i = 1, 2, 3, … N

R_s = Reliability of the series system (i.e., the probability of the system functioning at time t)

Figure 13.2 depitcs the series system.

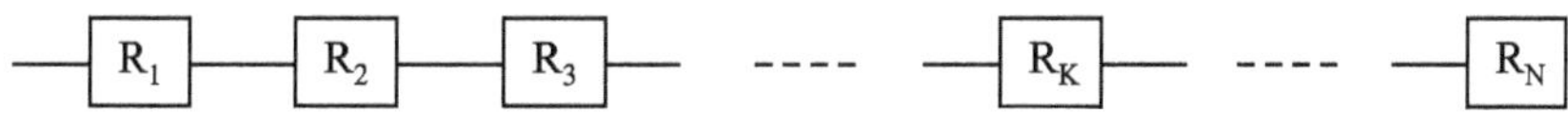

Figure 13.2 Series system of N

The reliability of the above system

$$R_s = R_1 \times R_2 \times R_3 \times \ldots \times R_K \times \ldots R_N$$

It can be discerned that the system fails if any of the sub-components fail.

i.e., $R_s = 0$, if R_1 or R_2 or R_3 or … R_K … $R_N = 0$.

Conversely, if even one element's (components) probability of functioning is improved the whole system's reliability is improved.

ILLUSTRATION 13.1:

The probability of functioning of each element in a series system of four components is given as $R_1 = 0.6$; $R_2 = 0.7$; $R_3 = 0.8$; $R_4 = 0.9$. Find the overall reliability of the system. What will be the change in the system reliability if the reliability of the third component is (a) increased to 0.9 (b) decreased by 0.1?

Solution.

The above system can be represented diagrammatically as follows.

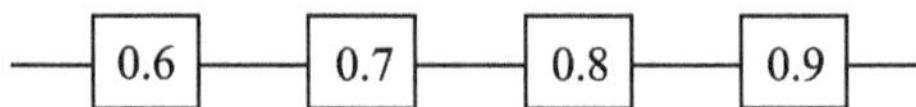

Figure 13.3 Series system of four components

The system's reliability can be calculated as, $R_s = R_1 \times R_2 \times R_3 \times R_4$

$$R_s = 0.6 \times 0.7 \times 0.8 \times 0.9 = 0.3024.$$

(a) If the reliability of the third component is increased to 0.9, then the system reliability is given by

$$R_a = 0.6 \times 0.7 \times 0.9 \times 0.9 = 0.3402$$

i.e., overall reliability increased by 0.0378.

(b) If the reliability of the third component is decreased by 0.1, i.e., to 0.7, then the system reliability is given by

$$R_b = 0.6 \times 0.7 \times 0.7 \times 0.9 = 0.2646$$

i.e., overall reliability decreased by 0.0378.

Observations

- Although the reliability of one of the components is as high as 0.9, the overall probability of the system is as low as 0.3024.
- The components with lower reliability connected in series will lower the overall reliability of the system.
- If the components of certain reliability are connected in series, the overall reliability of the system is less than that of their individual reliabilities.
- If the probability of an element is improved from 0.8 to 0.9, the system probability improved to 0.3402, i.e., an increment of 0.0378 only, i.e., not appreciably improved. It means that it is difficult to achieve high reliability with the basic series system.
- If the probability of an element is reduced from 0.8 to 0.7, the system probability is decreased to 0.2646, i.e., a decrement 0.0378 only (same as increment).
- It is difficult to achieve high reliability with the series system.

13.3.2 Parallel System

Suppose, a system is composed of two sub-systems say, A and B are connected in series as shown in Figure13.4. Let the reliabilities of the two components be R_A and R_B. Now let us compute the reliability say, R of the overall system.

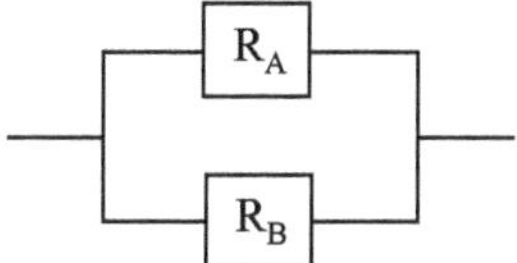

Figure 13.4 Series system of two components

Now, if the probability of failing the sub-systems A and B are denoted by F(A) and F(B), then we have

$$R_A = 1 - F(A) \quad \text{or} \quad F(A) = 1 - R_A$$

and
$$R_B = 1 - F(B) \quad \text{or} \quad F(B) = 1 - R_B$$

We know that there are three possibilities for the components of system not functioning,

(a) Sub-system 'A' fails

(b) Sub-system 'B' fails

(c) Both sub-system 'A' and 'B' fail

In first and second cases, the system will not completely fail, but only in the third case, the system completely fails. Thus, the reliability is affected if and only if both the components fail.

Thus, the probability of A failing, i.e., $F(A) = 1 - R_A$

And the probability of B failing, i.e., $F(B) = 1 - R_B$

The probability of both failing is $F(A) \cdot F(B) = [1 - R_A] \cdot [1 - R_B]$

Thus, the probability of system not functioning $[1 - R_P] = [1 - R_A] \cdot [1 - R_B]$

So, the overall reliability of the system $R_P = 1 - [1 - R_A] \cdot [1 - R_B]$

Now let us generalise the above concept for N components,

A parallel system is shown below which has 'N' number of elements connected together in Figure 13.5.

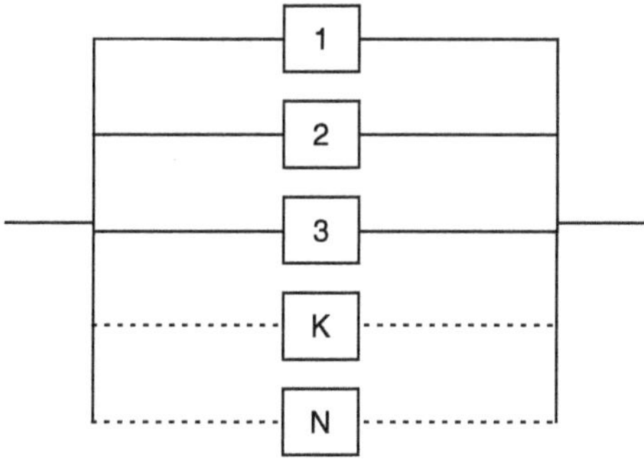

Figure 13.5 Parallel system of N components

In parallel system, if any one component (element) fails, still the system functions. Unless all the components fail simultaneously, the system does function (of course with less power (or capacity).

Let R_K = Probability of any element 'K' functioning at a given time, t

Then $1 - R_K$ = Probability of element 'K' failing in the system.

Also R_i = Reliability of the ith system where i = 1, 2, 3, … N

R_P = Reliability of the parallel system

(i.e., the probability of the system functioning at time t)

The probability of all the elements failing together

$$= (1 - R_1) \times (1 - R_2) \times (1 - R_3) \times \ldots (1 - R_K) \times \ldots \times (1 - R_N)$$

Therefore, the probability of a parallel system functioning at a given time t or reliability of the total system of parallel components is,

$$R_P = 1 - (1 - R_1) \times (1 - R_2) \times (1 - R_3) \times \ldots (1 - R_K) \times \ldots \times (1 - R_N)$$

ILLUSTRATION 13.2:

The reliability of four components connected in a parallel set up are given as $R_1 = 0.6$; $R_2 = 0.7$; $R_3 = 0.8$; $R_4 = 0.9$. Find the overall reliability of the system. What will be the change in the system reliability if the reliability of the third component is (a) increased to 0.9 (b) decreased to 0.7?

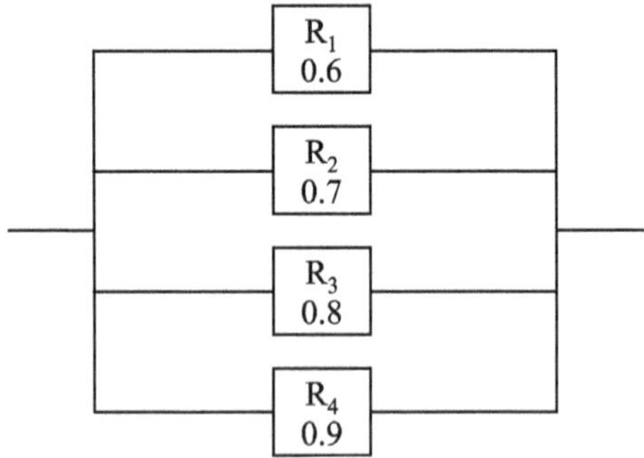

Figure 13.6 Parallel system of four components

The reliability of the system functioning at any time t is given by,

$$R_P = 1 - [(1 - R_1) \times (1 - R_2)\,(1 - R_3)\,(1 - R_4)]$$
$$= 1 - [(1 - 0.6)(1 - 0.7)\,(1 - 0.8)\,(1 - 0.9)]$$
$$= 1 - [0.4 \times 0.3 \times 0.2X0.1]$$
$$= 1 - 0.0024 = 0.9976$$

(a) If the reliability of the third component is increased to 0.9, then the system reliability is given by

$$R_a = 1 - 0.4 \times 0.3 \times 0.1 \times 0.1 = 1 - 0.0012 = 0.9988$$

i.e., overall reliability increased by 0.0012

(b) If the reliability of the third component is decreased to 0.6, then, the system reliability is given by

$$R_b = 1 - 0.4 \times 0.3 \times 0.3 \times 0.1 = 1 - 0.0036 = 0.9964$$

i.e., overall reliability is decreased by 0.0012.

Observations

- Although the reliability of one of the components is as low as 0.6, the overall probability of the system is as high as 0.9976.
- The components with high reliability connected in parallel will improve the overall reliability of the system.
- If the components of certain reliability are connected in parallel, the overall reliability of the system is greater than that of their individual reliabilities.
- If the reliability of an element is improved from 0.8 to 0.9, the system probability improved to 0.9988, i.e., an increment of 0.0012, i.e., not considerably increased.
- If the probability of an element is reduced from 0.8 to 0.7, the system probability decreased to 0.9964, i.e., a decrement 0.0012 only (same as increment), i.e., not considerably decreased.
- It is possible to achieve high reliability with the parallel system.

13.3.3 Combined (Series and Parallel) System

It is a combination of series and parallel systems. In such cases of a system having elements connected both in parallel and series, it is better to convert into one of the systems. Two cases are illustrated as shown below in the Figures 13.7 and 13.8.

Case-1

Here in this case, the parallel system is converted into equivalent series system so that the whole system turns to series type only.

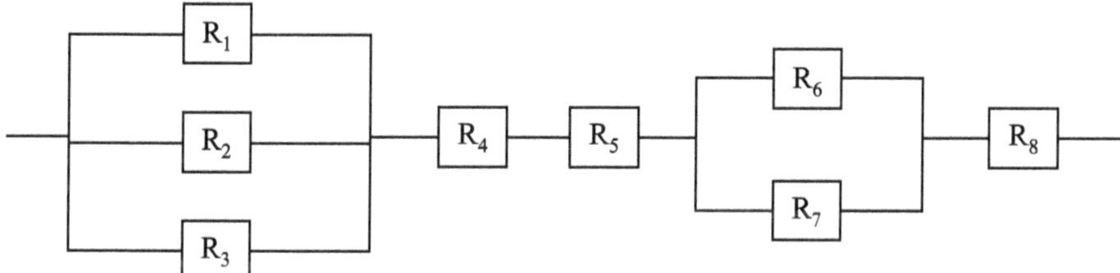

Figure 13.7 Combined system: Case-I

In the above system, we consider that we have a series system with (a parallel sub-system of R_1, R_2, R_3), R_4, R_5, (a parallel sub-system of R_6, R_7) and R_8, where first, we shall calculate the reliability of the sub-systems as follows.

$$\{1 - (1 - R_1)(1 - R_2)(1 - R_3)\} = R_x, \quad \text{say and}$$

$$\{1 - (1 - R_6)(1 - R_7\} = R_y, \quad \text{say}$$

Then, the overall system reliability is given by

$$R = R_x\, R_4.R_5.R_y\,.R_8$$
$$= \{1 - (1 - R_1)(1 - R_2)(1 - R_3)\}\, R_4.R_5.\{1 - (1 - R_6)(1 - R_7\}.R_8$$

Case-2

Consider the case of combined system as shown in Figure 13.8. In this case, it is better to convert the components into equivalent parallel system. Here consider the three sub-systems as parallel. The top sub-system and bottom sub-systems have the components in series. Thus, we consider as a parallel system of (a series sub-system of R_1, R_2) R_3 and a series subsystem of R_4, R_5, R_6.

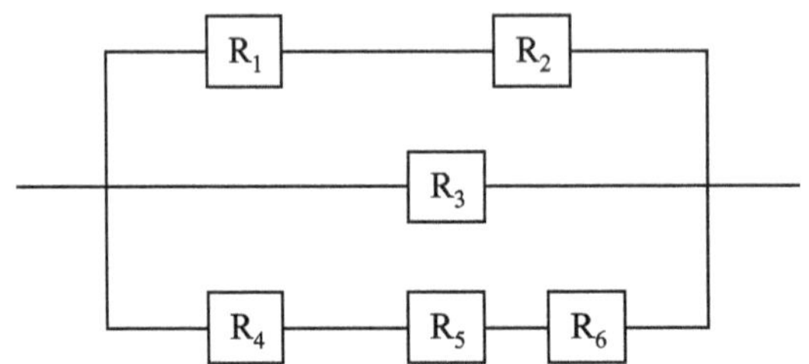

Figure 13.8 Combined system: Case-2

Thus, the overall reliability can be computed as

For series sub-system of R_1, R_2, it is, say, $R_x = R_1 \times R_2$

For series sub-system of R_4, R_5, R_6, it is, say, $R_y = R_4 \times R_5 \times R_6$

The overall system reliability is $R = 1 - (1 - R_x) \cdot (1 - R_3) \cdot (1 - R_y)$

$$= 1 - (1 - R_1 \times R_2) \cdot (1 - R_3) \cdot (1 - R_4 \times R_5 \times R_6)$$

ILLUSTRATION 13.3:

Consider a system with the combined (series and parallsel connections) system as shown in Figure 13.9. The reliability of each of the element is given in the figure itself. Find the overall reliability of the total system.

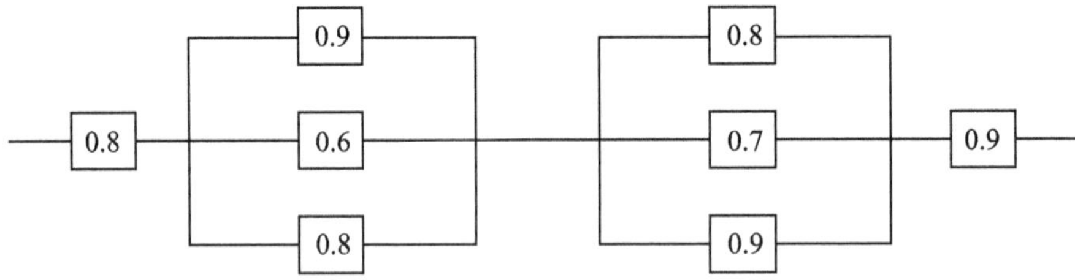

Figure 13.9 Combined system

Solution.

The reliability of the first parallel system

$$R_{p1} = 1 - (1 - 0.9)(1 - 0.6)(1 - 0.8) = 0.992$$

The reliability of the second parallel system

$$R_{p2} = 1 - (1 - 0.8)(1 - 0.7)(1 - 0.9) = 0.994$$

The above parallel-series system can be reduced to equivalent single series system as given below (see Figure 13.10).

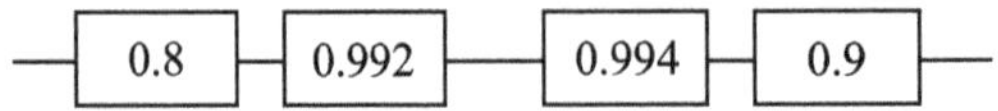

Figure 13.10 Combined system converted to series system

The reliability of this system is given by

$$R = 0.8 \times 0.992 \times 0.994 \times 0.9 = 0.709955$$

This is greater than the value of the weakest element in the system, i.e., 0.6.

Conclusion: The reliability of a system (having greater than components) can be improved by connecting the components in parallel.

If two weak components can be joined in parallel with each other, their joint (equivalent) reliability is higher than their individual probability of functioning.

ILLUSTRATION 13.4:

Costly components are expected to function more reliably, as they are manufactured precisely to the required dimensions, better quality materials and appropriate technology with no (or very few) defects. Suppose that a product is to be designed with two components 'a' and 'b' whose reliability and corresponding cost are given in the following table. If the components can be combined in any orientation (either series or parallel), find all the possible reliabilities of the system and corresponding costs. Give your remarks for designing an optimal combination.

Sub-component	*Cost for Reliability*		
Reliability	0.8	0.9	0.95
Cost of each 'a' in Rupees	50	100	150
Cost of each 'b' in Rupees	60	90	120

Solution.

There are nine (3^2) possible combinations with the two components in three different reliabilities. Further in each combination, we have either in series or parallel connections. Thus, we get eighteen (18) cases as given below.

Case 1 a(0.95) and b(0.95) in parallel

The reliability of the product,

$$R_p = 1 - (1 - 0.95) \times (1 - 0.95) = 0.9975$$

Cost of product = ₹ 150 + ₹ 120 = ₹ 270.

Case 2 a(0.95) and b(0.95) in series

The reliability of the product,

$$R_s = 0.95 \times 0.95 = 0.9095$$

Cost of product = ₹ 150 + ₹ 120 = ₹ 270.

Case 3 a(0.9) and b(0.9) in parallel

The reliability of the product,

$$R_p = 1 - (1 - 0.9) \times (1 - 0.9) = 0.99$$

Cost of product = ₹ 100 + ₹ 90 = ₹ 190.

Case 4 a(0.9) and b(0.9) in series

The reliability of the product,

$$R_s = 0.9 \times 0.9 = 0.81$$

Cost of product = ₹ 100 + ₹ 90 = ₹ 190.

Case 5 a(0.8) and b(0.8) in parallel

The reliability of the product,

$$R_p = 1 - (1 - 0.8) \times (1 - 0.8) = 0.96$$

Cost of product = ₹ 50 + ₹ 60 = ₹ 110.

Case 6 a(0.8) and b(0.8) in series

The reliability of the product,

$$R_s = 0.8 \times 0.8 = 0.64$$

Cost of product = ₹ 50 + ₹ 60 = ₹ 110.

Case 7 a(0.95) and b(0.9) in parallel

The reliability of the product,

$$R_p = 1 - (1 - 0.95) \times (1 - 0.9) = 0.995$$

Cost of product = ₹ 150 + ₹ 90 = ₹ 240.

Case 8 a(0.95) and b(0.9) in series

The reliability of the product,

$$R_s = 0.95 \times 0.9 = 0.855$$

Cost of product = ₹ 150 + ₹ 90 = ₹ 240.

Case 9 a(0.95) and b(0.8) in parallel

The reliability of the product,

$$R_p = 1 - (1 - 0.95) \times (1 - 0.8) = 0.99$$

Cost of product = ₹ 150 + ₹ 60 = ₹ 210.

Case 10 a(0.95) and b(0.8) in series

The reliability of the product,

$$R_s = 0.95 \times 0.8 = 0.76$$

Cost of product = ₹ 150 + ₹ 60 = ₹ 210.

Case 11 a(0.9) and b(0.95) in parallel

The reliability of the product,

$$R_p = 1 - (1 - 0.9) \times (1 - 0.95) = 0.995$$

Cost of product = ₹ 100 + ₹ 120 = ₹ 220.

Case 12 a(0.9) and b(0.95) in series

The reliability of the product,

$$R_s = 0.9 \times 0.95 = 0.855$$

Cost of product = ₹ 100 + ₹ 120 = ₹ 220.

Case 13 a(0.9) and b(0.8) in parallel

The reliability of the product,

$$R_p = 1 - (1 - 0.9) \times (1 - 0.8) = 0.98$$

Cost of product = ₹ 100 + ₹ 60 = ₹ 160.

Case 14 a(0.9) and b(0.95) in series

The reliability of the product,

$$R_s = 0.9 \times 0.8 = 0.72$$

Cost of product = ₹ 100 + ₹ 60 = ₹ 160.

Case 15 a(0.8) and b(0.95) in parallel

The reliability of the product,

$$R_p = 1 - (1 - 0.8) \times (1 - 0.95) = 0.99$$

Cost of product = ₹ 50 + ₹ 120 = ₹ 170.

Case 16 a(0.8) and b(0.95) in series

The reliability of the product,

$$R_s = 0.8 \times 0.95 = 0.76$$

Cost of product = ₹ 50 + ₹ 120 = ₹ 170.

Case 17 a(0.8) and b(0.9) in parallel

The reliability of the product,

$$R_p = 1 - (1 - 0.8) \times (1 - 0.9) = 0.98$$

Cost of product = ₹ 50 + ₹ 90 = ₹ 140.

Case 18 a(0.8) and b(0.9) in series

The reliability of the product,

$$R_s = 0.8 \times 0.9 = 0.72$$

Cost of product = ₹ 50 + ₹ 90 = ₹ 140.

A product can be selected on two criteria (i) cost of the product, (ii) reliability requirement. These are compiled and shown in the following table.

Case	*Reliability of*			*Parallel/Series*	*Cost*
	Component 'a'	*Component 'b'*	*System*		
1.	**0.95**	**0.95**	**0.9975**	**Parallel**	**270**
2.	0.95	0.95	0.9095	Series	270
3.	0.9	0.9	0.99	Parallel	190
4.	0.9	0.9	0.81	Series	190
5.	**0.8**	**0.8**	**0.96**	**Parallel**	**110**
6.	0.8	0.8	0.64	Series	110
7.	0.95	0.9	0.995	Parallel	240
8.	0.95	0.9	0.855	Series	240
9.	0.95	0.8	0.99	Parallel	210
10.	0.95	0.8	0.76	Series	210
11.	0.9	0.95	0.995	Parallel	220
12.	0.9	0.95	0.855	Series	220
13.	0.9	0.8	0.98	Parallel	160
14.	0.9	0.8	0.72	Series	160

15.	**0.8**	**0.95**	**0.99**	**Parallel**	**170**
16.	0.8	0.95	0.76	Series	170
17.	0.8	0.9	0.98	Parallel	140
18.	0.8	0.9	0.72	Series	140

Say, if the reliability of the component should be highest (0.9975) no matter how high the cost (₹ 270) is, then we can go for case 1 (i.e., 0.95 reliability for both 'a' and 'b' connected in parallel) while case 5 (0.8 reliability for both 'a' and 'b' connected in parallel) is better if we choose the lowest cost (₹ 110) for reasonably good reliability such as 0.96. However, it seems most optimal values (system reliability is 0.99 at the cost of ₹ 170) can be obtained in case 15 (i.e., the components 'a' and 'b' with the reliabilities 0.8 and 0.95, respectively connected in parallel). Another most important point to observe in the above table is that series connection cannot give higher reliability, therefore, we can ignore those nine (9) cases, i.e., all even numbered cases of the above table.

BUSTLE - 13.1

Divide the following into parts or sub-units and arrange them as a block diagram of these sub-units in series or parallel or combinations of sub-systems. Design the reliability requirements for the sub-units so that the total may run within range 0.9 to 0.95 reliability.

(a) A ceiling fan

___

___

(b) A bike

___

___

Test Your Understanding 13.1

1. Derive the expression for the overall reliability of a circuit containing 'k' components connected in series.
2. Derive the expression for the overall reliability of a circuit containing 'k' components connected in parallel.
3. Describe the procedure to evaluate overall reliability for a circuit with combination (series and parallel) of components.
4. How do you improve or reduce the reliability of the equipment using the series and parallel connections? Illustrate with an example.
5. Find the reliability of the system with four components connected in (a) series and (b) parallel each of having reliability 0.9.
6. Find the reliability of a system containing three sub-systems of 0.8 reliable each connected in (a) series and (b) parallel.
7. Three components of failure probabilities 0.1, 0.2, and 0.3 are to be connected to get maximum possible reliability. Design the most suitable circuit.

8. Find the reliability of the following combination of the components whose reliabilities are shown in the blocks of the diagram.

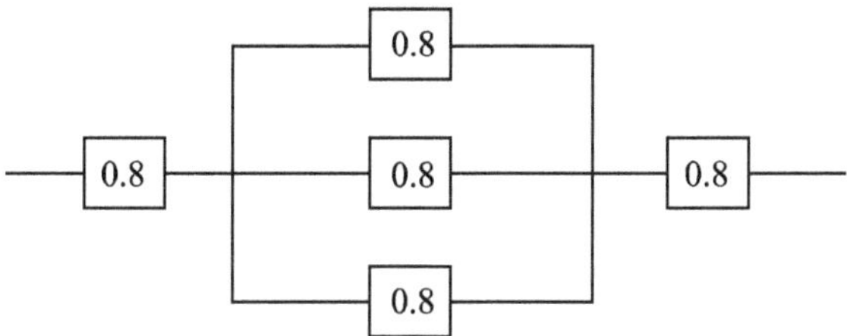

9. Find the reliability of the following combination of the components whose reliabilities are shown in the blocks of the diagram.

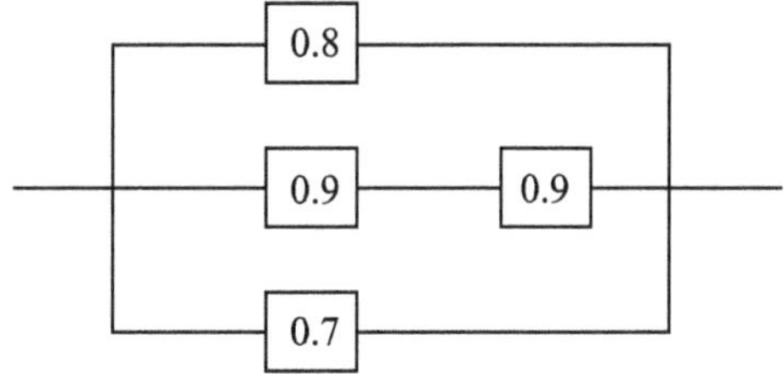

10. Calculate the reliability of the following systems in which the probability of functioning of each component is given in the figure itself.

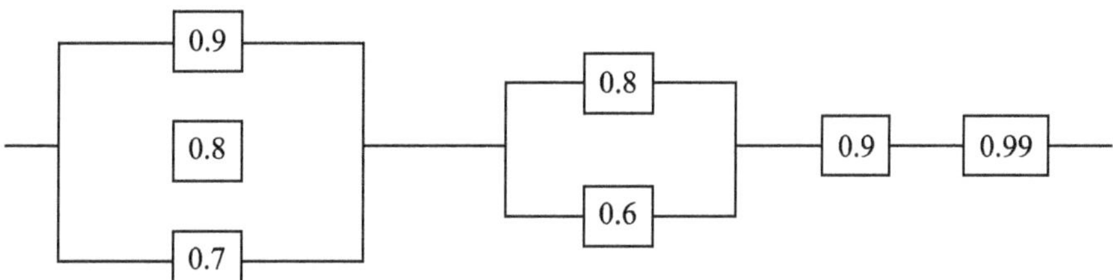

11. The reliabilities and corresponding costs of two sub-components x and y of a system are given in the following table. Find all possible system reliabilities and their costs. Give your comments to estimate the most optimal reliability for the system composed of two components x and y. Find the combination of components such that (a) reliability should not be less than 85% (b) cost should not be more than ₹ 150/- with

Sub component	*Cost for Reliability*		
	0.8	0.9	0.95
A	50	90	150
B	70	90	120

Answer: (a) x(0.95) and y(0.90) or x(0.95) and y(0.95) (b) x(0.9) and y(0.8)

13.4 MODULAR DESIGN OR MODULARISATION

The concepts of series and parallel systems of reliability discussed above have raised the scope for modular design concept. The modular design concept gives us, a range of choice in choosing a product(s). Suppose for a personal computer, two types of keyboards, three types of printers, three types of scanners, and three video monitors are available (all these are compatible to the personal computer). Then, we can have $2 \times 3 \times 3 \times 3 = 54$ varieties to choose from. Thus, modular design helps in creating a product in more variety through various combinations of its sub-systems. The design of various modules once stabilised (standardised), its production method and process are standardised.

13.4.1 Advantages of Modular Design

The modular design concept offers the following additional benefits:

- During usage, problems are easier to diagnose.
- Modules are easier to service.
- Production efficiency increases.
- Easy transferability and adaptability.
- Replacement of components is made easy because of availability, and compatibility of the components is simplified due to interchangeability.
- More variety to satisfy different taste of different customers.
- Production cost reduces.
- Increases the competitive spirit in the manufacturers (or vendors) of the components and thus increases the quality and reliability of the parts.
- Materials planning and inventory control can be simplified, especially, in finished goods inventory.

13.4.2 Redundancy

This is a technique that utilises two parallel elements to reduce the probability of failure.

$$\text{Reliability} = 1 - F^2 \text{ for two parallel elements.}$$

13.4.3 Group Redundancy

In group redundancy, each channel of elements which performs a function is provided with a duplication channel. That is, there is a normal channel and a standby (or a redundant) channel.

ILLUSTRATION 13.5:

Two group systems R_1 and R_2 have reliability 0.5 and 0.8, respectively. Determine the reliability under the following arrangements:

(i) R_1 and R_2 are cascaded.

(ii) R_1 and R_2 are cascaded and only R_1 is redundant.

(iii) R_1 and R_2 are cascaded and only R_2 is redundant.

(iv) R_1 and R_2 are cascaded and but are redundant.

(v) R_1 and R_2 are cascaded and the cascaded combination (say R_3) has group redundancy.

Solution.

Given that

Reliability $R_1 = 0.5$

Reliability $R_2 = 0.8$

(i) **R_1 and R_2 are cascaded.**

The reliability of the configuration

$$= R_1 \times R_2$$
$$= 0.8 \times 0.5$$
$$= \mathbf{0.4.}$$

(ii) **R_1 and R_2 are cascaded and only R_1 is redundant.**

Reliability of R_2

$$= 0.8$$

Reliability of redundant R_1

$$= 1 - F^2$$
$$= 1 - 0.5 \times 0.5$$
$$= 0.75$$

The reliability of (cascaded) combination

$$= 0.8 \times 0.75$$
$$= \mathbf{0.600}$$

(iii) **R_1 and R_2 are cascaded and only R_2 is redundant.**

Reliability of $R_1 = 0.5$

Reliability of redundant R_2

$$= 1 - F^2$$
$$= 1 - 0.8 \times 0.8$$
$$= 1 - 0.64 = 0.36$$

The reliability of cascaded combination

Reliability of combination

$$= 0.5 \times 0.36$$
$$= \mathbf{0.180}$$

(iv) **R_1 and R_2 are cascaded and both are redundant.**

Reliability of redundant R_1

$$= 1 - 0.2 \times 0.2$$
$$= \mathbf{0.96}$$

Reliability of whole system

$$= 0.75 \times 0.96$$
$$= \mathbf{0.72.}$$

(v) **R_1 and R_2 are cascaded and the cascaded system say R_3 has group redundancy.**

Reliability of cascaded combination of R_1 and R_2

$$= 0.8 \times 0.5$$
$$= \mathbf{0.40}$$

Reliability of group redundancy

$$= 1 - (1 - 0.40)^2$$
$$= 1 - 0.6 \times 0.6$$
$$= \mathbf{0.64}$$

13.5 DISTINCTION BETWEEN QUALITY AND RELIABILITY

Quality of a product is the degree of conformance to a set or specified standard for it. Quality can be assured for efficiency, accuracy of dimension, accuracy of surface finish, aesthetic standards, etc. For example, a component is specified with a dimension of say 10.0±0.2 mm. The component is manufactured and the dimension is then measured. If the actual dimension lies between 9.8 mm and 10.2 mm, it is said that the component is of good quality otherwise, of poor quality. Suppose you want a surface to be very smooth (1 micron roughness value). If the surface generated is not very smooth (i.e., it does not conform to the specified or desired degree), it is said to have poor quality. On the contrary, if your requirements are a rough surface, then a component of rough surface is said to be of good quality.

In simple words, *reliability* is *'trustworthiness'* particularly, in terms of functioning of the system. Reliability can be defined in different ways in different contexts. A few definitions are given below:

In the context of equipment health, *"reliability of an equipment is the ability to function against failure".*

or

In the context of production, *"reliability of a machine is the ability to produce continuously good items".*

or

In the context of measurement, inspection, and quality control, *"reliability of measuring device is the ability to measure dimensions with equal accuracy or precision, etc."*

13.6 DEFINITIONS AND TERMS USED IN RELIABILITY CALCULATIONS

The definitions and mathematical relationships often used in reliability calculations are given below.

(a) **Reliability:** Reliability is defined as probability that a product, a component, an equipment, or a system will perform its intended function for a specific period of time or life time under specified operating conditions or sum total of reliabilities of all components in prescribed operating conditions.

OR

It is also equal to one minus probability of failure

$$R(t) = 1 - F(t)$$

(b) **Time To Repair (TTR):** It is the time taken to restore the equipment to effective working condition.

(c) **Mean Time To Repair (MTTR):** It is the time required to locate a failure and to repair (or to restore the condition of the equipment).

(d) **Time Between Failures (TBF)**: It is the time difference between one failure to the next failure.

(e) **Mean Time Between Failures (MTBF):** It is the mean of the time interval between consecutive failures of equipment. Therefore,

MTBF = Total operating time/Number of failures

(f) **Time To Fail (TTF):** It is the total failures of the machine, i.e., the time taken after failure to working.

(g) **Mean Time To Fail (MTTF):** It is the expected value of mean time to failure, which is expressed as

$$\text{MTTF} = \int R(t)dt \quad \text{where } R(t) \text{ is reliability function on 't'}$$

(h) **Time Between Maintenance (TBM):** It is the time gap between two consecutive maintenances (may be planned or unplanned).

(i) **Mean Time Between Maintenance (MTBM):** It is the average of all the durations between the end of one maintenance to the start of the next, may it be planned or unplanned. In other words, it is the average time that the equipment is utilised for the operation.

(j) **Failure Rate:** It is the number of failure per unit time. Therefore,

Failure rate (λ) = 1/MTBF

(k) **Probability of Failure or Unreliability:** Probability of failing is equal to one minus the reliability of a system. It is also called the probability of not functioning or unreliability.

$$\text{Probability of failure } [F(t)] = n.t/T$$
$$= 1 - \text{Reliability of the system}$$
$$= 1 - e^{-nt/T}$$

Where n = number of elements in a system (or device)

t = Cumulative operating time

T = MTBF

(l) **Reliability Improvement Factor (RIF):** It is the ratio of probability of failure of a system to that of improved system.

RIF = Probability of failure before improvement/Probability of failure after improvement

(m) **Maintainability:** Maintainability is defined as the ease with which machine can be put into operation. It is a factor of design.

(n) **Downtime:** It is the total time during which the system is not in acceptable operating condition. Downtime can further be subdivided into a number of categories.

Actual repair time is that portion of downtime during which one or more technicians are working on the system to affect a repair. This time includes preparation time, fault location time, fault correction time, and checkout for the system.

(o) **Operating Time:** It is the time during which system is operating in a manner acceptable to the operator.

(p) **Logistic Time:** It is that portion of down time during which repair is delayed slowly because it is necessary to wait for manpower and/or for replacement part or subdivision of the system.

(q) **Administration Time:** It is that portion of downtime not included under active repair time and logistic time; this includes both necessary administrative activities and unnecessary wasted time.

ILLUSTRATION 13.6:

The cumulative operating time is found to be 50 hours in a system consisting of 100 components each having mean time between failures (MTBF) of 5000 hours. Find the reliability of the system.

Solution.

Given that

Number of components (n) = 50

Cumulative operating time = 100 hrs.

MTBF = T = 10, 000

We know that

Probability of failure [F(t)] = 1 – Reliability of the system = $1 - e^{-nt/T}$

Therefore,

reliability of the system = $1 - F(t) = e^{-nt/T}$

$= e^{-50*100/5000}$

= 0.905

ILLUSTRATION 13.7:

The reliability of a system is estimated at 0.8. On application of condition monitoring techniques, the reliability of a system is found to be improved to 0.9. Find reliability improvement factor (RIF) and give your comments on its probability of failures.

Solution.

The probability of failure before application of CM techniques (say F_1) is

= 1 – 0.8 = 0.2

The probability of failure after application of CM techniques (say F_2) is

= 1 – 0.9 = 0.1

The reliability improvement factor (RIF)

= Fail probability before applying CM Techniques/Fail Probability after applying CM

$= F_1/F_2$

= 0.2/0.1 = 2

Hence, the system is improving (the improvement is twofold, i.e., doubled).

BUSTLE - 13.2

In a factory having at least 50 machines, choose few machines and collect the failure data of these machines for the last one year from the logbooks of maintenance department. Calculate the MTBF (or MTTF for non-repairable type), MTTR, MTBM, and MDT.

____

____

____

____

Test Your Understanding 13.2

1. How does the modularisation help in reliability design of the equipment? Explain with an example.
2. What do you understand by the term 'modular design' or 'modularisation'? List its advantages.
3. Distinguish between quality and reliability. How can you interlink them?
4. Define and explain the terms MTBF, MTBM, MTTF, MTTR, and MDT.
5. Define and explain the terms operating time, downtime, logistic time, and administration time.
6. Define and explain the terms FR, RIF, unreliability, redundancy, and group redundancy.
7. The mean time between failures (MTBF) of 100 components is 10000 hours. The cumulative operating time is found to be 100 hours in the system. Find the reliability of the system. [Ans: 0.905]
8. The reliability of a system is estimated at 0.75. After renovation, the reliability of the system is found to be increased by 0.05. Find reliability improvement factor (RIF). [Ans: 1.25]

13.7 AVAILABILITY

The capacity of a system to perform its intended function when called upon to do so is often referred to as availability of operational readiness. In the simplest two state case, where equipment or system is either working at full capacity or not working the availability is commonly defined as the probability of the system operating satisfactory and is represented as

Availability = Total up time/(Total up time + Downtime)

Mathematically, it can be expressed as the ratio of mean time between failures (i.e., operating time) to the total available time of the system. Therefore,

Availability = MTBF/ (MTBF + MTTR)

13.7.1 Classification of Availability

There are three forms of availability based on the classification of downtime viz. working time, standby time, and lost time in routing preventive maintenance.

(a) **Intrinsic Availability (A_i)**

This is the probability that system is operating satisfactorily at any point of time, when under stated conditions, where the time considered is operating time and active repair time.

Design accuracy: It is the probability that the system will successfully accomplish its mission given the system is operating within design specifications.

$$A_i = \text{Operating time/(Operate time + Actual repair time)}$$
$$= \text{MTBF/(MTBF + MTTR)}$$

(b) **Achieved Availability (A_a)**

It is inclusive version of the inherent availability and includes factors influencing the availability of a system which are directly related with the designed maintainability and reliability characteristics of the system. This includes corrective and preventive maintenance and the mean maintenance time. Administrative or supply delay times are not included as downtime for the calculation of availability.

A_a= Mean time between maintenance/Total elapsed time exclusive of administrative and logistic delays

(c) **Operational Availability (A_0)**

It differs from the achieved availability in the sense that in case of operational availability as the total operational time, the total downtime including administrative delays is also taken into account.

13.7.2 Factors Affecting Availability and/or Reliability

Availability and hence reliability is affected by many aspects. However, the following are the more often registered factors that affect the reliability of the equipment. Hence, attention should be paid on these only which make the product or system reliable.

- Design
- Material
- Manufacturing method
- Complexity of product
- Working environment
- Testing and inspection
- Maintenance policies
- Large number of components with many sub-assemblies

ILLUSTRATION 13.8:

A complex electronic system consisting electronic gadgets has MTBF 100 hours and MTTR of 120 minutes. Find the availability.

Solution.

Given MTBF = 100 hrs.

MTTR = 120 minutes = 2 hrs.

Availability = MTBF/(MTBF + MTTR)

100/(100 + 2) = 100/102 = 0.9804

Comment: If the reliability of the system is increased from 80% to 90%, the probability of failure is reduced to half. Consequently, the system has become twice reliable after application of condition monitoring techniques as compared to initial situation.

ILLUSTRATION 13.9:

An old electronic system consists five vacuum tubes whose MTBF is 10,000 hrs. Now, when the system is fitted with 20 Integrated Chips (ICs), the MTBF is increased to 80,000 hours. Find the effect of the ICs on reliability of the system. If the transistors are used instead of ICs, we require 40 transistors. Evaluate if the transistors can be used.

Solution.

(i) The probability of failure in initial condition with vacuum tubes

$$P_1 = n_1.t/T_1$$
$$= 5.t/10000 = t/2000$$

(ii) The probability of failure with ICs

$$P_2 = n_2.t/T_2$$
$$= 20.t/80000 = t/4000$$

Reliability improvement factor (RIF)

$$= P_1/P_2$$

$$= (t/2000)/(t/4000) = 2$$

With application of ICs, the reliability of the system has improved (because the probability of failure has reduced). Hence, it is better to use ICs.

(iii) The probability of failure with 40 transistors

$$P_3 = n_3.t/T_3$$
$$= 40.t/(80000)$$
$$= t/(2000)$$

The reliability improvement factor is

$$= P_1/P_3$$
$$= (t/2000)/(t/2000) = 1$$

Thus, we observe that with increase in number of transistors (i.e., elements in general), the reliability of the system has reduced. Hence, the number of parts or components should be as less as possible for higher reliability.

ILLUSTRATION 13.10:

An electronically controlled system has 100 elements and each has MTBF of 10,000 hours. Its cumulative operating time is one hour. Calculate the following:

(i) The probability of failure

(ii) The probability of failure if the elements are grouped as a set of 10 each in redundant manner.

(iii) The reliability improvement factor (RIF)

Solution.

(i) Given,

$$n = 100 \text{ elements}$$
$$t = 1 \text{ hr}$$
$$T = 10{,}000 \text{ hrs.}$$

The probability of failure,

$$P_{f1} = n.t/T$$
$$= 100 \times 1/10{,}000$$
$$= 0.001$$

(ii) The group redundancy case

$$P_{f2} = N.n$$
$$= (t/T)^2$$
$$= 100 \times 10 \times (1/104)^2$$
$$= 10^{-5}$$

(iii) The reliability improvement factor

$$RIF = p_{f1}/p_{f2}$$
$$= 0.01/10^{-5} = 10^3 \text{ or } 1000$$

13.8 MAINTAINABILITY

It is defined as *the probability that a system can be restored to predetermined level of quality (or condition) in a specified period, when the maintenance is done with prescribed procedure and resources.*

Better maintainability of an equipment means,

- The downtime of the system is low,
- Time to carry out preventive maintenance is low, and
- Time taken in fault-analysis, repair, replacement, and subsequent operations are low.

The following terms are used in reference to maintainability:

(a) **Mean Time Between Maintenance (MTBF):** It indicates the frequency of maintenance.

(b) **Mean Time To Repair (MTTR):** It is mean time to require for repair of equipment. It does not include wait time.

13.8.1 Factors Affecting Maintainability

- Training of maintenance crew (poorly trained crew will take unduly more time)
- Maintenance procedure (complicated procedure delay maintenance period)
- Maintenance guidance and monitoring (lack of proper guidance and monitoring will increase the maintainability)
- Availability of maintenance manual (unavailability of proper maintenance manual delay the time)
- Safety requirements (higher safety requirements delay the process)
- Standardisation prescribed and to be achieved (standardisation needs of high degree will take more time to achieve)
- Accessibility (of equipment, tolls, gadgets necessary to carry out maintenance)
- Expertise required
- Special items (of extraordinary quality and specifications) essential for replacements

BUSTLE - 13.3

Calculate the availability and other relevant maintenance characteristics for the data you have collected in Bustle - 13.2.

____

____

____

____

Test Your Understanding 13.3

1. What is 'availability'? Explain. Classify and explain different types of availabilities of equipment.
2. List the factors affecting the availability of the equipment.
3. Define the term maintainability. What are the factors that affect maintainability?
4. Explain the terms with suitable examples.
   (a) Operating Availability
   (b) Intrinsic Availability
   (c) Achieved Availability
5. Suggest some measures to improve the availability and maintainability of given equipment.

13.9 OVERALL EQUIPMENT EFFECTIVENESS (OEE)

The key figure used with respect to these factors is equipment effectiveness, defined as follows:

Overall equipment effectiveness is the effective utilisation of the equipment on the whole so that the desired productivity is achieved at a certain quality level and is mathematically expressed as the product of equipment availability, performance efficiency, and the rate of quality.

Overall Equipment Effectiveness (OEE)

$$= \text{M/C Availability} \times \text{Performance Efficiency} \times \text{Rate of Quality}$$

or

$$= A.\eta.RoQ$$

Where,

$$\text{Equipment Availability (A)} = \frac{\text{Available Time}}{\text{Planned Time}} = \frac{\text{Planned Time} - \text{Downtime}}{\text{Planned Time}}$$

$$\text{Performance Efficiency } (\eta) = \frac{\text{Theoretical Cycle Time}}{\text{Actual Cycle Time}}$$

$$\text{Rate of Quality (RoQ)} = \frac{\text{Accepted Quantity}}{\text{Produced Quantity}}$$

Thus,

$$\text{OEE} = \frac{\text{Planned Time} - \text{Downtime}}{\text{Planned Time}} \times \frac{\text{Theoretical Cycle Time}}{\text{Actual Cycle Time}} \times \frac{\text{Accepted Quantity}}{\text{Produced Quantity}}$$

13.9.1 Performance Efficiency

Performance efficiency is product of operating speed rate and the net operating rate. Operating speed rate of equipment refers to discrepancy between the ideal speed (based on equipment capacity as desired) and its actual operating speed.

$$\text{Operating Speed Rate} = \frac{\text{Theoretical Cycle Time}}{\text{Actual Cycle Time}}$$

The net operating rate measures the maintenance of given speed over a given period.

$$\text{Net Operating Rate} = \frac{\text{Actual Processing Time}}{\text{Operation Time}}$$

$$= \text{Processed Amount} \times \frac{\text{Actual Cycle Time}}{\text{Operation Time}}$$

Therefore,

Actual Cycle Time = Net Operating Rate × Operating Time/ Processed Amount

ILLUSTRATION 13.11:

Five presses of certain workshop are planned in three shifts per day. Any machine is kept idle during the lunch hour (one hour) and in tea breaks (15 minutes × two times per shift). The machines are found to be under breakdown for 50, 20, 28, 40, and 45 hours, respectively in a certain month from the logbooks of Plant Engineering. The production department observes that the cycle time of operations are 12, 10, 9, 12, and 10 seconds while these are found as 10, 10, 8, 9, and 8 seconds from the record books of the Industrial Engineering department. The Quality Control department reported that the numbers of defective parts produced by these machines are 150, 220, 125, 140, and 145 respectively among the quantities 2000, 1500, 1500, 1800, and 1500 produced. The plant manager has to predict and decide the order of these machines for maintenance. He assumes that the machine is to be taken for maintenance immediately if overall equipment effectiveness (OEE) falls below 65% and 5 days hence for every increment of 1% above 65%. Schedule the machines for maintenance. Consider 25 working days in the month.

Solution.

Given that

Number of working days = 25

Number of shifts per day = 3

Number of hours per shift = 8

Total Time = Number of working days × Number of shifts per day × Number of hrs per shift

= 25 × 3 × 8 = 600 hrs

Idle time per shift = 1 hr (Lunch break) + 2 × 15 min (Tea breaks)

= 1.5 hrs or 90 min.

Idle time per day = Idle time per shift × Number of shifts = 1.5 × 3

Total idle time = Idle time per shift × Number of shifts × Number of working days

= 1.5 × 3 × 25 = 112.5 hrs

So,

Planned Time = Total Time – Idle Time = 600 – 112.5 = 487.5 hrs

The calculations for each machine and the solution are shown in the following tabular form.

The Formulae used are:

Availability (A) = (Planned Time – Down Time)/Planned Time

Performance Efficiency (η) = Theoretical Cycle Time/Actual Cycle Time

Rate of Quality (RoQ) = Accepted Production Quantity/Total Production Quantity

Overall Equipment Effectiveness (OEE) = A × η × RoQ

Abbreviations used: TT: Total Time; PT: Planned Time; DT: Downtime
TCT: Theoretical Cycle Time; ACT: Actual Cycle Time
TP: Total Produced Quantity; Def: Defective Quantity
Acc: Accepted Production Quantity; RoQ: Rate of Quality
OEE: Overall Equipment Effectiveness

M/c	TT	PT	DT	Availability	TCT	ACT	η	TP	Def	Acc	RoQ	OEE	Rank
Press 1	600	487.5	50	0.897435897	10	12	0.833333	2000	150	1850	0.925	0.6918	III
Press 2	600	487.5	20	0.958974359	10	10	1	1500	220	1280	0.853333	0.8183	I
Press 3	600	457.5	28	0.942564103	8	9	0.888889	1500	125	1375	0.916667	0.768	II
Press 4	600	487.5	40	0.917948718	9	12	0.75	1800	140	1660	0.922222	0.6349	V
Press 5	600	487.5	45	0.907692308	8	10	0.8	1500	145	1355	0.903333	0.656	IV

From the above table, the following conclusion can be derived:

Press 4 is to be taken up for maintenance immediately.

Press 5 is to be taken up within the first 5 days.

Press 1 is to be taken up after 20 days from starting day.

Press 3 is to taken up between 55th to 60th day from the starting day.

Press 2 is to be planned in between 80th to 85th day from the starting day.

ILLUSTRATION 13.12:

A manufacturing firm wants to sell a machine which is run for 5 years. The finance manager estimated at 10% straight-line depreciation per annum. But the maintenance manager claims that the machine was not continuously run and is in very good condition and hence it should not be considered as 5 years old. He advises to consider the overall equipment effectiveness to decide the selling price. The production records of the machine in the past 6 months reveal the information that 20, 30, 25, 30, 25, and 20 defectives were found in the productions of 400, 500, 500, 600, 400, and 500 parts produced, respectively. The theoretical cycle time of the operation is 20 seconds while it is found to be 25 seconds in practice. The planned hours of production in the past six months were 200, 250, 200, 200, 250, and 200 hours of which it was under breakdown for 20, 10, 15, 10, 50, and 5 hours respectively. Check whether the maintenance manager's claim is considerable.

Solution.

Total Planned Time in Six Months = 200 + 250 + 200 + 200 + 250 + 200 = 1300

Total Down Time = 20 + 10 + 15 + 10 + 50 + 5 = 110

% Availability = (Planned Time – Downtime)/Planned Time

= (1300 – 110)/1300 = 91.5%

Performance Efficiency = Theoretical Cycle Time/Actual Cycle Time

= 20/25 = 80%

Total Production = 400 + 500 + 500 + 600 + 400 + 500 = 2900

Defectives Produced = 20 + 30 + 25 + 30 + 25 + 20 = 150

Rate of Quality = (2900 – 150)/2900 = 94.8%

OEE = 0.915 × 0.8 × 0.948 = 0.694 or 69.4%

By straight line depreciation, the selling price reduces to 50% of its cost while by OEE it comes to 69.4% (or 70%). Hence, the maintenance manager's claim is justified.

13.9.2 Evaluation of Maintenance Performance [Maintenance Performance Indices (MPI)]

The performance of maintenance department can be assessed in many ways since it is connected to many factors (refer section 13.10.3). In fact, it becomes very difficult at times and may raise conflicts also. Of course, the terms like MTTF, MTTR, MTBF, MDT, etc., are the most common methods to evaluate the performance of the maintenance and the efficiency of its personnel. Some people do not accept or do not get satisfied with these analyses. They claim that they did the best and throw the blame on the equipment inefficiency or its operators' faults. Though acceptable in some cases, the blames of the production and maintenance departments on each other can be answered by the following ratios or indices. These are often referred to as maintenance performance indices or maintenance performance indicators (MPI). These can also be used to decide the incentives for the maintenance workmen.

1. Effectiveness of Maintenance Planning

$$= \frac{\text{Downtime due to scheduled maintenance}}{\text{Downtime due to total maintenance work}}$$

2. Frequency of Breakdown $= \dfrac{\text{Number of breakdowns}}{\text{Available machine hour}}$

3. Equipment Utilisation Factor $= \dfrac{\text{Equipment operating time}}{\text{Total available time}}$

   Equipment Capacity Utilisation $= \dfrac{\text{Equipment operating time}}{\text{Total available time}} \times 100$

4. Equipment Availability = Operation time/(Operation time + Maintenance time)

   Percent Availability $= \dfrac{(\text{Total available hours} - \text{Downtime}) \times 100}{\text{Total available hours}}$

5. Maintenance Cost Index $= \dfrac{\text{Total maintenance costs}}{\text{Annual sales}}$

6. Overhaul Expenditure Ratio $= \dfrac{\text{Overhaul expenditure}}{\text{Value of equipment}}$

7. Spare Parts Ratio $= \dfrac{\text{Value of spares consumed}}{\text{Value of equipment}}$

8. Breakdown Cost $= \dfrac{\text{Cost of total breakdown repair}}{\text{Total number of breakdown}}$

9. Maintenance Man Power Efficiency $= \dfrac{\text{Actual total man hours spent on maintenance job}}{\text{Total planned man hours for maintenance work}}$

10. Maintenance Overtime Ratio $= \dfrac{\text{Overtime hours used for maintenance}}{\text{Total maintenance hours}}$

The following aspects are also useful in evaluating maintenance performance in certain situations:

- Trend analysis of spare parts consumptions
- Obsolete items value
- Design imperfections
- Cost overrun frequency
- Overtime requirements
- Maintenance expenses per man-hour
- Trend of downtime
- Replacement requirements
- Maintenance expenses per kilowatt of power consumed
- Ease of repair
- Trend in skill requirement

13.9.3 Maximising Equipment Effectiveness

The rate of equipment effectiveness is actually the operating rate or availability. We know that,

$$\text{Availability} = \frac{\text{Operating Time}}{\text{Load Time}} = \frac{\text{Loading Time} - \text{Downtime}}{\text{Loading Time}}$$

Here, loading time is available time per day and is derived by subtracting planned downtime from total available time per day. Planned downtime is the official time scheduled in the production plan for maintenance activities. Thus, if this can be reduced, the numerator can be increased and thence the equipment effectiveness can be improved.

13.9.4 Factors Affecting Maintenance Performance Indices

The maintenance performance indices are affected by many reasons. And it is very difficult to evaluate in many occasions. For example, the problems that are encountered in evaluating the maintenance manpower efficiency index in a hypothetical industry are enumerated here below. In the same manner, one can enlist such practical problems in assessment of other MPIs.

Case Objective: The assessment of number of maintenance personnel or evaluating the maintenance manpower efficiency index.

Practical Problems Noticed

- For decentralised system, every section/department may need 2 or 3 maintenance persons, for day-to-day inspection and minor corrective actions. Additionally, the central organisation may need 10 or 12 people. They are deputed to the departments on requirement basis. These persons may be utilised for other services or transferred or influenced by hire/fire decisions when the work in their allocated departments are subjected to changes. Thus, this term may affect the maintenance manpower efficiency ratio (of the department) given above.

- For centralised system, all persons remain at the central office; they are given jobs (i.e., sent to various departments) either on emergency demand or planned basis. Here each team performance cannot be assessed and the efficient worker as well as the inefficient worker may receive the same incentive which can affect the morale of the former one.
- Smaller organisations keep a small team of maintenance personnel. They may hire experts/ skilled persons from outside on need. This saves the expenditure on maintenance of the organisation. The special maintenance jobs are even contracted to outside party. It becomes difficult in such situations to calculate the efficiency index.
- To prevent any loss of production, the regular maintenance work is scheduled on weekly off day. The assembly line of an automobile industry is generally run in two shifts only or in general shift. In such cases, the preventive maintenance or regular check-up is done during off time.
- The planned maintenance for different sections/departments is scheduled on different days. Say, if the engine section is on Monday, then maintenance of crank-shaft section is scheduled on Tuesday. In this way, the same maintenance crew is engaged for the job in different shops. There may be one or two specialised maintenance man for each section or machine. This saves cost, by employing fewer maintenance people.
- Maintenance is a specialised job. For maintaining water supply system, a set of persons are employed. For electrical maintenance, a different set of persons are employed. For blast furnace or cold rolling mill, a third set of persons are engaged. For measuring instruments, optical, mechanical, electrical, etc., a different set of persons are required. Their natures of jobs largely differ.

BUSTLE - 13.4

1. *Collect relevant data and calculate the maintenance performance indices for the organisation/institution where you are working/studying or the one with which you are familiar.*
2. *Suppose you are the maintenance foreman. When a critical machine is under breakdown, the spares are found to be missing. They may not be available in the local market and even if available, they are very costly. What action plan would you suggest? How would you convince your superiors?*

____

____

____

____

Test Your Understanding 13.4

1. What is meant by the term 'overall equipment effectiveness (OEE)'? Explain the terms involved in evaluating OEE of a machine.
2. What do you understand by the term 'performance efficiency'? How do you evaluate? Illustrate with an example.
3. Explain various maintenance performance indices (MPIs). Also explain their importance.
4. Discuss the factors that affect the maintenance performance indices.
5. Define and explain the terms OEE, MPI, RoQ, maintenance cost index, and breakdown frequency.

6. An entrepreneur of press tools manufacturing plans to purchase the machines for his unit. Due to financial constraints, he decides to take new as well as some old machines. He buys two machines namely, one milling and one horizontal-boring machine, which are 3 years old. The finance manager estimates at 10% straight-line depreciation per annum. But the maintenance manager claims that even if the machine has been continuously run it is in very good condition and hence it should not be considered as 3 years old. He advises to consider the overall equipment effectiveness to decide the buying price. The production records of the machines in the past 9 months reveal the information that 12, 3, 5, 10, 2, 7, 2, 8, and 13 for milling and 2, 1, 0, 4, 0, 1, 1, 2, and 2 defectives were found in the productions of 1000, 1100, 800, 1200, 1400, 1200, 1000, 850, and 1150 parts produced, respectively. The theoretical cycle time of the operation is 20 minutes while it is found to be 15% of additional time in practice. The planned hours of production in the past nine months were 500 hours for each machine but the machines were under breakdown for 4, 10, 10, 6, 5, 4, 15, 12, and 5 hours for milling machine and 2, 7, 5, 4, 4, 8, 0, 10, and 9 hours for boring machine, respectively. Check whether maintenance manager's claim is considerable.
7. A firm wants to sell a machine which is run for 4 years. The finance manager estimates at 10% straight-line depreciation per annum. According to the maintenance manager's claim, the machine is in very good condition and hence it should not be considered as 4 years old. The following information is available from the production records of the machine in the past 6 months:

Month:	*Jan*	*Feb*	*Mar*	*Apr*	*May*	*Jun*
Produced Quantity:	2000	1800	1500	2500	1500	2200
Number of Defectives:	200	150	120	300	100	200
Planned Hours:	250	200	120	250	180	250
Breakdown Hours:	20	25	10	20	10	25

The theoretical cycle time of the operation is 20 seconds while it is found to be 25 seconds in practice. Use overall equipment effectiveness to check maintenance manager's claim.

8. Four CNC lathes in a workshop are planned in two shifts per day. Any machine is not kept idle. The machines are found to be under breakdown for five times with downtimes as 50, 10, 12, 18, and 10 hours respectively in a certain month from the logbooks of Plant Engineering. The Production department observes the cycle time of operations are 12, 10, 15, 10, and 10 seconds while these are found to be 10, 10, 10, 9, and 8 seconds from the record books of the Industrial Engineering department for five different jobs loaded on these machines. The Quality Control department reported that the numbers of defective parts produced by these machines are 105, 120, 125, 150, and 145, respectively among the quantities 2000, 2500, 1500, 1000, and 1500 produced. The plant manager has to predict and decide the order of these machines for maintenance. He assumes that the machine is to be taken for maintenance immediately if overall equipment effectiveness (OEE) falls below 65% and 5 days hence for every increment of 1% above 65%. Schedule the machines for maintenance. Consider 25 working days in the month.

13.10 CONCLUSION

The models of reliability, availability, maintainability, redundancy, and evaluation of overall equipment effectiveness (OEE) are the most significant issues of a maintenance department to judge whether a system and its maintenance is functioning well or not. The maintenance performance indices are used in several industries today to assess whether their system is effectively and efficiently maintained or not. These calculations in one form or the other play an important role in the engineering process and equipment and also are necessary in evaluating life cycle costing, cost benefit analysis, operating, and capability studies.

13.11 RECAP

This chapter focuses on reliability improvement by various combinations of the components such as series system, parallel system, and combined (series and parallel) system so that the overall reliability of the systems can be improved or designed as needed. The advantages of modular design or modularisation are explored and then quality and reliability are distinguished. Availability of the machine is another facet that this chapter covers. Different types of availability such as intrinsic availability (A_i), achieved availability (A_a), operational availability (A_o), and the factors affecting availability and hence the reliability are discussed. The effect of redundancy and group redundancy on reliability is narrated through the numerical problems. The details of maintainability and the factors affecting maintainability are very significant issues in reliability studies. Brief notes of these have been given a place in this chapter. One can evaluate the efficiency and effectiveness of the maintenance department through the maintenance performance indices (MPIs). In view of this, various MPIs are defined. The evaluation of overall equipment effectiveness (OEE), maintenance performance, and efficiency are also depicted through numerical problems.

13.12 KEY CONCEPTS

1. **Reliability in Series System:** $R_s = R_1 \times R_2 \times R_3 \times \ldots \times R_K \times \ldots R_N$
2. **Reliability in Parallel System:** $R_P = 1 - (1 - R_1)(1 - R_2)(1 - R_3) \ldots (1 - R_K) \ldots (1 - R_N)$
3. **Modularisation:** The design of various modules (sub-units) to standardise its production method and process and optimise the reliability.
4. **Redundancy:** This is a technique that utilises two parallel elements to reduce the probability of failure. Reliability = $1 - F^2$ for two parallel elements.
5. **Group Redundancy:** Each channel of elements which performs a function, is provided with a duplication channel or a standby (or a redundant) channel.
6. **Reliability:** The probability that a product, a component, an equipment or a system will perform its intended function for a specific period of time or life time under specified operating conditions or sum total of reliabilities of all components in prescribed operating conditions. $R(t) = 1 - F(t)$
7. **Mean Time To Repair (MTTR):** The time required to locate a failure and to repair (or to restore the condition of the equipment).
8. **Mean Time Between Failures (MTBF):** The mean of the time interval between consecutive failures of equipment. MTBF = Total operating time/Number of failures
9. **Mean Time To Fail (MTTF):** The expected value of mean time to failure, which is expressed as $MTTF = \int R(t)\,dt$ where R(t) is reliability function on t.
10. **Mean Time Between Maintenance (MTBM):** The average of all the durations between end of one maintenance to the start of next, may it be planned or unplanned.
11. **Failure Rate:** The number of failure per unit time. Failure rate (λ) = 1/MTBF
12. **Unreliability:** Probability of failing or not functioning. $1 - R(t)$ or $1 - e^{-nt/T}$
13. **Reliability Improvement Factor (RIF):** The ratio of probability of failure of a system to that of improved system. RIF = Probability of failure before improvement/Probability of failure after improvement.

14. **Operating Time:** The time during which system is operating in a manner acceptable to the operator.
15. **Downtime:** The total time during which the system is not in acceptable operating condition.
16. **Logistic Time:** The portion of downtime during which repair is delayed slowly because of necessary waiting for manpower and/or for replacement part or subdivision of the system.
17. **Administration Time:** The portion of downtime not included under active repair time and logistic time; this includes both necessary administrative activities and unnecessary wasted time.
18. **Availability:** The capacity of a system to perform its intended function when called upon to do so. Availability = Total Up Time/(Total Up Time + Downtime) = MTBF/(MTBF + MTTR) or Availability = Available Time/Planned Time = (Planned Time – Downtime)/ Planned Time
19. **Intrinsic Availability (A_i):** The probability that a system is operating satisfactorily at any point of time, when under stated conditions, where the time considered is operating time and active repair time. A_i = Operating time/(Operate time + Actual repair time) = MTBF/(MTBF + MTTR)
20. **Achieved Availability (A_a):** The inclusive version of the inherent availability and includes factors influencing the availability. This includes corrective and preventive maintenance and the mean maintenance time but not administrative or supply delay times. A_a = Mean time between maintenance/Total elapsed time exclusive of administrative and logistic delays
21. **Operational Availability (A_0):** It differs from the achieved availability in the sense that in case of operational availability as the total operational time, the total down time including administrative delays is also taken into account.
22. **Maintainability:** It is defined as the probability that a system can be restored to predetermined level of quality (or condition) in a specified period, when the maintenance is done with prescribed procedure and resources.
23. **Overall Equipment Effectiveness (OEE):** Overall equipment effectiveness is the effective utilisation of the equipment on the whole so that the desired productivity is achieved at a certain quality level. OEE = M/C Availability × Performance Efficiency × Rate of Quality = A.η.RoQ
24. **Rate of Quality (RoQ):** Accepted Quantity/Produced Quantity
25. **Performance Efficiency:** Performance efficiency is product of operating speed rate and the net operating rate. Operating speed rate of equipment refers to discrepancy between the ideal speed (based on equipment capacity as desired) and its actual operating speed.

    η = Theoretical Cycle Time/Actual Cycle Time

REVIEW QUESTIONS

1. Explain the change in system reliability when the components connected in series are changed to parallel. Illustrate with an example.
2. Discuss, with the help of an example, the change in system reliability when the components connected in parallel are changed to series.

3. What is modularisation? What are its merits?
4. Distinguish between
   (a) MTBF and MTTF
   (b) MTBM and MTTR
5. Write short notes on
   (a) Unreliability
   (b) Redundancy
6. Define and explain the terms reliability, availability, and maintainability. Discuss on what factors each of these terms depend.
7. What is meant by 'availability'? Explain different kinds of availabilities that we come across in reliability analysis. On what factors do they depend?

CHAPTER 14

Total Productive Maintenance and Kaizen

PREAMBLE

This chapter focuses on total productive maintenance. TPM aims at maximising the equipment effectiveness like tero-technology. The eight pillars of total productive maintenance and equipment effectiveness in maintenance are discussed. This chapter also covers the TPM template, expected contributions by application of TPM, and the barriers of adopting the TPM in the organisation. It also explains the concept of zero defects, zero maintenance, and zero accidents. This chapter will also help understand and implement Kaizen with right outlook.

CHAPTER OUTLINE

14.1 INTRODUCTION

Total productive maintenance (TPM), as defined by Seiichi Nakajima (a Japanese scientist), is an innovative approach to maintenance that optimises equipment effectiveness, eliminates breakdowns, improves productivity, and promotes autonomous maintenance by operators, through day-to-day activities involving total work force. TPM is a philosophy, a culture, and a new attitude towards maintenance. The total effectiveness is an indicator that is derived through equipment availability, performance efficiency, and the rate of quality product. An organisation that believes in total maintenance system should be able to set up activities directed towards appropriate preventive/predictive maintenance, maintainability improvement, and maintenance prevention. TPM involves wide range of activities involving proper operation, failure analysis, cost benefit analysis, design improvement, participation of operators in maintenance, positive attitude, and culture across organisation. Above all, it needs strong support and belief of top management in TPM programme and the total participation of all employees in TPM.

14.2 TPM ENVIRONMENT

Total productive maintenance (TPM) is the Japanese approach to maximise the effectiveness of the equipment or facility that is used in the industry. Total productive maintenance is an innovative and non-traditional approach to plant maintenance. It is complementary with total quality management (TQM), just in time (JIT), total employee involvement (TEI) and continuous performance improvement (CPI) and other world fame strategies like tero-technology and logistics which lie in the pursuit of economic life cycle Cost (LCC). The concept of preventive maintenance (PM) paved the way to new needs of the industry in the direction of its maintainability and method improvement (MI). This is made possible by predictive maintenance philosophy with condition and instrumentation to predict failure and plan maintenance strategy. Latest is maintenance prevention (MP) which incorporates equipment design stage that results in equipment's easy maintenance. The new concept of total productive maintenance (TPM) makes use of these techniques to achieve and maintain optimal condition for production, quality yield, and safety. TPM advocates overall effectiveness of eEquipment (OEE) through availability, performance, and quantity. TPM aims at reducing major losses in the system through zero defects, zero breakdowns, and zero accidents.

(The evolution of each of the maintenance philosophies has already been discussed at length in Chapter 3.)

14.3 RELATIONSHIP BETWEEN TPM, TERO-TECHNOLOGY, AND PHYSICAL ASSETS MANAGEMENT (PAM)

Tero-technology (developed in U.K.), Physical Assets Management and Logistics (in USA), and TPM (in Japan) have the same goals. Tero-technology as per BSI is a combination of management, financial, engineering, and other practices applied to physical assets in pursuit of economic life cycle costs (LCC). Its practice is concerned with the specification and design for reliability and maintainability of plant and machinery, equipment, buildings, and structures with their installation, commissioning, maintenance modification, and replacement with feedback of information on design performance and costs.

TPM aims to maximise equipment effectiveness. In fact, this is same as tero-technology's goal of attaining an economic life cycle cost.

Logistics is an old military term referring to support to front line through procurement, storage, transportation and maintenance of manufactured goods and systems. In current methods, logistics consists of LCC, reliability engineering and maintenance engineering. Thus, the goal of TPM, tero-technology, and logistics is same as goal of LCC and they differ in terms of pre-use target location and responsibility.

BUSTLE - 14.1

Assemble any five of your friends and any officials of Plant Engineering department (if possible) and raise a discussion on the following questions assuming a hypothetical organisation (such as a laptop manufacturing company or any other company you know).

1. *What is TPM?*
2. *Why TPM is to be adopted?*
3. *How TPM is to be implemented?*

____

____

____

____

Test Your Understanding 14.1

1. What do you understand by TPM? Explain the concept.
2. Explain the relation between TPM, tero-technology, and physical assets management.

14.4 TPM: AN OVERVIEW

"What is Total Productive Maintenance (TPM)?"

Suppose you have to answer the above question in one or two words or sentences, then what would be your answer? Is it a maintenance technique? Is it a method of maintenance? Is it a set of principles? Is it standard norm? Is it a theory? Are there any hard and fast rules to follow? Perhaps, none of these are right ways to answer because it is something more.

Then what is TPM? Well, TPM is a culture. TPM is the philosophy of maintenance culture that stands on the pillars of belief and works out by total employee involvement with respecting the promises for high quality productivity. It aims at attaining high quality, high productivity, zero defects, zero breakdowns, and zero accidents. To understand the TPM philosophy, we should be aware of its background and the need to implement.

There is an old story, which you might have learnt from your grandmother in your childhood. The story goes on as follows.

There was a fisherman who brought some fish one day and kept them under the sun. All the fish got dried up except one among them. The fisherman asked the fish why it had not dried up. The fish answered that the grass had shaded on it. Then the fisherman asked the grass why it came in the way. "The cow has not eaten me, what shall I do?" replied the grass. The fisherman then asked the cow why it did not eat. The cow replied that it was the fault of the cowboy who had not taken her to that grass. Then the fisherman asked the cowboy why he had not taken the cow to graze the grass. The cowboy blamed the ant that had bitten him due to which he could not do his duty. The fisherman asked the ant "Hello ant, why did you bite the cowboy?" "Why not! he has spoiled my house," replied the ant.

Of course, the story may not be a realistic one and very much hypothetical. However, this scene in the story can be observed everyday in almost every industry. In review meetings of an industry about why the targeted production is not achieved, the discussions often go in the following direction.

The production manager puts the blame on the poor condition of the machines that work as an impediment in reaching the target. The plant engineer blames the purchase department for bringing spurious spares. The purchaser forwards it to the store manager for not specifying correctly. Store manager passes it on to designer or planning department. And this goes on.

The attitudinal and behavioural environment of such employees is witnessed; it is known as ***throw out of wall syndrome***. (In short, it is referred to as TOW syndrome in this book.) This TOW syndrome to an industry is analogous to acquired immune deficiency syndrome (AIDS) to human beings. The TOW syndrome is even more dangerous than AIDS.

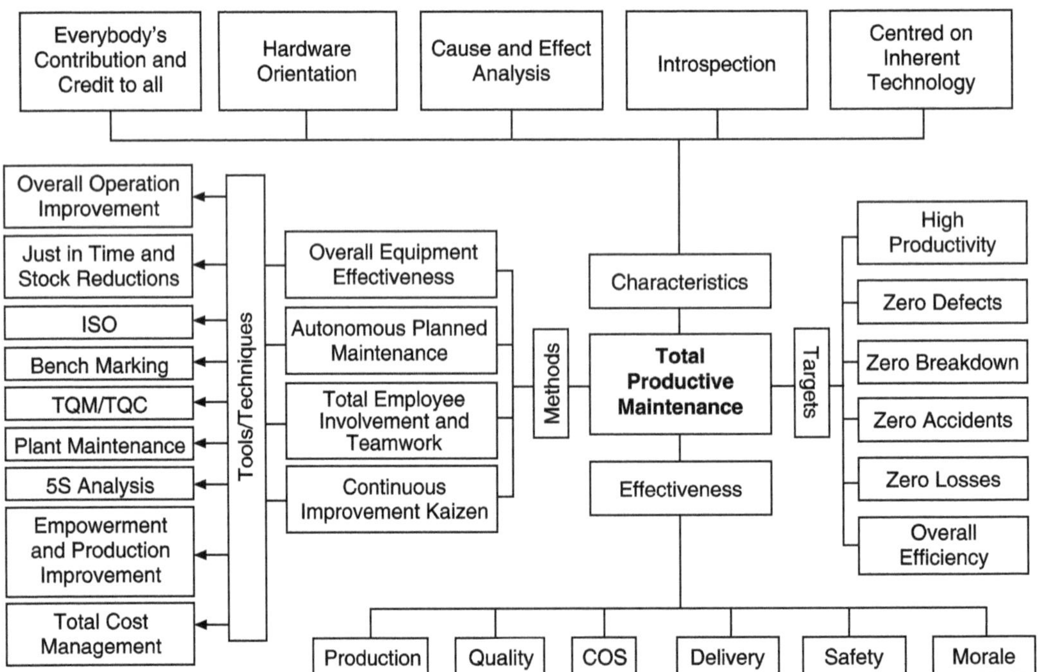

Figure 14.1 Management by TPM

Thence, there is a need for trouble-free and defect-free work atmosphere where there will be introspection at every stage of every person about his/her work for others and to the organisation rather than blaming somebody.

If we can imagine an environment where everybody's motto is to achieve the targeted production, for which the machinery is to be thoroughly maintained for high production throughout their full length of life right from design to discard, and there is no blaming of one over the other; ... such *workers' paradise* is said to have a culture of TPM. TPM takes into account of the design, manufacture, performance, life, cost, and quality of output of the machine and even up to discard as shown in Figure 14.1.

14.4.1 Objectives of TPM

(i) To maximise overall equipment effectiveness (OEE), i.e., availability, performance, and output.

(ii) To achieve zero defect.

(iii) To achieve zero breakdown.

(iv) To realise zero accident.

(v) To attain zero losses/waste.

(vi) To achieve total involvement, i.e., participation of all departments and sections, including planning, purchase, and user of the equipment (i.e., the production department).

(vii) To improve the overall general and congenial environment and working conditions of the plant.

(viii) To boost the morale of the employees by inculcating the sense of pride in them.

(ix) To strive for high productivity.

(x) To provide high quality in equipment and hence the products.

(xi) To maximise overall plant efficiency.

14.4.2 Approaches of TPM

The organisation that observes TPM runs on four tracks. These can be compared as the head lights or torches that show the path for the vehicle (organisation) on the road of TPM. These four approaches are given below.

- Overall equipment effectiveness (OEE)
- Autonomous planned maintenance (APM)
- Total employee involvement (TEI) and teamwork
- Kaizen or continuous improvement (CI)

These four approaches are simultaneously applied through various tools and techniques. These are discussed in the later section of this chapter. Some of the important tools and techniques connected with these approaches are listed below:

- Overall operation improvement
- Just in time and stock reductions
- ISO
- TQM/TQC
- Benchmarking

- Plant maintenance
- 5S analysis
- Empowerment and production improvement
- Total cost management

14.4.3 Characteristics of TPM Philosophy

When the organisation implements TPM, the following characteristics are found in the organisation.

(a) *Centred on Inherent Technology:* The TPM philosophy insists on the inherent technology and recommends to eliminating the defect at grass root level.

(b) *Hardware Orientation:* TPM promises defect-free environment with hard orientation.

(c) *Everybody's Contribution:* Everybody contributes to the success of TPM. Everyone in the organisation is part and parcel of the programme. Every individual owns wilful responsibility.

(d) *Credit to All:* Every individual of the organisation will enjoy the fruit of the success.

(e) *Introspection through Cause and Effect Analysis:* Each person at every stage of the job checks by own-self if it is useful or does cause any damage. The causes and effects of the failure, if any that are likely to occur are also studied and efforts are made to prevent the failure.

(f) *Respect to the Promises:* Everybody respects the system and its culture. The promises are kept up with utmost value. Everybody will see that the words (promises) and deeds (achievements) exactly match.

14.4.4 Effectiveness of TPM

TPM results in high feelings among the employees. They take pride of being a part of the system. However, this can occur when it is effectively implemented observed from macro-level to micro-level. The effectiveness of the implementation of TPM is known from its economic efficiency or profitability or productivity which in turn can be measured in several dimensions. The major dimensions through which the effectiveness of TPM is measured are given below.

(a) Production (b) Quality (c) Cost
(d) Delivery (e) Safety (f) Morale

BUSTLE - 14.2

1. *Suppose you are the plant engineer and the production manager is trying to throw the blame on you. He complains every time that he could not achieve the production target due to poor maintenance, low machine performance, and less availability of equipment. What quantitative and qualitative measures would you take to counteract?*
2. *Suppose you are the production engineer and you are unable to meet the production targets due to poor maintenance, low machine performance, and less availability of equipment. What quantitative and qualitative evidences would you present before management to act upon the plant engineer?*

Test Your Understanding 14.2

1. Explain the philosophy of TPM.
2. Enumerate the characteristics of TPM philosophy.
3. What are the approaches of achieving TPM? Explain.
4. List the objectives or targets of TPM and explain.
5. Write a detailed note on 'management by TPM'.

14.5 CONCEPT OF TOTAL PRODUCTIVE MAINTENANCE (TPM)

This TPM philosophy is the brain child of Dr. Sujuki (Japanese), which deals with maintenance management. The basic concept of the total productive maintenance (TPM) can be understood by the word *'total'*. The word 'total' covers every nook and corner of the organisation. However, for understanding the concept of total productive maintenance (TPM), let us investigate the term *totality* through the following questions.

1. What does 'total' mean? In what sense the term 'total' is used?
2. What are the means and ways (or approach) of achieving totality?
3. How is this totality achieved?

The answer to the first question can be derived as "to achieve the *total effectiveness* in equipment maintenance", i.e., overall equipment effectiveness (OEE). One good indicator of measuring this effectiveness is profitability. Thus, it means that everyone of the organisation should strive for achieving the economic efficiency, i.e., profitability at every point.

The second question, i.e., "what are the means and ways of achieving totality?" is answered by taking the totality approach of plant maintenance. The *total plant maintenance* includes preventive maintenance, maintenance prevention, improving maintainability, and so forth.

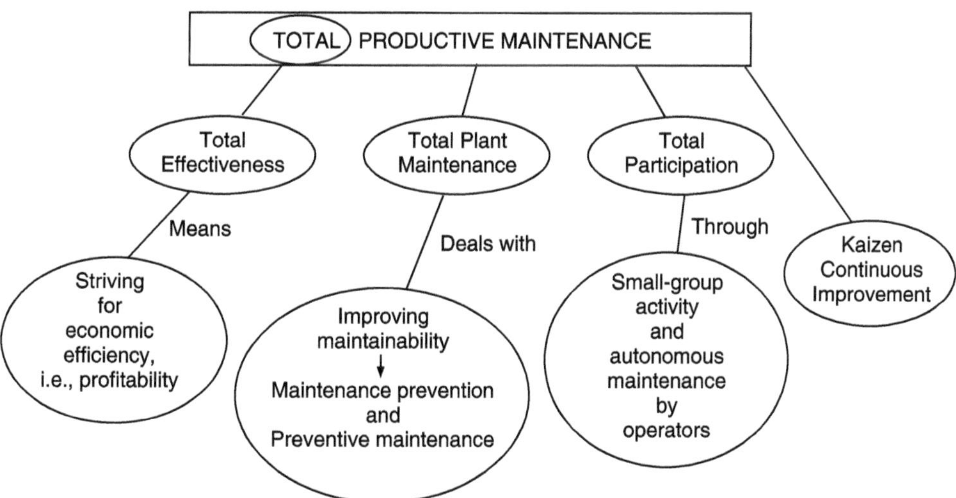

Figure 14.2 Concept of totality in TPM

But how is this totality achieved? The answer of this question can be obtained satisfactorily by making everyone a part in the system. The totality cannot be achieved by one individual or even a group but can be achieved if everybody is involved. The 'total participation' or otherwise called *total employee involvement (TEI)* is sought through various techniques such as small group activity (SGA) and autonomous maintenance by the operators.

The term maintenance in TPM speaks not only about the maintenance but also about the maintenance of system continuous improvement.

Thus, the concept of 'totality' in total productive maintenance can be understood by achieving the total effectiveness by total plant maintenance by total participation.

14.6 PREVENTIVE MAINTENANCE, CORRECTIVE MAINTENANCE, AND MAINTENANCE PREVENTION FOR CONTROLLING DETERIORATION OF AN EQUIPMENT

The significance of preventive maintenance in controlling the deterioration of the equipment and enhancing the life has already been explained in Chapter 3. However, it is viewed through a different perspective in this chapter.

Analogy between Preventive Medicine and Preventive Maintenance

The importance of preventive and corrective maintenance can be better explained by taking an analogy with the ailments and treatment done by doctors. Disease is analogous to equipment failure or malfunctioning; preventive maintenance (PM) is analogous to preventive medicine; and a physical repair work can be comparable to treatment and surgery, etc. A hypothetical relationship is shown in Figure 14.3.

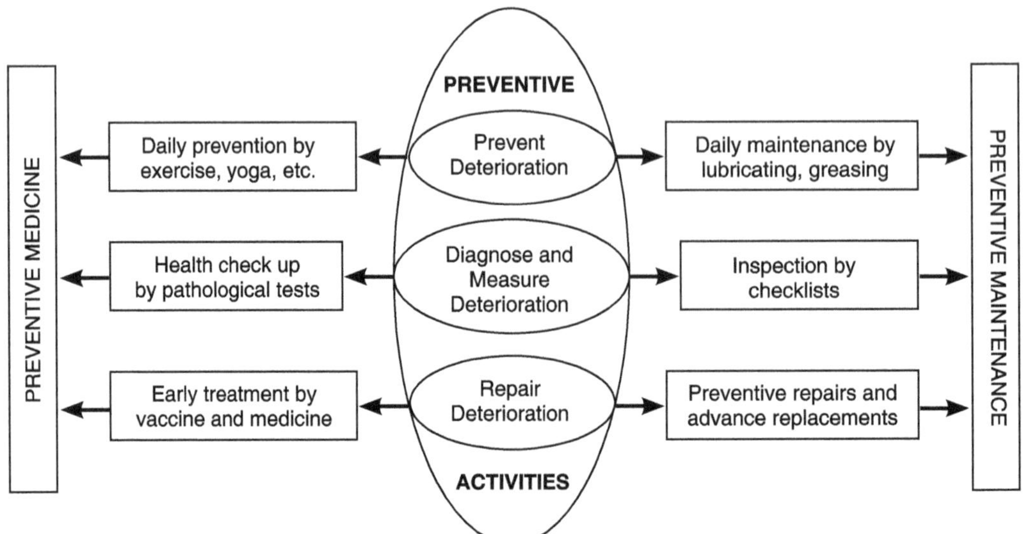

Figure 14.3 Anology between preventive medicine and preventive maintenance

Periodic check-up and by diligently lubricating, cleaning, and performing adjustments and conducting inspection deterioration can prevent and avert potential equipment failures (disease). Thus, PM decreases number of breakdowns and inevitably increases life span.

It is cheaper to repair equipment on preventive basis than to wait until it completely deteriorates. At that stage, the cost of restoring equipment is exorbitant.

When dust and dirt adhere to moving parts and sliding surfaces of machines are scratched, causing deterioration. And when lubrication is neglected, excessive friction or burning can result wasting energy.

Life Span Characteristics and Breakdown Counter Measures

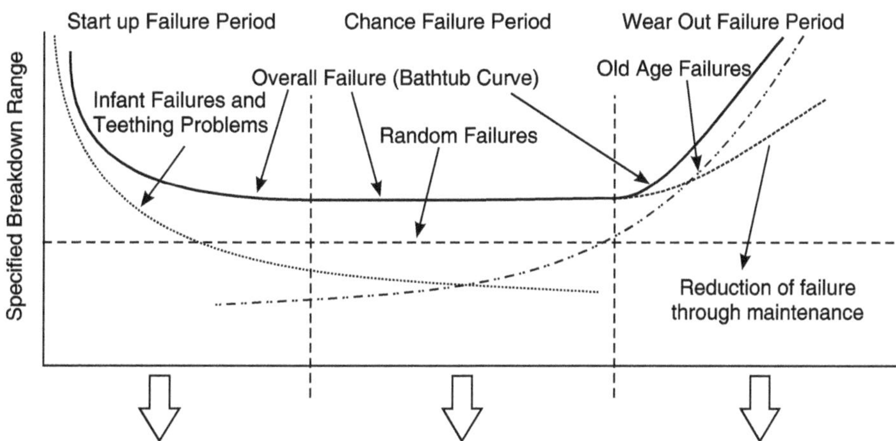

Category	*Start up Failure*	*Chance Failure*	*Wear Out Failure*
Cause	Design/manufacturing errors	Operational errors	Wear out
	Trail seen at acceptance and start up control	Proper operation	Preventive and maintainability improvement
	Maintenance Prevention		

Preventive maintenance alone cannot eliminate breakdowns. Figure 14.4 clearly shows life span characteristic curve or machine life cycle (MLC) or bathtub curve of equipment. (The MLC curve has been discussed in detail in Chapter 7.) When the equipment is new there will be heavy failure rate (early failure period) which eventually drops and levels off. The failure rate stabilises to a certain level for a long period of time (accidental failure period). Finally, equipment approaches the end of useful life, the failure rate increases once again (wear-out failure period).

The causes of early failure are design and maintenance errors. So to combat them, the design department must conduct test runs at early stage. Furthermore, maintainability improvement should be pursued to discover and treat weakness in design and manufacturing.

Accidental failures are caused primarily by operator errors. Hence, the most efficient counter measure is to ensure that operators use equipment properly. Wear-out failures can be reduced and lifespan extends by preventive maintenance and maintainability improvement (through changes in design).

Maintenance prevention is an effective counter measure for all three types of breakdowns. A maintenance-free design of equipment must be incorporated at planning stage to prevent early period, accidental, and wear-out failures.

14.7 THE EIGHT PILLARS OF TPM

The philosophy of the TPM stands on the eight factors as shown in Figure 14.5 that depicts the eight pillars of total productive maintenance.All these when tackled in the right proportions, the objective of TPM will be fully realised.

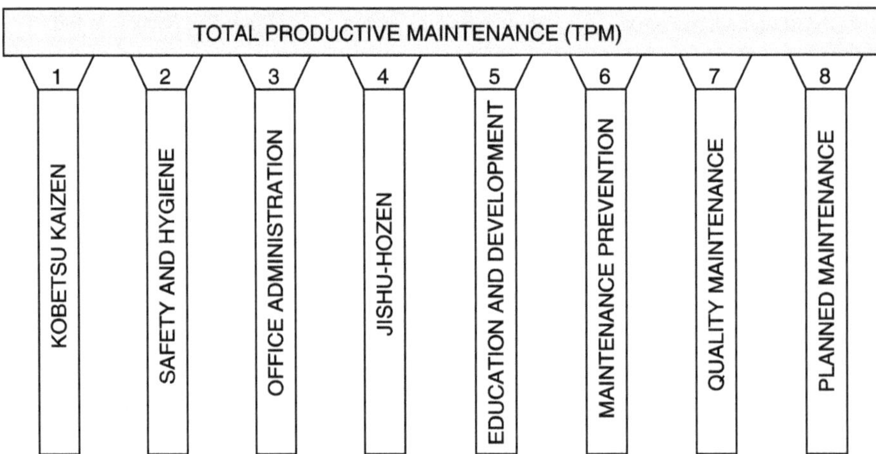

Figure 14.5 The eight pillars of TPM

The eight pillars are:

(a) Kobetsu Kaizen
(b) Safety and hygiene
(c) Office administration
(d) Jishu-Hozen
(e) Education (training) and development
(f) Maintenance prevention
(g) Quality maintenance
(h) Planned maintenance

Brief explanations about these eight pillars are given below.

(a) **Kobetsu Kaizen**

Kobetsu Kaizen, popularly known as **Kai-zen** (Kai means change – zen means good), which means continuous improvement (change for betterment). It is the concept of eliminating various losses that occur in the production and productivity due to human negligence or ignorance in maintaining equipment and quality of material, tools, and other auxiliary items. A detailed discussion on various types of Kaizen is given at the end this chapter.

The objective of Kaizen can be stated concisely as to maximise the overall effectiveness of equipment, process, and the plant by elimination of all losses through small group activities.

The various losses which need to be eliminated are categorised into four groups as given below.

Group 1

- Yield loss.
- Energy loss.
- Losses due to improper design of die, jigs and fixture, etc., or their application.

Group 2

- Managerial inefficiency.
- The factors that impede human work efficiency.
- Improper monitoring of work and workers.
- Improper arrangement of materials, tools, machinery, and instructions.
- Lack of automation.

Group 3

- Machine breakdowns.
- The factors affecting equipment efficiency and thus resulting in low utilisation and low production.
- Failure of workers to arrive in time and collect instructions, materials, and tools.
- Idling of the machine due to non-availability of material, tools, or temporary absence of workmen.
- Losses due to non-optimal speed, feed, poor knowledge of the operator, or poor condition of the machinery.
- Late starting.
- Reworking of defective elements.
- Setting of tool, work piece, and inspections.

Group 4

- Loss to restore basic conditions of equipment.
- Losses due to scheduled downtime.
- Improve 'why-why' analysis system to get correct information about reasons of breakdown and poor performance.
- Failure analysis system.
- Training of all maintenance work personnel on basic modules of maintenance.
- Supporting machinery in workshops and handling equipment, operations for full implementation of the principle underlying in Jishu-Hozen in inspection, lubrication, minor adjustments, cleaning, etc.

(b) **Safety and Hygiene (Autonomous System)**

Objective

- To achieve pleasant working environment.
- To achieve zero accident.
- To provide safe working conditions and workplace.

(A detailed discussion is made on this topic in Chapter 15.)

Brief explanation

The intention of including this pillar in the TPM is that the worker can work more if the environment as well as his/her health is in good condition. For maintaining such conditions, the workplace must be kept clean, safe, and hygienic.

How to achieve (Tips)

A few tips to achieving are given below:

- Keep the environment clean, neat and tidy.
- Use dustbins.
- Think always as 'safety first'.

(c) **Office Administration**

Objective

The objective of this pillar is to keep the workmen free from the tensions of administrative problems.

Explanation

The implementation of TPM should also include the administrative part that indirectly influences the job. The planning work, logistics and spare parts management, information flow, job security, and wage administration, etc., can indirectly affect the work of the operators of maintenance, which is not desirable.

How to achieve (Tips)

A few tips to achieving are given below:

- Pay the wage regularly, correctly, and on time.
- Keep the right tools available at the right time in the right fashion.
- Keep cordial relations.
- Resolve conflicts if any, immediately when noticed.
- Build team work.
- Treat all equally.
- Use scientific management principles righteously.
- Get the right spares at the right time in correct quantity and quality.
- Use scientific base for scheduling and planning of the jobs.

(d) **Jishu-Hozen**

Jishu-Hozen can be defined as carrying out initial or basic maintenance of equipment by the operators themselves and informing of the potential of major breakdown.

The concept of Jishu-Hozen lies in the fact that *'a mother is the best judge regarding the health of her kid than an outsider'*. Therefore, the person who works on certain machine knows better about its health and performance. Hence, the basic maintenance work is assigned to the operators of that machinery.

Objective

The objective of Jishu-Hozen is to achieve zero breakdown (equipment) and zero defects (product) through the autonomous maintenance activities.

How to achieve (Tips)

Establish standard method and time for cleaning, lubricating and tightening of nuts and other loose parts so that minimum time is consumed in these activities.

- Initial cleaning.
- Inspection.
- Cleaning, lubrication, greasing, tightening, etc.
- Systematise autonomous maintenance.
- Taking counter measure at the source of the problem.
- Practice full autonomous management.

(e) **Training, Education, and Development**

Objectives

- To disseminate maintenance and engineering knowhow.
- To eliminate equipment failure, product defects and accidents resulting due to ignorance and lack of skill.
- To train each and every maintenance personnel in their respective jobs.
- To develop multi-skill workforce.

How to achieve (Tips)

These can be achieved through:

- Identifying training requirements for each individual in various skills.
- Arranging skill enhancement training programmes.
- Assessing the effectiveness of training schemes.
- Apprising the maintenance personnel of the various reasons for failure and breakdowns, defects in products, low quality of products, and reasons for accidents.
- Assigning engineering and maintenance people to handle training, emphasising practical knowledge about real machine past failures.
- Focusing on the relationship between failures and improper operating procedures and poor maintenance.

(f) **Maintenance Prevention**

This maintenance consists of preventive maintenance and maintenance prevention techniques to improve mean time between failures (MTBF).

Objectives

- To improve operating time and hence the availability.
- To eliminate equipment failure, product defects, and accidents by technological upgradation.
- To strive for preventive maintenance than to prevent the failures.
- To reduce losses due to breakdown, setting up, stops, rejections, slow down, and yield loss.
- To enhance proprietary technology.

How to achieve (Tips)

To improve operating time

- the six major losses are reduced and
- the propriety technology is to be updated (or improved).

Six major losses

- Loss due to equipment failures
- Loss due to set up adjustments
- Loss due to defectives
- Loss due to minor stops
- Reduced speed losses
- Yield loss

Proprietary Technology: This involves design of production equipment, which provides for

- Extent of automation
- Maintenance free operation
- Easy start up
- Life cycle costing
- Energy usage

(g) **Quality Maintenance**

Objectives

- To build a reliable production line (zero failures).
- To build quality through equipment.
- To ensure that no product defects are caused by faulty equipment.

Methodology

To achieve these objectives, TPM uses MTBF and MQM (man-quality-machine) management and also dissects operating mistakes to determine, if additional training or use of fail-safe devices.

The operational standard of TPM quality management is "*do not accept, manufacture, and send a failure or defect*".

Workplace Organisation and 5S Frame Work

TPM organisation of workplace along five dimensions is known as 5S (in Japanese the 5S are Seiri, Seiton, Seiso, Seiketsu, and Shitsuke) framework.

- Organisation
- Orderliness
- Attention to details
- Cleanliness
- Discipline

The focal idea of 5S workplace organisation is identifying all the subtle abnormalities that might cause failures or defects if not discovered and treated.

Practising 5S on regular basis encourages the formation of an improvement cycle beginning with reducing variations caused by abnormalities and then by creating stability and uniformity in manufacturing process and finally enhancing process analysis, which in turn leads to further improvements. These are discussed in detail at the end of this unit.

The purpose of TPM licensing system is to expand an operator's capabilities by developing new skills and promoting operational stability. The system also encourages acquisition of skills necessary to perform multiple jobs, creating a multi-functional workforce.

The usual management as suggested by Dr. W. Edward Deming, plan–do–check–act (P–D–C–A) cycle of workplace organisation is shown in Figure 14.6.

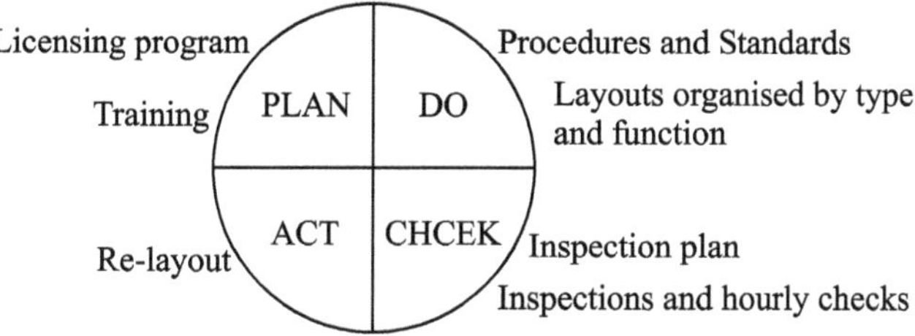

Figure 14.6 Deming's P–D–C–A cycle

(h) **Planned Maintenance**

Objectives

- To maximise performance of the equipment.
- To minimise cost in maintenance.
- To achieve zero equipment breakdown.

Methodology

- Establish predictive maintenance system.
- Set up preventive maintenance system.
- Institute corrective maintenance system.
- Establish training system for maintenance crew and supervisors.
- Detailed activity planning for planned maintenance.

BUSTLE - 14.3

Examine the possibility of application of eight pillar of TPM in your organisation or institution?

Test Your Understanding 14.3

1. Write a detailed note on role of preventive maintenance in TPM.
2. Explain the 5S frame for workplace organisation.
3. Preventive maintenance to the equipment is like preventive medicine to human beings. Explain.
4. *Do not accept, manufacture and send a failure or defect.* Critically analyse the TPM slogan.
5. Distinguish between productive and predictive maintenance.
6. Discuss the methods, tools, and techniques to achieve the TPM.
7. Write notes on the following in light of TPM.
   (i) Kobetsu Kaizen (ii) Jishu-Hozen
   (iii) Safety and Hygiene (iv) 5S Framework

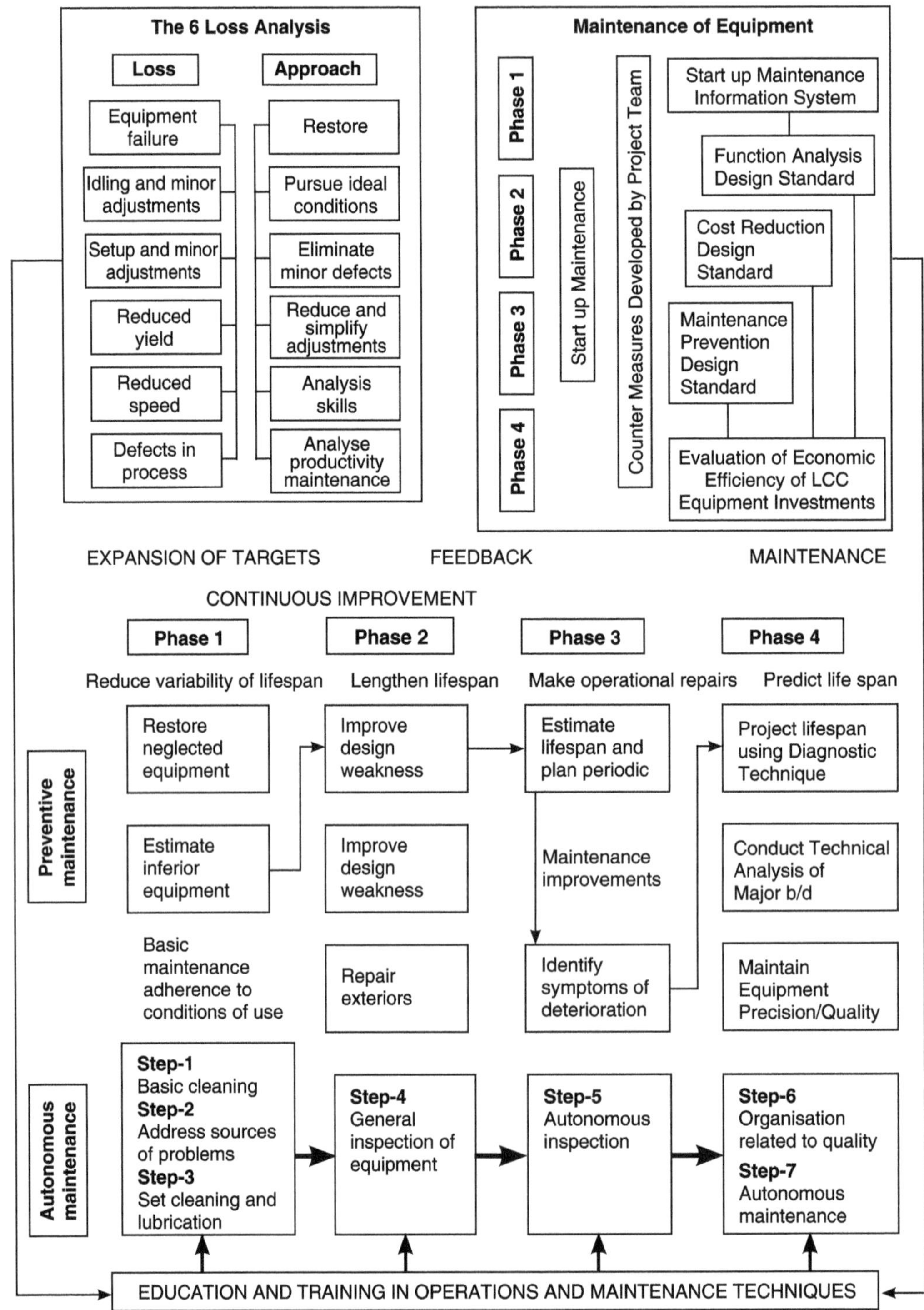

Figure 14.7 The TPM template

14.8 TPM TEMPLATE

The template of TPM is given in Figure 14.7, in which all the information regarding the TPM is greatly simplified and summarised.

14.9 BARRIERS TO IMPLEMENT TPM

The relevance of TPM and the need to implement in Indian industries is of special interest in capital scarce economy to sustain local and international markets and to face ever-growing competition because of recent globalisation and liberalisation. It is desirable to mould the mechanism of implementation in India to suit local conditions succeed a consensus style of decision making, etc. Indian industry suffers traditionally with factors like inadequate compensatory package, below average level of workers awareness, segmented hierarchical kind of setup, old maintenance practices, etc. Most of the Japanese concepts including TPM provide to satisfy people's higher order needs and it can be observed that TPM promotes achievement motivation. Workshops will become important in this regard in addition to giving technical inputs to improve maintenance skills.

TPM will be more successful if it is tried

- in those industries, where reasonable pay packages are offered.
- where there is required degree of job security.
- by setting up the attitudes of people towards accepting and appreciating the concept of TPM, through meetings.

Companies that plan to implement TPM should aim for zero breakdowns and zero defects. Every company that adopts TPM approach must adopt three imperatives.

- The quality and functioning of equipment must change.
- The equipment operators must change their way of thinking about equipment.
- As a result, workplace itself must change drastically.

TPM will not be successful in companies that will not make change in cultures that decision making responsibility are pushed to lower levels of hierarchy. Road to successful TPM implementation raises some tough questions like

- Do you trust and believe that operators will adjust their machines?
- What is the cost of our current approach to maintenance?
- Why should maintenance assume management responsibilities?
- How does the company reward operators?

The TPM can only succeed if companies realise that path to progress and success lies through integrating maintenance fully into total business system of company, i.e., to integrate the technical features of the equipment with personnel, social, and organisational features.

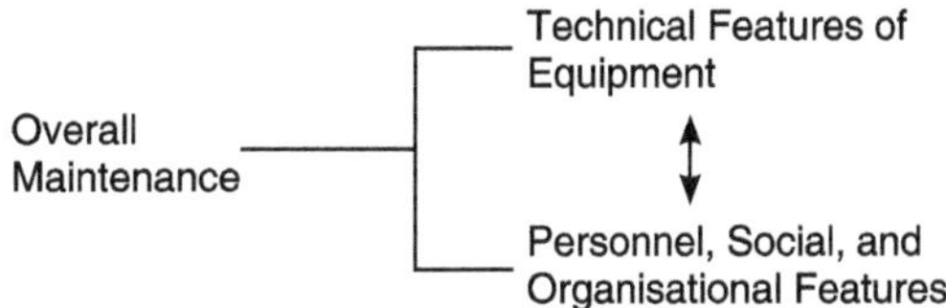

Thus, it can be seen that overall maintenance effectiveness is of technical features of system being used as well as the culture of the organisation.

Predictive Maintenance

The objective of predictive maintenance is early prediction of failures by judicious application of condition monitoring techniques and instrumentation. The objective of such a philosophy is to predict failures in advance, so that maintenance actions can be planned to take place at a time convenient to man and machine, and unplanned interruption of production system are minimised. But this requires high technological competence, versatile knowledge of the equipment, and effective information system.

14.10 COMPARISON BETWEEN TPM AND TQM

Total quality and total productivity are strategic management concepts combining quality and productivity. Total quality's aim of delivering the highest quality at lowest price is unachievable unless total productivity is achieved. The quality of work at each step is dependent upon man, methods, machines, materials, and environment which management establishes. Men have to be skilled, have positive attitude to quality and productivity. They must feel involved and motivated to give their best.

Total Quality Management (TQM)

The total Quality Management (TQM) approach of an organisation is centred on quality. It is based on the participation of all its members and aiming at long term success through customer satisfaction and benefits to members of the organisation and society. TQM has four segments –

- Total quality control (TQC)
- Total waste control (TWC)
- Total employee involvement (TEI)
- Continuous improvement (Kaizen)

Total quality control (TQC) is wide spread throughout the Japanese industry as Company Wide Quality Control (CWQC). The quality circles were evolved as part of TQC.

Total Productive Maintenance (TPM)

The total productive maintenance (TPM) approach of an organisation is centred on productivity through maintenance. It is based on the participation of all its members and aiming at long-term success through effective maintenance of equipment and benefits to members of organisation and society. TPM has four segments:

- Overall equipment effectiveness (OEE)
- Autonomous planned maintenance (APM)
- Total employee involvement (TEI) and teamwork
- Kaizen or continuous improvement (CI)

Comparison between TPM and TQC

Similarities:

The concepts of TPM and TQM have some similarities. These are

- Both philosophies aim at bringing out change in organisational culture.

- Both the concepts aim at improving and elevating the reputation of the organisation.
- Kaizen, the concept of continuous improvement, is an important segment in both the programmes.
- Both the concepts are achieved by total employee involvement (TEI), i.e., every employee is responsible and part of it.
- In both situations, the culture of the organisation will become congenial and cooperative.

In simple sentence, we can say that both TPM and TQC are aimed at corporate improvement. However, there are slight differences.

Differences:

The differences of TPM and TQM are represented in the following tabular form.

S.No.	*TPM*	*TQM*
1.	The subject of management is equipment in TPM.	The subject of management is quality in case of TQC.
2.	TPM aims at realisation of proper form of equipment and workplace.	TQC aims at management that implements systematisation and standardisation.
3.	TPM imparts training in maintenance engineering and maintenance skills.	TQC imparts personnel training in QC techniques.
4.	TPM aims at small group activities as integral part in how workplace is organised.	TQC aims at voluntary QC aids and quality circles.
5.	The goal of TPM is elimination of loss and waste due to equipment its maintenance.	The goal of TQC is elimination of loss and waste due to quality.
6.	TPM uses condition-based maintenance, preventive maintenance, and maintenance prevention.	TQC uses process inspection and proof mechanisms.
7.	The tools for TPM are machine failure records and MTBF.	TQC tools are statistical process control/control charts.
8.	TPM recommends zero breakdown (ZBD) programme.	TQM recommends zero defect (ZD) programme.

It is observed that TPM makes use of some of techniques of JIT and TQC but relates them specifically to equipment management. For example:

- Life cycle planning used to predict the maintenance needs of equipment throughout its lifetime.
- Design of virtually maintenance-free equipment man-quality-machine (MQM) management in order to ensure defect-free products, analyses interface between maintenance, product quality, machine parts and links, specific maintenance procedures, machine actions.

As with JIT and TQC processes, TPM process is driven by employee involvement at all company levels. TPM transfers primary responsibility for machine maintenance as close to action as possible to the machine operators. The goal of this responsibility transfer is to develop highly coordinated approach to productive maintenance, where everyone works together as a part of improvement process. Another key component of TPM is education and training. To fulfil their newer possibilities, operators are given in-house instruction for correct operating practices and simple repairs.

BUSTLE - 14.4

The success of both the philosophies TQM and TPM is by and large dependent on the two segments viz. Total Employee Involvement (TEI) and Kaizen. Therefore, the top management tries to transfer the responsibilities down to operator's level. List these responsibilities that are to be transferred. If there is any opposition, how would you tackle the situation?

Test Your Understanding 14.4

1. Enumerate the merits of TPM in light of effectiveness of the system performance.
2. Distinguish between TPM and TQM.
3. Draw the TPM template and explain its features.
4. List the barriers of adapting TPM.

14.11 ZERO BREAKDOWN

Physical condition of equipment deteriorates due to several reasons, one of them is usage. The deterioration can be visible (external) or invisible (internal). If the deterioration or decay or defects are not noticed and arrested on time, the equipment breaks down. Therefore, the plant engineer

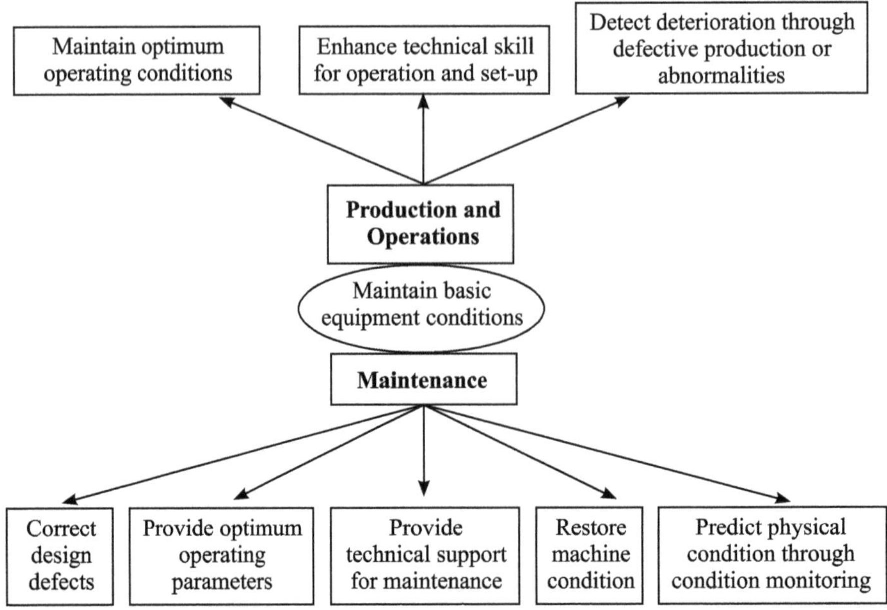

Figure 14.8 Role of maintenance and operations departments in materialising zero breakdowns

is required to keep keen observation of the equipment conditions regularly so as to forestall the occurrence of any breakdown. To achieve zero breakdown (i.e., no breakdown) the important steps are listed below:

- Rectify design deficiency as soon as it is identified.
- Rectify the deterioration so as to bring the condition nearest to its original state.
- Adhere to optimum operating conditions as prescribed by OEM in the manual.
- Maintain the basic equipment conditions by proper and timely cleaning, lubricating, bolt tightening, etc.
- Improve maintaining and operating skills.

Figure 14.8 depicts the role of maintenance and operations departments in achieving zero breakdowns. However, it is believed that the zero breakdowns can be achieved by two facet approach, i.e., a part of the job by maintenance and another part by production.

Functions of Maintenance Personnel

The functions of maintenance personnel under zero breakdown programme are as follows:

- Provide technical support to the production department for its autonomous maintenance activities.
- Provide optimum operating parameters and clear cut instructions to the production department for safe functioning of the machinery.

  For example:

  (i) Suppose the maximum temperature at furnace is given. Instruct that if the temperature exceeds a given value, the machine should be put off.

  (ii) Provide the information regarding the maximum pressure that a compressor should develop.7
- If deteriorated, restore normal condition through inspection, overhauling, and/or replacement.
- Enhance maintenance skills for routine check-up and minor maintenance and repair work.
- Predict status of the physical condition through condition monitoring techniques and take preventive steps if there are any symptoms of occurrence of breakdown or faulty production.
- Rectify minor design defects if possible, otherwise warn the production manager of the fault so that the machine may not be operated or may be operated carefully, which will otherwise produce defective products.

Functions of Production Personnel

The functions of production personnel under zero breakdown programme are as follows:

- Basic maintenance functions such as cleaning, checking of lubricant, water, grease, tightness of nut, screw, etc.
- Detect deterioration in functional performances through wear debris size and/or size distribution, product quality, noise level, smoke level, temperature of cooling water, consumption of lubricating oil, grease, etc.
- Maintain operating conditions as specified in equipment manual given by OEM or as suggested by maintenance department.
- Enhance technical skill for set-up, operation, adjustments, and visual inspection.

14.11.1 Five Counter Measures for Zero Breakdown

Ideally, breakdowns can be eliminated by Maintenance Prevention (MP) or adoption of maintenance free design. Defects that go undetected and untreated are called hidden defects. To eliminate failures, we must expose hidden defects and treat equipment before it breaks down. The following five steps help eliminate failures:

- Maintaining well regulated basic conditions (cleaning, lubricating, and bolting).
- Adhering to proper operating procedures.
- Restoring deterioration.
- Rectify weakness in design.
- Improving operation and maintenance skills.

The interrelationship maintenance and operation skills through the five zero breakdown counter measures are shown in Figure 14.9.

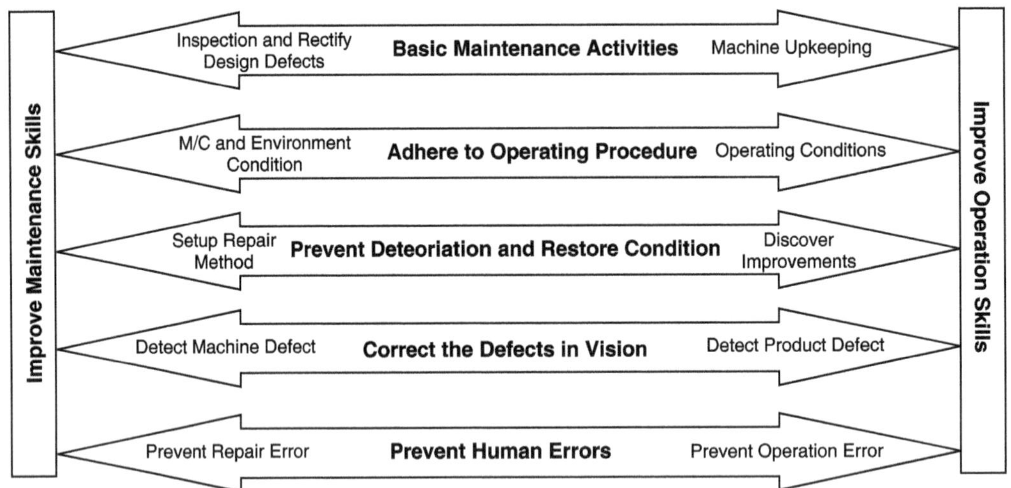

Figure 14.9 Five counter measures of zero breakdown

The above figure shows the relationships between these five counter measures. Everyone involved in equipment operation or maintenance must work to eliminate failures. The figure also depicts division of labour between operation and maintenance departments in pursuing the goal of zero breakdowns.

14.11.2 Zero Defect (ZD) and Zero Breakdown (ZBD)

Philip Crosby, the creator of *zero defect* (ZD), defined quality as conformity to requirements and called for the company wide implementation of ZD or quality measurement. His four principles of quality management are given below. The goal of zero defects is to create a means of promoting perception on essential elements in the pursuit of quality.

- The definition of quality is confirmation to requirements.
- The system of quality prevention.
- The performance standard is zero defects.
- The measurement of quality is the price of non-conformance.

Zero defects and productive maintenance have a common philosophy. While zero defects strive to prevent defects, zero breakdown maintenance has emphasised the importance of preventing breakdowns. Since equipment failure is a type of defect, both ZD and ZBD, in effect are preventive systems aimed at eliminating defects. In TPM operators, these are the inspectors responsible for quality assurance. ZD is considered a significant factor in the success of just in time (JIT) production system as well.

14.12 EXPECTED CONTRIBUTION BY APPLICATION OF TPM

(a) Increase in

- Labour efficiency and productivity.
- Output of the equipment and machinery.
- Innovative ideas generation.
- Reliability in meeting supply schedule.
- Functional and physical life of the equipment.
- Confidence and morale of operators, maintenance personnel.
- Value addition to the operators and maintenance crew.
- Amicable relationship between production and maintenance group.

(b) Reduction in

- Failures and breakdowns.
- Wastage of energy.
- Accidents.
- Stock of raw-material and semi-finished components stock.
- Wastage and rework.
- Customer complaints.
- Maintenance cost.

Test Your Understanding 14.4

1. Explain the terms zero defects, zero breakdown, zero accidents with reference to TPM.
2. Explain the five breakdown counter measures.
3. What are the contributions of application of TPM?

14.13 KAIZEN—THE CONTINUOUS IMPROVEMENT

Definition or Meaning of Kaizen

Kai means change; zen means good. *Kobetsu Kaizen* is popularly known as *Kai-zen.* As discussed earlier (in this chapter under section 14.5) Kaizen, one of the 8 pillars of TPM, means 'change for good' or a 'good change'. This concept is elaborated in this section.

The objective of the Kaizen in Plant Engineering may be set to maximise the overall effectiveness of equipment, process, and the plant by elimination of all losses through small group activities. To achieve this objective, one has to first understand the nature and features of Kaizen and its applicability to plant maintenance. These are discussed at length here under.

14.13.1 Salient Features

- It is a continuous improvement tool but not a single-time job.
- It targets excellence.
- It coordinates between various functions on/off the shop floor.
- Its outlook is global production system.
- It is a step-by-step improvement or change for betterment involving employees at grass-root level.
- It is done by a team (team approach). The team members belong to the same organisation and even to the same section or department where the problem exists.
- It is bottom-up approach.
- It does not need substantial investment.
- The improvement is experienced in a very short span of time.

14.13.2 Need for Kaizen

Some areas consume a lot of operator's time in shop floor. In fact, these times can be avoided or reduced. A few areas are identified and suitable questions are framed for applying Kaizen. If these questions are answered without any bias, we can realise the need for implementation of Kaizen. And Kaizen can promise the best possible results.

- Can we process or inspect the work piece/job/equipment while transporting?
- Can we avoid lifting or carrying heavy loads?
- Can multiple handling be avoided?
- Can we modernise or mechanise the operation instead of carrying out manually?
- Can we make some arrangement so as to pick up parts easily?
- Can setting time be reduced?
- Can we combine two or more tools/operations in to a single tool/operation?
- Can the arrangement be made for easily filling up, topping up, and drainage of coolant?
- Can chips removal and handling be made easy?
- Can all the metal guards and covers be replaced by transparent one?
- Can the control switches and emergency stops be provided nearest to the operator?
- Can we arrange material and tools in front of the operator (Shadow box) in the sequence of work?
- Can we reduce an operator's to and fro transportation of material?
- Can suitable handles or lifting arrangement be provided on jigs and fixtures to handle easily?
- Can we avoid the movement required for non-productive elements such as looking, selecting, or searching?

- Can we minimise an operator's waiting time (or enhance machine availability) by line balancing, supplying spares just in time, minimising machine downtime?

14.14 TYPES OF KAIZEN

Kaizen can be applied with different focus at different situations. However, it is the tool that is more relevant to Plant Engineering and Maintenance, particularly to those who are on the way implementing the TPM culture. Whatsoever, the form and outlook of application of Kaizen may be, it becomes inherently or apparently the part and parcel of job of the maintenance department.

Various types of Kaizen (different facets of Kaizen) are:

- Standard Work Kaizen
- Set-Up Reduction Kaizen
- TPM Focus Kaizen
- Quality Focus Kaizen
- Kanban Kaizen
- 5S Kaizen

14.14.1 Standard Work Kaizen

The standard work Kaizen does not mean to be applied on a shop floor only. It can be applied to any area. However, it is more relevant to maintenance department since it is not only applicable to the operations and processes but also to layout or equipment. Thus, suitably it can be applied in maintenance department also for its operations.

Standard work Kaizen consists of the following four components process Kaizen, operation Kaizen, layout Kaizen, and equipment Kaizen (POLE):

(a) **Process Kaizen**: Improvements done to eliminate the waste in processing, inadequate technology, or design can be grouped under the process Kaizen.

Following are the few examples to attempt for the process Kaizen:

- Optimisation of tool-life.
- Reduction of unnecessary over machining by better planning.
- Change the machining sequence to reduce cutting time or machining cycle time.
- Eliminating or reducing an unproductive time of stroke of the hammer or press or tool for example by quick return mechanism.

(b) **Operation Kaizen**: In this philosophy, efforts are made to reduce cycle time.

Following are the few examples to attempt for the operation Kaizen:

- Change/improving work procedures.
- Change in the object position *vis-a-vis* operator or processing equipment.
- Combining or overlapping more than one operation.
- Better management of tools inventory.
- Modification of materials handling system.

(c) **Layout Kaizen**: Design a layout in such a way that it is flexible and can be changed to respond to changes in process flow. Layout that can make the job easier and safer for the worker removes drudgery from operator's job and makes them more productive. Layout Kaizen should focus to assure ease and safety in

- Flow of objects.
- Flow of people.
- Flow of information.

How to Do?

After a decision is taken to change the layout, a full scale cardboard cut-out has to be made for each machine/equipment. As per the changed layout, these cut-outs can be laid out in a vacant space so that any interference to man or material besides maintainability can be foreseen. Layout can be changed in any idle shift or off days without hampering production. Attention must be paid to the following points while conducting layout Kaizen:

- Possibility of standardising operations.
- Safety and hygiene of operators.
- Delays in flow and counter flow of material.
- Flexible manpower planning.
- Built-in process quality.
- Maintainability of equipment and machines.
- No bottlenecks to house-keeping, proper collection of chips, and their disposals.
- Prevent flying of chips using chips guards.
- Flexible hydraulic hoses and electrical cables between machine and panels.
- Use of single piece conveyors between the machines, wherever required and no trolleys, tables, etc., for keeping work in process (WIP).

(d) **Equipment Kaizen**: Under equipment Kaizen, the following exemplary points need to be considered:

- Can we reduce the cost or miniaturise to suit production volume within the target time?
- Is the cycle time to machines/equipment in a given line matching to facilitate one piece flow?
- Can we reduce the set up change time?
- Can machine controls and emergency stops be placed nearest to the operator?
- Are the oil filling or checking points easily visible and approachable?
- Can coolant be easily topped up/filled/drained?
- Can we improve quality of tools to increase intervals between tool changes?
- Can we replace all metal guards by transparent one?
- Can we install one touch clamps to replace manual clamping or automatic ejection device?

14.14.2 Set-up Reduction Kaizen

Set-up time is the total length of time between the last good part of previous run to the first good part of the next run; during this time a machine or equipment is down and not producing parts.

This can be divided into two parts:

- **Internal Set-up:** The portion of the set-up that can only be done while the machine is shut down.

- **External Set-up:** Parts of the set-up that can be performed while the machine is producing the parts.

Steps for Set-up Reduction Kaizen:

- Review objectives with team and keep team members informed of all the activities for which their support and cooperation is needed.
- Video tape the existing set-up and define each element.
- Separate the internal and external elements of the set-up.
- Post a set-up time bar chart with target time indicated.
- Convert internal set-up to external set-up.
- Improve all the elemental operations, starting with internal.
- Develop and train operators in new work sequence.

14.14.3 TPM Focus Kaizen

Equipment downtime is a significant area to control if any firm is thinking of progress. Such relevant problems are addressed through TPM, whose objectives are:

- To maximise overall equipment effectiveness (OEE) by intensive effort to remove hidden causes.
- To improve the equipment reliability, availability, and maintainability.

The following are the steps in TPM focus Kaizen:

- Establish the basis, such as MTBF. Advice the team of what to look for during initial cleaning.
- Before beginning the job, maintenance and plant safety is notified to ensure that equipment is safely put off.
- Identify, record, and repair minor defects like oil leaks, choked orifices, worn guides/gears, missing guards/covers, etc.
- Record and tag defects that will require maintenance assistance to repair.
- Mark lubrication and inspection points, prepare and display standard work for daily inspection and maintenance.
- Provide maintenance by skilled group.
- Collect data for calculation of MTBF and post results.

14.14.4 Quality Focus Kaizen

The following are the steps for quality focus Kaizen:

- Introduce and review objectives with team members and study the data.
- Discuss the problems with operators concerned on site where the quality problem exists by performing 5W+1H (what, when, where, why, who, and how).
- List the expected levels of quality for the items under Kaizen and examine the defects/failures keenly and define the effect or symptoms of quality problems.
- Look for the signals or physical conditions in the processes that can cause these quality problems.
- Conclude a statement (cause and effect) that exactly defines the problems and construct a cause and effect diagram after brainstorming and reaching a consensus.

- Collect data on identified causes and select the best solution after brainstorming for possible solutions.
- Implement the solution, monitor and document the results, and follow up to ensure permanent implementation of solution.

14.14.5 Kanban Kaizen

A Kanban is a tag or card attached to a product to facilitate its proper movement. It is used as an authority to supply material or start production.

A Kanban system uses Kanbans as the principal source of information which travels backwards as the material travels forward. Kanban system is the basis of just in time or pull system production as it ensures the exact type, quantity, and quality of material and at the time needed at the point of use.

A Kanban system, therefore, ensures:

- Pull production (no production or conveyance is done until the next process needs the material).
- Levelled production.
- Quality (no defects are sent to the following process).
- Controlled inventory, only specified quantity as per the bin size, will be supplied at defined intervals.

14.14.6 5S Kaizen

The 5S Kaizen is composed of 5 terms originally framed in the Japanese language. These are explained in paragraphs to follow.

The 5S steps of house-keeping, with the Japanese names (starting with S), their equivalent English names (starting with S) and their meanings (equivalent names starting with C) are as follows:

S.No.	*Japanese 'S'*	*Equivalent English 'S'*	*Meaning/Equivalent 'C'*
1.	Seiri	Sort	Clear Out
2.	Seiton	Straighten	Configure
3.	Seiso	Scrub	Clean and Check
4.	Seiketsu	Systematise	Conform and Continue
5.	Shitsuke	Standardise	Custom and Practice

(a) **Seiri (Sort/Clear Out)**

Under this principle, first all items are classified into two categories in terms of necessary and unnecessary. And then the latter is discarded or removed. Usually, a ceiling on the number of necessary items should be established otherwise everything may look like necessary in one way or the other. In usual situations, the department is found with full of unused machines, jigs, dies and tools, rejects, work in process, used/unused materials, supplies and spare parts, sleeves, containers, desks, work benches, files of documents, carts, racks, pallets, and many other materials. As a rule, anything that will not be used within the next 30 days should be removed.

How to Do?

- Seiri begins with a red tag campaign.
- Select an area for Seiri.
- Designate appropriate 5S team and delegate suitable authority.
- Team members go to the site with a handful of red tags and place them on the item they believe are unnecessary.
- If red tags are found on items actually required by the site managers, they must demonstrate the necessity for such items.
- The specific material that is not required in the next thirty days but may be needed in the near future, are moved to their appropriate places such as the warehouses.

(b) **Seiton (Straighten/Configure)**

Seiton means configuring the items marked as 'useful' and arranging them so as to minimise search time and effort. The minimum number of needed item can be of no use if they are stored too far from the work station or in a place where they cannot be found.

How to Do?

- Each item must have a designated address, name, and volume.
- Floor space for bins containing WIP or supplies must be marked by painting.
- Maximum allowable number is indicated clearly.
- In a nut shell, each item should have its own address and conversely each space should have its designated address.
- Each wall can be numbered.
- The gang ways should be marked clearly.

(c) **Seiso (Scrub/Clean and Check)**

Seiso is cleaning and also checking. Clean the working environment, including machines and tools, floors, walls, and other areas of the shop floor, because when the machine is covered with oil, soot, and dirt, it is difficult to identify any problem that may be developing.

It is understood that most machines go under breakdown due to vibration, or wear debris or introduction of foreign particles or due to inadequate lubrication. Seiso is a great learning experience for the operators of the machine, since they can make many useful innovations regarding the cleaning of the machines. Naked electrical wires may be covered safely while cleaning the machine.

(d) **Seiketsu (Systematise/Conform and Continue)**

Seiketsu is interpreted in two ways. One is that to keep each individual safe, clean, neat, and tidy by wearing proper and safe working dresses, using safety glasses, gloves, and shoes as well as maintaining a clean, healthy working environment. Another interpretation is to adhere to work on Seiri, Seiton, and Seiso continually and every day, i.e., systematically observe the above three as a part of the job in the system.

For instance, it is easy to go through the process of Seiri once a while and make some improvements, but without an effort to continue such activities, the situation will soon be back to where it has started.

(e) **Shitsuke (Standardise/Custom and Practice)**

Plant managers must determine who should be involved and the frequency, i.e., how often the 5S team should visit the entire factory. These should take place in such a way that the 5S programme should not disturb the regular work and should not cost in terms of money and time. Thus, a standard procedure, duration, frequency, responsible, and authorised persons, etc., for this purpose are to be assigned. The people concerned should be trained properly about the 5S system and about the significance of house-keeping, cleanliness, orderliness, and traceability of the items.

14.15 EXPECTED CONTRIBUTIONS BY APPLICATION OF KAIZEN

(a) **Optimising Manpower**

The excess manpower employed can be identified and optimised through proper load balance.

(b) **Reducing Work-in-Process Inventory**

- Optimal stock of material for the process can be identified.
- The lot size can be reduced to small.
- Reconsideration of decisions regarding volume of finished product is possible.
- Material handling can be reduced and hence rejections or reworking can be reduced.
- Small and frequent transport runs can be avoided.

(c) **Quality can be Enhanced**

- A system can be built up that makes it easy to understand the causes.
- Measures can be taken to stop the process immediately whenever defect occurs.

(d) **Enhancing Production Capacity**

- Providing simple and in-house developed machine to minimise manual work.
- Finding out bottlenecks in process or machine.

(e) **Advantages due to Re-layout**

- No delays in flow of material.
- Possibility of better working space and condition.
- Possibility of standardising operations.
- Quality built into process.
- Consideration on proper space for movement of trolleys, forklifts, etc.
- Ability to realise flexible manpower line.
- Attentive maintenance.
- Safety and hygiene.

(f) **Installation Visual Control on Line**

- Display of standard worksheet at the stations.
- Pre-warning to the operators on approach.
- Provides rhythm on the line.
- Makes line supervisor alert regarding delays in operation.
- Production control board to display actual production figures on hourly basis and reason for shortfall in planned schedule along with counter measure taken.

14.16 TOTAL MAINTENANCE PLANNING

Total maintenance planning includes the planning of the following (A to Z) aspects:

A. Lubrication
B. Overhauling
C. Breakdown or corrective maintenance
D. Safety
E. Predictive maintenance
F. Maintenance record keeping
G. Preventive maintenance including shutdown maintenance
H. Reconditioning
I. Kaizen activities
J. Modifications or design out maintenance
K. Inspection
L. Spare parts planning and spare parts consumption pattern
M. Refurbishing
N. Equipment replacement
O. Maintenance plan formulation
P. Standardisation (of inspection technique, equipment, material, specification, etc.)
Q. Materials requirement planning
R. Obsolescence (of equipment, technique)
S. Scrap and disposal system
T. Spare parts manufacture
U. Level of training (of maintenance crew, inspection personnel, etc.)
V. Liaison with production department
W. Liaison with support organisation
X. Cost control effort
Y. Utilities and services (such as compressed air, power generator, water, etc.)
Z. Documentation (of history of maintenance of each equipment, breakdown records, type of breakdown, frequency of breakdown, repair time, replacement requirements, etc.)

BUSTLE - 14.4

Examine the possibility of application of Kaizen in your organisation/institution.

____

____

____

____

Test Your Understanding 14.5

1. Define and explain the concept of Kaizen.
2. Describe the impact of application of Kaizen on process, operation, layout, and equipment.
3. Discuss various points of focus of Kaizen application.
4. Enumerate the aspects in total maintenance planning.
5. Write short notes on
   (a) Standard Work Kaizen
   (b) TPM Focus Kaizen
   (c) Quality Focus Kaizen
   (d) Kanban Kaizen
   (e) 5S Kaizen
6. Write short notes on
   (a) Seiri
   (b) Seiton
   (c) Seiso
   (d) Seiketsu
   (e) Shitsuke
7. Describe the role and focus of Kaizen on 5S.
8. Explain the role and focus of Kaizen on quality.
9. Enumerate the contributions of application of Kaizen.
10. What are the roles and responsibilities of Kaizen team?
11. Kaizen and TPM are part and parcel to each other. Discuss your views on the statement with some practical examples.

14.17 CONCLUSION

The increased use of TPM and condition monitoring techniques will indicate that the nature of management information systems will change. It is necessary to collect information about equipment performance, convert information through analysis needed for people concerned for decision making at different levels of management.

It can be finally stated that organisations adopting high level of mechanisation and automation of production processes should adopt a maintenance strategy that takes into account technical characteristics of system (such as the reliability-centred maintenance (RCM) studied in Chapter 13) and the related human aspects (with TPM) and integrate them to achieve the best economic results in maintenance. A closer look at the industry indicates predictive maintenance strategy in replacing concept of failure-based and time-based maintenance philosophies. Online monitoring systems have proved their utility in case of large capacity complex machines and they are found to be cost effective. While TPM is a proven system for high productivity, cost effectiveness, and defect-free culture. In addition, the Japanese concepts such as JIT, Kaizen, 5S approach, etc., if blended with the system, enhance the magnificence of the TPM culture.

14.18 RECAP

TPM is a culture, a philosophy, and a new attitude towards the maintenance in the plant. It involves wide range of activities in proper operation, failure analysis, design improvement, cost benefit analysis, etc. TPM aims at maximising the equipment effectiveness like tero-technology. The eight pillars of total productive maintenance and equipment effectiveness in maintenance are discussed. This chapter also covers the TPM template, expected contributions by application of TPM and the barriers of adopting the TPM in the organisation. The predictive and productive maintenance and TPM and TQM are differentiated with good clarity. Zero defects and zero breakdowns of maintenance and applications in reality in the plant are also dealt. The Japanese concept Kaizen in its various focuses of applications (standard work Kaizen, TPM focus, quality focus, 5S focus, etc.) is explained because of its relevance to total productive maintenance.

14.19 KEY CONCEPTS

1. **TPM:** The maintenance philosophy and a culture of maintaining the equipment in good condition from its design to discard considering all the factors.
2. **Just In Time (JIT)**: This technique is developed by Japanese that means to act in time when problem arises.
3. **Maintenance Prevention:** Avoiding the occurrence of maintenance.
4. **Zero Breakdown:** Elimination of breakdowns by maintenance prevention (MP) or adoption of maintenance free design, etc.
5. **Jishu-Hozen:** Carrying out initial or basic maintenance of equipment by the operators themselves and informing the potential of major breakdown.
6. **Kaizen:** Kaizen means change for good or a good change (continuous improvement).
7. **Standard Work Kaizen:** Continuous improvement by eliminating the waste in the shop floor.
8. **Process Kaizen:** Improvements done to eliminate the waste in process, inadequate technology or design.
9. **Operation Kaizen (Method Improvement):** In this philosophy, efforts are made to reduce cycle time. Adopting the scientific approach in maintenance of the plant.
10. **Layout Kaizen.** Design a layout and re-layout or change or improve so that it is flexible and can be changed to respond to changes in process flow.
11. **Equipment Kaizen.** Modify or improve the equipment productivity, efficiency, and utility.
12. **Set-up Reduction Kaizen:** Improve all the elemental operations and develop/train operators in new work sequence.
13. **TPM Kaizen:** To maximise OEE by intensive effort to remove hidden causes and to improve the equipment reliability, availability, and maintainability.
14. **Quality Focus Kaizen:** Define the problems and construct a cause and effect diagram; after brainstorming and reach a consensus, select the best solution.
15. **Kanban Kaizen:** A Kanban is a tag attached to a product to facilitate smooth and proper material (inventory) movement oriented at wastage reduction.

16. **5S Kaizen:** The 5S (Seiri, Seiton, Seiso, Seiketsu, and Shitsuke) steps aimed at good house-keeping, waste reduction, safety and hygiene, etc.
17. **Seiri/Sort/Clear Out:** Classify items into two necessary and unnecessary discard or remove the latter.
18. **Seiton/Straighten/Configure:** Seiton means configuring the items marked as 'useful' and arranging them so as to minimise search time and effort.
19. **Seiso/Scrub/Clean and Check:** Seiso is cleaning and also checking.
20. **Seiketsu/Systematise/Conform and Continue:** Adoption of the procedures and methods in a proper way in the organisation at either macro or micro level to keep each individual safe and hygiene by proper and safe working and maintaining a clean and healthy working environment. And also to adhere to work on Seiri, Seiton, and Seiso continually and every day.
21. **Shitsuke/Standardise/Custom and Practice:** Adoption of the systems in the organisation and making them universally applicable (standard) for the smooth functioning of the organisation. By standardising procedure, duration, frequency, responsible and authorized persons, etc.

REVIEW QUESTIONS

1. What is total productive maintenance (TPM)? What are its objectives?
2. What are the approaches to achieve TPM? Explain each briefly.
3. List and briefly explain the eight pillars of TPM.
4. Enumerate the merits of TPM. What are the barriers to implementation of TPM?
5. Discuss the meaning and concept of the terms zero defects, zero breakdowns, and zero accidents.
6. What is Kaizen? Explain briefly different methods or approaches of Kaizen.
7. Examine the applicability of various features of Kaizen to Plant Engineering Maintenance.
8. Explain the 5S Kaizen in detail.
9. Explicate the role and focus of Kaizen on TPM.

CHAPTER 15

Industrial Safety

PREAMBLE

This chapter focuses on the importance of safety in industry and the role of safety manager during discharging the duties of safety management entrusted with three-fold responsibilities, i.e., before, during, and after the accident. It analyses the causes and prevention of the occurrences of accidents. Various legislative provisions relevant to industrial safety are discussed in detail. The topics discussed in this chapter should be of utmost importance to the employees of plant engineering due to the risk associated task they involve in on a daily basis.

CHAPTER OUTLINE

15.1 INTRODUCTION

Accidents may cause injuries, which sometimes result in temporary/permanent and/or total disablement or death. Thousands of industrial accidents occur every year claiming large number of lives of workers all over the world in addition to the loss of assets. In India, average number of casualties due to industrial accidents is nearly one lac every year out of which nearly 1200 causes death. Statistics show that 98% accidents are avoidable and only 2% accidents cannot be prevented. Seriousness of this problem can be understood by these figures.

15.2 SAFETY MANAGEMENT

The role of safety manager during discharging the duties of safety management is entrusted with three-fold responsibilities. These are

(a) Pre-accident responsibility, i.e., activities before the accident occurs
(b) During accident responsibility, i.e., activities during the occurrence of accident
(c) Post-accident responsibility, i.e., activities after the accident occurs

15.2.1 Pre-accident Role

The safety manager should take all precautionary measures to forestall or prevent the occurrence of accidents. These are accomplished through

(a) Inspection of the working premises on regular basis and identifying the areas those are prone to accidents.
(b) Taking necessary steps to prevent accidents in the accident prone areas such as providing machine guards, fencing, and safety equipment.
(c) Classifying the total working areas into three types, hazardous area, chancy area, and normal area based on the factors that are prevailing in the respective areas (and past records if available).
(d) Taking acute care on the hazardous areas and considerable care on chancy areas. However, one has to remember the words of Murphy, "if a failure can occur, it will". Therefore, sufficient care must be taken on all areas as required.
(e) Educating and training all the people, including workmen, supervisors, and officers, working in the hazardous and chancy areas about the possibilities of occurrence of accidents and inculcate the safety consciousness in them.
(f) Conducting training programmes on safety to bring the seriousness and alertness while doing their jobs.
(g) Pasting the posters, writing safety slogans on the walls and prominent places so that they will be reminded.
(h) Conducting contests on safety slogans, posters, and cartoons, etc., and creating interest in safety.
(i) Keep first aid box available in their vicinity for emergencies.
(j) Keep fire extinguishers and other safety equipment and train the employees about their usage.

15.2.2 Role During Accident Occurrence

The safety manager has to act fast and intelligently during this point of time. As far as possible, the safety manager should not allow this situation to arise. However, it is not possible to eliminate such situations. To act in such situations, the safety manager should have courage, bravery and guts along with initiative. This activity has to be carried out though

(a) Preparing a rescue team in advance for this purpose and sufficiently train them with mock tests.
(b) Rushing up to the accident spot within no time and alerting the rescue team and activate.
(c) Keeping all rescue equipment such as fire extinguishers, ropes, water resource, first-aid box, etc., at the nearest spots of suspected areas for 'ready to use'.
(d) Keeping all phone numbers, contacting resources, and communicating equipment, and using them immediately.
(e) Informing the situation to the superiors or officers concerned for their suggestions or actions.
(f) Contacting with fire-engine (101*), ambulance (108*), stretchers and medical facility (104*), and police (100*), etc., as and when required. *(*The numbers in the parentheses are the toll free phone numbers in India.)*
(g) Carrying out first-aid on the injured or affected or victims of the accident.
(h) Preparing the first investigation report (FIR).
(i) Making alternate arrangements for the work to run uninterrupted immediately after the accident.
(j) Preparing a detailed report by inquiring with eye witnesses after the situation is pacified.

15.2.3 Post Accident Role

The safety manager plays a vital role even after the accident has occurred. The top officials are keen about the after effects and the information how it has occurred. Quite often, the managements try to escape from the scene to avoid the compensations and legal implications and complications. The safety manager's role, therefore, becomes crucial to bring a balance between the two sides. He has to do justice to the victims and at the same time safeguard the management. Generally, the activities during this situation are

(a) Form a committee with experts and officers concerned to investigate by conducting a detailed inquiry with eye witnesses and relevant persons.
(b) Prepare the reports and submit to the top management for necessary arrangements.
(c) See that proper medical assistance is available to the victims.
(d) Console the victims and their family or dependents.
(e) Helping the victims to recover to their normal state.
(f) See that satisfactory 'compensation' is received by the victim or his dependents for his/her loss.
(g) Alternate arrangement for the work immediately after the accident and during their recovery period.
(h) Recommend for the improvements that are to be taken up in the area and steps not to recur.

(i) In case of death, the funeral expenses, evaluating the possibility of employing one of his/her family members, etc.

15.3 SAFETY EDUCATION AND TRAINING

The training should be imparted in safety measures to the workers. This can be accomplished by conducting vestibule schools. Further, it should be reminded regularly through safety posters, safety films, safety slogan contests and suggestions. These are useful to increase the interest of employees in accident prevention. This interest in turn creates alertness and safety consciousness. The purpose of this training is to induce care in using dangerous tools in hazardous environment to carry out risky operations.

FIRST AID

Although sufficient safety measures are taken in factories to minimise accidents, they cannot be totally avoided and, hence, the employers must provide proper first aid facilities for common accidents such as fire, heat stroke, cuts, etc., in every factory.

A first aid box must be provided in the charge of a responsible person, who must be always available in working hours and he should be trained in first aid. Some firms have elaborate arrangements of stretcher service, ambulance arrangements, surgeries, etc., in the work dispensary.

The workers as human being are entitled to every consideration at the hands of the management and it is for the administration to institute adequate precautionary and prompt first aid towards its workers, and hence it is essential that the management should be fully equipped to offer first aid to the workers.

Contents of First Aid Box

The specified contents of first aid box in workshops employing more than 60 persons (as per Factory Act) are:

1. Twenty four small sterilised dressings.
2. Twelve medium size sterilised dressings.
3. Twelve large size sterilised dressings.
4. 10 gms packets sterilised cotton wool.
5. One snake bite lancet.
6. One pair of scissors.
7. Twelve large size burn dressings.
8. Two (25 gms) bottles of potassium permanganate crystals.
9. One (100 gms) bottle containing 2% alcoholic solution (spirit).
10. One (100 gms) bottle of antiseptic lotion/powder having the dose and mode of administration indicated on label.
11. One copy of first aid leaflet.
12. Twelve rolled bandages 10 cm wide.
13. Twelve rolled bandages 5 cm wide.
14. Two rolls of adhesive plaster.
15. Six triangular bandages.
16. Two packets of safety pins.
17. A supply of suitable splits.
18. One tweeser set.
19. Eye drops.
20. Tincture of iodine.
21. A bottle (500 ml) of hydrogen peroxide.

BUSTLE - 15.1

Recall an accident you have seen in the recent past and prepare a first investigation report (FIR) on it. Further, construct a cause and effect diagram assuming various situations.

____

____

____

____

Test Your Understanding 15.1

1. Discuss the role of safety manager in an industry.
2. What are the responsibilities of safety manager under the following occurrences?
   a) Before the occurrence of an accident
   b) During the occurrence of accident
   c) After the occurrence of accident
3. Mention important precautions with reference to industrial safety, which should be followed during the industrial establishment.
4. What is the first aid box? What are its contents?

15.4 ACCIDENTS

The cost of an accident cannot be expressed in terms of money only. It also affects the individual and his family. Its social aspects are very tragic because for a family, an accident may mean a descent from independence to economic dependence. They will have to accept the financial assistance from a charitable agency and which affects the social relationship of the family. After the accident, the mental attitude also changes.

Definition

"Accident is defined as any sudden or unusual occurrence that interferes with the orderly progress of activity."

15.4.1 Adverse Effects of Accidents

In present days, grave attention is being paid in this matter, because it has been clearly understood that 'accidents cause heavy losses'. The losses may be direct and/or indirect.

(a) **Direct losses**

Generally, these losses are measured in terms of money. These are often the losses to the employer in the following forms:

- Compensation paid to the worker.
- Medical expenses incurred on the worker.
- Loss due to damage of assets.
- Money spent to reinstate the situation.

(b) **Indirect losses**

These indirect losses arise from the following sources:

- Loss of time of the injured person.
- Loss of time of his fellow workers, who stop work at the time of accident to help the injured. Even the worker who may not help would also stop the work to show sympathy or at least for curiosity.
- Loss of time of supervisors/foremen
  - (i) in assisting injured worker;
  - (ii) in investigating and preparing a report of accident;
  - (iii) in making alternative arrangement;
  - (iv) in selecting and training the new worker to fill the vacancy in case of death.
- Loss due to damage caused to machines.
- Loss due to lowered efficiency of the worker when he returns after recovery.
- Loss due to reduced efficiency of other workers due to fall in their morale.
- Loss to the injured worker. Injured worker suffers the following losses:
  - (i) Loss to his income.
  - (ii) Loss due to medical expenditure or for recovery.
  - (iii) Pain felt by worker, which cannot be compensated.
  - (iv) Disablement if any.

15.4.2 Causes of Accidents

To forestall or at least to minimise the accidents, it is necessary to know about the cause of accidents. Some general causes leading to accidents are given below:

(a) **Hazardous machines or areas** are commonly prone to accidents. These accidents occur from boilers, pressure vessels, prime movers, transmission system, and high speed rotary parts, etc.

(b) **Unsafe working practices or physical conditions**, which include improper guards, improper illumination, improper ventilation, unsafe clothing.

(c) **Precautions on moving or falling objects**.

(d) **Personal factors** such as lack of knowledge, physical weakness.

(e) **Violation of commonly accepted safe procedure**. These include
  - Working at unsafe speed
  - Loading machines beyond capacity
  - Not using safety devices
  - Adopting unsafe procedure

(f) **Electrical causes.** Some of the important causes are:
  - Working at unsafe speed
  - Not providing proper protecting devices.
  - Not obeying proper instructions and not following safety precautions.
  - Failure to use insulated pliers, screwdrivers, rubber gloves, etc.

(g) **Exposure to harmful substances** such as toxic gases, fumes, dust, vapours, mist, and aerosols.

15.4.3 Preventive Measures

(a) **Safety Guards:** By providing proper safeguards to the machines, accidents can be prevented. Some guards are built into a permanent casing, while some are attached afterwards.

(b) **Fencing:** Machines or their parts should be fenced when it is not possible to provide safeguards.

(c) **Proper Condition:** All boilers and other pressure vessels must be kept in proper condition. Safety valves, pressure gauges and water gauges, etc., must be examined thoroughly at regular intervals.

(d) **Sound Construction:** There must be sound compensation of hoists, cranes, and lifts, etc. They must be tested periodically.

(e) **Physical Conditions:** Sufficient illumination and ventilation should be provided. Floor should be free from oiliness. Floor should be kept clean.

(f) **Protective Equipment:** Safety measures include special clothing for the protection of body, such as gloves, apron, goggles, etc. Loose clothing may be a source of danger.

(g) **No Repair While Running:** Repair work on machines should not be done when it is running.

(h) **Orderliness:** All the tools should be kept at their proper places.

(i) **Use Appropriate Equipment for the Purpose:** For example chips should not be removed by hand but brushes should be used.

(j) **Training:** Workers should be trained about correct procedures and they should be educated about safety precautions. Constant warning, publicity, and placards carrying slogans (as 'Safety-First', Danger, 440 volts, etc.) are also helpful to reduce accidents.

(k) **Fire Hazards:** To avoid the danger, inflammable materials should be kept away from general storage at a safe distance (minimum 50 ft or 15.25 m). Fire extinguishers should be kept at suitable places.

(l) **Prevention of Electric Accidents:** To prevent electric accidents, the following measures should be taken:

- Electrical insulation should be periodically tested.
- Use proper tools for testing and repairing.
- Repair work should be done after switching the power off.
- Use the safety equipment such as insulated tools and rubber gloves, etc.

15.4.4 Types of Industrial Accidents

Industrial accidents may be classified under two general classes:

(a) **Machinery Accidents:** Usually, these accidents are caused due to inadequate safeguards, bad machine condition, over loading, improper operating or handling of machines. Machinery accidents can be reduced by providing safety guards on moving parts such as belts, gears, etc. Further, fencing is to be made at hazardous areas. Workmen are to be insisted to wear or use the safety equipment such as goggles, helmet, nose mask, ear plugs, aprons, shoes, gloves, face masks, pads, etc., and are to be thoroughly trained about their usage.

(b) **Non-machinery Accidents:** These accidents are commonly caused due to personal reasons such as age, physical weakness, inexperience and carelessness or from the plant conditions such as poor ventilation and illumination.

Non-machinery accidents are generally caused by different factors; some of them are listed here.

Factors Responsible for Accidents

The following are the common factors responsible for occurrence of an accident.

(a) **Age:** Young persons due to their ignorance, immaturity, physical weakness, etc., cannot withstand or counteract to the accidents. It has been seen that accidents are more frequent with younger persons (below 15 years of age).

(b) **Physical Condition:** Minor illnesses like cold, cough, sore throat, headache, etc., are also responsible for accidents to a large extent. These small frequent illnesses deviate the concentration of the job on one hand and lower the general health on the other.

(c) **Mental Conditions:** If a worker works wilfully, he takes sufficient care on himself, the organisation, and the environment. Further, if the mental condition adversely affects the work. Good working conditions along with motivating, inspiring, congenial supervision produce a good effect on the workers' psychology in addition to greater efficiency. In such conditions, the worker will always be ready to offer his services and co-operation. It is essential for the success of any industry that its workers should extend good co-ordination.

(d) **Experience:** Experience obviously teaches the lessons. Rate of accidents for less experienced workers is more than those of more experienced workers.

(e) **Rate of Production:** This factor should also be considered while analysing the cause of accidents. Generally, the number of accident increases with the increase in production but the proportion of accidents tends to decrease with the increasing production. This means, the rate of change in accidents per man-hour is less than the rate of change in production per man-hour.

(f) **Working Conditions Affecting Health:** Working conditions also have their impact on the work. When a worker is allowed to work in good working conditions, then his efficiency increases. Bad environment or working conditions may ultimately lead to:

- Physiological fatigue;
- Mental fatigue, i.e., feeling of boredom; and
- Decreased efficiency resulting in reduced output. In earlier days, no attention was paid on the working conditions like illumination, humidity, air ventilation, temperature, etc. But its importance is now being realised (Elton Mayo's Hawthorne experiments are probably first to study in this direction).

(g) **Fatigue:** Experiments have shown that suitably arranged rest pauses reduce the number of accidents to a large extent. As these reduce the fatigue, and hence the accident which might occur due to excess fatigue is forestalled. Researches have shown that when only one break (for lunch) is provided, then accidents tend to increase with each successive hour of work in the morning and reaching a maximum approximately at 11 am then reduces in the noon. Number of accidents again increase to reach maximum value towards the later part of the afternoon but a slight drop in the last hour (probably it occurs due to the fact that in this hour, speed of work decreases and worker feels relaxed mentally that shortly after he will be free).

(h) **Climatic Conditions:** Researches have shown that accidents are minimum at a temperature of 67.5 °F (nearly 20 °C). At higher temperatures, rate of occurrence of accidents is more and after 24 °C rate of occurrence of accidents increases substantially.

(i) **Illumination:** Poor illumination raises accident frequency. In daylight, accident frequency is less as compared to the nights' artificial illumination. Dim illumination reduces the speed of work and results in strain on eyes and lead to more accidents. Light should come from the right direction and of desired illumination. In artificial light, glare is most common defect, which harms the eyes. It also produces strain and headache. Spoilage of work also increases due to glare.

(j) **Ventilation:** A badly ventilated working atmosphere will cause discomfort and fatigue. It reduces interest in the work and hence the efficiency. Proper ventilation takes away the heat of human body, furnaces, boiler, and other equipment, thus, reducing the effect of heat to some extent. Proper ventilation also reduces dampness. Fans arranged in a systematic way helps to achieve this objective to a considerable extent. Sometimes, fans placed in wrong location or direction may move the air through furnace, hot parts of machines, etc., thus, transmitting the heat to the workers, which they would have not received otherwise.

(k) **Hours of Work:** Working hours should be distributed uniformly over the week. A worker should get at least one weekly holiday so that he can enjoy on that day and relieve his fatigue and boredom and thence he may return on duty as fresh in the next week. As discussed above, rest pauses also lowers mental fatigue on the worker. It is better to allow at least 5 minutes break in one working hour and one lunch/dinner break. Duration of rest may vary slightly based on the nature of work and working conditions. *(About the working hours, it is further discussed in detail again in this chapter under Factories Act).*

(l) **Noise and Vibrations:** Too much noise and vibrations produce mental fatigue and reduce the efficiency of the worker. Of course, noise cannot be completely stopped particularly, at running machinery; it can be reduced by covering the source of noise by using baffles, silencers, and soundproof materials, etc. Noise and vibrations can also be controlled to some extent by proper maintenance, checking, lubrication, proper functions, etc. Its reduction is very essential as it distracts attention and concentration on the work and sometimes too much noise even affects the hearing capacity.

(m) **Plant and Shop Layout:** The systematic layout helps reducing accidents, movements of the products, etc. If the shop layout is such that it looks pleasant then worker may take keen interest in his work. Thus, layout should be in such a way that material handling be economical and safe, and avoid overcrowding. Passages for movement must be quite safe and sufficient enough. It should also be planned to get sufficient natural light in proper direction. Factory should be kept clean, doors and windows have to be properly coloured and walls should be whitewashed so that atmosphere in the factory looks cheerful. A well designed factory brings pleasure in every worker to feel proud in working and take utmost interest in his work.

15.4.6 General Precautions (Dos and Don'ts)

The following are some of the common precautions, which should be observed from the viewpoint of industrial safety so as to prevent accidents:

Dos

- Always be alert and in the right physical and mental condition.
- While doing the job, always wear appropriate clothes, safety glasses, gloves, footwear, hard hat, etc., as per the job requirement.
- Always keep guards and covers in position especially while operating a machine.
- Keep hands away from moving parts such as fan, V-belt, gears, drive shafts, etc.
- Ensure that you are fully aware of the equipment and its operation before operating.

- See that the equipment is carefully inspected at regular intervals.
- Stick to all the provisions laid down under various Acts (labour legislation).
- Follow precautions suggested by the manufacturer.
- Follow maintenance schedule.
- Keep the equipment and working area clean and tidy.
- Always run on safe speed as per the working conditions and the job requirement.
- Always use appropriate tools.
- The tools must be free from grease and oil, and properly maintained.

Don'ts

- Never check for leaks in a pressurised system with hand.
- Do not wear ties, loose hair, rings or watches, etc., which can be caught by moving parts of the equipment and may pull.
- Do not perform maintenance work while machine is running.
- Never put the machine on overload.
- Do not lean on machine.
- As far as possible, do not stand parallel to moving parts such as spindle, belt drive.
- Do not deviate from any instruction mentioned under the head 'preventive measures'.

BUSTLE - 15.2

1. *State the measures of safety you would introduce in any manufacturing organisation with which you are familiar.*
2. *Enumerate the safety precautions to be observed by a worker working on any one of the following machines:*

   (i) Surface grinding machine (ii) punch press (iii) electric welding machines.

Test Your Understanding 15.2

1. Write short notes on:

   (a) First aid (b) Effect of working conditions on safety (c) Causes of accidents
2. What is the importance of industrial safety? Explain various losses which may occur due to an accident in a factory.
3. Enumerate common causes of industrial accidents. Describe various measures to prevent them.
4. How safety education and training will help in industrial safety? Explain.
5. Explain different factors which cause/enhance the chances of accident.
6. What are the safety measures? Suggest different ways to minimise the accidents.

15.5 EFFORTS BY GOVERNMENT

To ensure industrial safety, the Government of India has made number of legislation like Indian Electricity Act, 1884, Indian Electricity Act (Revised), 1910, Indian Boilers Act, 1923, Workmen's Compensation Act, Petroleum Act, 1934, the Factories Act, 1948, Mines Act, 1952. The provisions under these Acts help governing the safety of personnel and equipment in industrial units in the country. However, legislation alone cannot ensure safety in industrial operations, unless there is an effective approach to prevention of accidents. These are possible only by adopting suitable control measures including safe designs of machines and processes, use of protection devices and personal protective equipment, effective safety procedures and practices as well as creation of self regulating system on the shop-floor.

To assure safety to workers and eliminating chances of damage to machinery and equipment, Indian Standards Institute (ISI) has done commendable job. Some of the principles laid down by ISI are listed below:

- Safety precautions to be taken during manufacturing operations.
- Requirements for effective maintenance of tools and equipments.
- Standards for proper layout, proper lighting, and ventilation of factory building.
- Guidance on safe welding and cutting, use of powered industrial trucks, belt conveyors, and fire fighting equipment.
- Standards and specifications of safe industrial operations and practices.
- Classification of hazardous chemicals and use of accident prevention tags and pictorial markings for handling and labelling of dangerous goods.
- Safety codes for handling acids and other chemicals.
- Safety requirements for personal protective equipment.
- Standards for fire safety in industrial buildings and safety procedures to be followed in electrical work and use of electrical appliances in hazardous area and explosive atmosphere.
- Specifications for protective clothing, safety helmets, face shields, and safety equipment's for eyes, lungs, hands, feet, and legs. These include eye and ear protectors, gas mask, gloves, safety boots and shoes for mines, and heavy metal industries, etc.

15.6 PROVISIONS UNDER FACTORIES ACT, 1948

15.6.1 Health Provisions

This act prescribes the following provisions for maintaining the health of the workers and reducing the possibilities of injuries to their body:

(a) **Cleanliness**

Every factory shall be kept clean and tidy and free from gases arising from any drain or other nuisance. For this, the following steps should be taken:

- The dust, waste, scrap, and refuse shall be removed daily by sweeping the floors and benches of workrooms, staircases, and passages.
- The floor of every work spot shall be cleaned at least once in every week by washing, using disinfectant where necessary. Oily or greasy floor must be avoided.

- Effective drainage system should be provided and maintained. Leakages should be arrested.
- All inside walls, partitions, ceiling of rooms and passages:
  - (i) Where they are painted or varnished, shall be repainted or re-varnished at least once in every five years.
  - (ii) Wherever needed, shall be whitewashed or colour washed at least once in every 14 months.
- Walls and roofs should be of such material and so designed that temperature shall not be exceeded but kept as low as practicable.
- Effective arrangement should be made to protect the workers from process producing excessively high temperature. Hot parts of the machinery shall be properly insulated.

(b) **Artificial Humidification**

- The water used for such humidification shall be clean and free from dirt.
- If there is an artificial humidification, it shall be of prescribed standard and created by the prescribed methods.

(c) **Overcrowding**

No room in the factory shall be overcrowded.

- For this, at least 350 cu. ft. or 9.9 m^3 of space for every work is necessary in the factories which were in existence before the date of commencement of this Act.
- At least 500 cu. ft. or 14.2 m^3 of space for every worker shall be provided in the factories built after the commencement of this Act.
- For this purpose any space, which is more than 14 ft. or 4.2 m above the level of the floor of the room, shall not be taken into account.

15.6.2 Safety Provisions under Factories Act, 1948

The following important provisions for safety are provided in the Act:

(a) **Fencing of Machinery**

The following items shall be properly fenced in every factory:

(a) Head race and tail race of every water turbine.

(b) Every moving part of a prime-mover, every flywheel

(c) The following items should be properly safeguarded;

- every part of an electric generator,
- a motor, or a rotary converter;
- every part of transmission machinery and
- moving parts of other machines.

(b) **Work on or near Machinery in Motion**

The work on or near moving machinery should be done by a specially trained adult male worker wearing tight fitting cloths. Neither woman nor young person shall be permitted to work, clean, lubricate, or adjust any part of a prime-mover or of any transmission machinery when it is in motion.

Devices for cutting off power from running machines in emergency should be provided.

(c) **Casing of Machinery**

Every set screw or key on any revolving shaft, spindle, wheel, or pinion shall be so sunk, encased or otherwise effectively guarded so as to prevent danger. Every toothed or friction gearing must be completely encased.

(d) **Hoists and Lifts**

In every factory:

- Every hoist and lift should be of sound mechanical construction with adequate strength. They have to be properly maintained and thoroughly checked regularly. It should be marked with safe maximum load and should not be overloaded.
- Hoists and lifts used for carrying persons should have at least 2 ropes or chains separately connected with the cage and balance weight and each rope/chain with its attachment should be capable of carrying the weight of the cage together with its maximum load.
- Every hoist way and lift way must be adequately protected by enclosures or fitted with gates.

(e) **Lifting Machines, Chains, Ropes, and Tackles**

All parts such as the working gears, lifting machines (crane, crab, winch, pulley block, etc.), chains, ropes or lifting tackles shall be of sound construction and with enough strength, well maintained and keenly examined at least once in a year or as specified by the manufacturer.

(f) **Revolving Machinery**

Every grinder, milling machine, or revolving machinery should be provided with the notice of the maximum safe working peripheral speed of every grind stone or abrasive wheel, mills, the speed of the shaft, or spindle upon which the wheel or cutter is mounted.

(g) **Excessive Weights**

No person has to be employed in any factory to lift, carry, or move any load so heavy that is likely to cause injury. State Government may regulate by prescribing the maximum weight which may be lifted or carried by adult (man/woman), adolescent, and child.

(h) **Protection of Eyes**

Screens or goggles must be provided to the persons engaged in the following manufacturing processes to protect the eyes:

- Welding or cutting the metals with electric or oxy-acetylene flame or any other such process.
- Fitting, riveting, removal of scales or dressing of metals or stones or any other such process involving risk of injury to the eyes from particles chip off in the course of process.

(i) **Precautions against Dangerous Fumes**

- Any person shall not be allowed to enter any hearth, chamber, pit, tank, pipe, flue, or other confined space in which dangerous fumes are likely to be present to such an extent as to involve risk to persons, unless it is provided with a manhole of adequate size to escape and that too if he is trained.
- Portable electric light above 24 volts should not be permitted to such confined space.Wearing suitable breathing apparatus and a belt securely attached to a rope, the free end of which is held by a person readily standing outside to rescue from the confined space.

(j) **Precautions against Fire**

- Every factory should provide means to escape in case of fire as prescribed.

- A free passage-way giving access to escape in case of fire must be maintained.
- Exit doors or gates from any room should not be locked or fastened so that they cannot be easily and immediately opened from the inside while any person is inside the room.
- Every window, door, or other exit affording means of escape in case of fire, other than the means of exit in ordinary use should be marked as “Emergency Exit” in all the languages so as to be understood by any worker.
- Effective means of warning in case of fire to every person has to be provided.

15.6.3 Welfare Provisions

(a) **Washing Facilities:** Adequate and suitable facilities for washing are to be provided separately for male and female workers and hygienically maintained for their use.

(b) **Changing Rooms:** Suitable arrangements should be made for changing dresses or cloths or other safety wear separately for males and females and the areas shall be marked clearly in all the languages so that every employee can understand.

(c) **Drying and Storing Rooms:** Suitable place for storing cloths not worn during working hours and for the drying of wet cloths should be marked.

(d) **Sitting Facilities:** Suitable arrangements for sitting shall be provided and maintained for all workers, who work in standing position.

(e) **First Aid Appliances:** First aid boxes or cupboards equipped with the prescribed contents shall be provided and maintained at readily accessible place during all working hours. At least one such box shall be provided for every 150 workers.

(f) **Canteens:** A hygienically maintained canteen or canteens shall be provided in every factory wherein more than 250 workers are employed.

(g) **Shelter, Rest Room, Lunch Room:** In every factory where more than 150 workers are employed, neat, adequate and suitable shelter, rest rooms and lunch rooms with provision of drinking water and dustbins, where the workers can take the food brought by them, shall be provided and maintained.

(h) **Crèches:** In every factory where more than 30 female workers are employed should provide and maintain adequately spaced, well-equipped crèche for the use of their children under the age of 6 years. A trained woman should be kept in-charge to take care of the children. Women having kids under the age of 3 years should be permitted to feed as and when required.

(i) **Welfare Facilities:** Welfare officers shall be appointed in every factory employing more than 500 workers. The government may prescribe the duties, the qualifications, and conditions of service and number of such officers.

15.6.4 Working Hours

(a) **Hours of Work:** No worker shall be required or allowed to work in factory for more than 48 hours in a week and for more than 9 hours in a day. No worker shall work for more than 5 hours before he/she has an interval of at least half an hour. The period of work on any day shall be so arranged that inclusive of rest intervals shall not spread over more than 10.5 hours. However, the chief inspector may increase this spread over time to 12 hours for reasons to be specified in writing.

(b) **Holidays:** No worker shall be required or allowed to work in a factory on Sunday (or a certain specified day), unless he/she has or will have a holiday for a whole day on one of the 3 days immediately before or after the Sunday (or the certain specified day). Thus, no worker shall work for more than 10 days continuously without a holiday.

(c) **Overtime wages:** If a worker works for more than 9 hours in a day or for more than 48 hours in a week, he/she shall be entitled to wages at the rate of twice his/her ordinary rate of wages, in respect of overtime work. Here, ordinary rate of wages means the basic wages plus such allowances as the worker is for the time being entitled to, but does not include a bonus.

(d) **Restriction on Double Employment:** No worker employed in a factory shall be required or allowed to work in any other factory.

(e) **Employment of Women:** This provision adds a further restriction to employment of female workers that no women shall be employed in any factory except between 6 am and 7 pm However, State Government may vary this limit, but variation shall not in any case authorise the employment of any woman between 10 pm and 5 am.

(f) **Employment of Young Persons:**

- A child, who has not completed 14 years, shall not be allowed to work in any factory.
- A child who has completed 14 years or an adolescent shall not be allowed to work unless a certificate of fitness is granted to him/her. This certificate shall be renewed every year.
- An adolescent, who has been granted a certificate of fitness to work in a factory, shall be deemed to be an adult. Further, an adolescent who has not attained the age of 17 years shall not be permitted to work between 10 pm and 7 am.
- No child (14 to 17 years) shall be permitted to work for more than 4½ hours in a day and he/she will also in any case not be allowed to work in the night (10 pm to 7 am).

For granting leave, the worker should apply 15 days before or immediately in case of illness. The act further provides that the number of times this leave taken during any year shall not exceed 3.

15.6.5 Dangerous Operations

If any operation carried out in the factory exposes any employee to a serious type of bodily injury, poisoning or disease, the State Government may declare that operation to be dangerous or hazardous and may ask for the provision of proper safeguard and, if necessary, may prohibit or restrict the use of any materials.

It also prohibits weak, adolescent, and children to be employed therein and to ask for the medical examination of workers employed.

15.6.6 Accidents and Diseases

- In a factory, accident resulting in death or causing any bodily injury that prevents the person from working for a period of 48 hours or more immediately following the accident, manager of the factory shall send intimation to the factory inspector within the prescribed time.
- If any worker suffers from any disease specified in the schedule, the manager of the factory shall send its intimation to the factory inspector within the prescribed time.
- If a medical practitioner attends to a person employed in a factory and suffering from any disease specified in the schedule, shall without delay send a report in writing to the chief inspector stating the name and address of the patient and the factory mentioning the name of disease.

15.6.7 Penalties/Punishments

(a) (i) On any breech or contravention of any of the provision of this Act or any rule made thereunder or any order in writing given thereunder, the occupier and the manager of the factory shall each be guilty of an offence and punishable with imprisonment upto 2 years or with a fine up to ₹ 1 lac or with both, if the contravention is continued after conviction, with a fine up to ₹ 1000 per day.

(ii) If he/she is again guilty of an offence involving in contravention of the same provision, he/she shall be punishable on a subsequent conviction with a fine up to ₹ 5000 in case of injury and ₹ 25,000 in case of death.

(b) Any worker contravenes any provision of this Act or rules or orders made thereunder imposing any duty or inability on workers, shall be punishable with fine up to ₹ 500.

(c) If a child works in two factories on any day, his/her parent or guardian or the person having custody of or control over him/her or obtaining any direct benefit from his/her wages, shall be punishable with a fine up to ₹ 1000.

BUSTLE - 15.3

1. *With reference to provisions of Factories Act, discuss the following:*
   (a) Power press safety measures to be taken while working.
   (b) Transmission machinery such as driving belts, gears, pulley flywheels, and shafts-safety measures to be taken while in motion.
   (c) Measures to be taken to safeguard the health of the workers.

____

____

____

____

Test Your Understanding 15.3

1. Explain various provisions under Factories Act, which helps in industrial safety.
2. List the health provisions mentioned under the Factories Act. Explain how they help in reducing the accidents.
3. What are the welfare provisions under Factories Act? Explain how they will affect the industrial safety.
4. Write short notes on:
   (a) Efforts by Government towards industrial safety, (b) Working hours described under Factories Act, (c) Penalties and punishments for breach of the Factories Act.

15.7 ELECTRICITY RULES

(a) **Protective Equipment and First Aid:** Fire buckets filled with clean and dry sand should be kept ready for immediate use for extinguishing fires. In addition, suitable fire extinguishers be provided at convenient points to deal with electric fires and kept in all generating stations, enclosed substation, and enclosed switch stations or any prone/suspected places.

First aid boxes must also be provided and maintained and be kept in the custody of responsible persons who shall be available during working hours.

(b) **Work:** No electrical installation work, including additions, alterations, repairs, and adjustments to existing installations, except such replacement of lamps, fans, switches, low voltage domestic appliances and fittings, shall be carried on behalf of any consumer or owner, except by an electrical contractor licensed on this behalf by the State Government and under the direct supervision of a person holding a certificate of competence issued or recognised by the State Government.

(c) **Periodical Inspection:** Every installation which is connected to the supply system of the supplier shall be periodically tested and inspected at interval of not exceeding 5 years.

15.7.1 Supply and Use of Energy

(a) **Testing of Installation and Connection:** On receiving an application for a new/additional supply of energy, the supplier shall inspect and test installation at the premises of applicant. If supplier notices is likely to constitute danger, he/she shall ask the applicant to make such modifications as necessary for safe installation. The supplier may refuse to connect the supply until the modifications are made.

Further, the consumer has to ensure that no person other than the supplier shall interfere with the service lines and the apparatus placed by the supplier on the premises of the consumer.

(b) **Switch Board Installation:**

- A clear space of at least 3 ft. (0.914 metre) in width must be left in front of the switch board.
- If there are bare connections at the back of the switch board, the space behind the board should be either less than 9 inches (23 cm) or more than 30 inches (76 cm) from the wall. In the latter case, there should be passage-way from either end of the switch board.

(c) **Variation of Voltage:** Variation in voltage should not be beyond 4% of the declared voltage for low/medium voltage and not beyond 12½% for high/extra-high voltages.

(d) **Variation of Frequency:** Frequency of an AC must not exceed 3% of the declared.

(e) **Sealing of Meters and Cut-outs.** Supplier may affix a seal to any cut-out or meters and it is the consumer's responsibility to see that seal is not broken.

15.7.2 Penalties

(a) **Breaking Seal:** A person breaking the seal shall be punishable with a fine up to ₹ 200 and the consumer, when he himself has not broken the seal, shall be punishable for every such breach with a fine up to ₹ 50.

(b) **Breach of Rules:** Any person, who commits breach of these rules, shall be punishable for every such breach with a fine up to ₹ 300.

[Some other provisions under Electricity Act are annexed for the reference of any interested.]

Connection with Earth

For Low and Medium Voltage

(a) Provisions for connection with earth of systems at low voltages (above 125 volts) and of systems at medium voltage are furnished here below:

- Neutral conductor of a 3-phase 4-wire system and the middle conductor of 2-phase 3-wire system shall be earthen both at the generating station and the sub-station.

- In a DC 3-wire system, the middle conductor shall be earthen at the generating stations only, and the current from the middle conductor to earth shall be continuously recorded by means of a recording ammeter and when current exceeds one thousandth part of the maximum current supply, insulation system should be checked immediately.
- When middle conductor is earthen by means of circuit breaker with a resistance connected in parallel, the resistance shall not exceed 10 ohms. The resistance shall be used only as protection for the ammeter.

(b) Frame of every generator, stationery motor and so far as is practicable portable motor, metallic parts of all transformers and any other apparatus used for regulating and controlling energy shall be earthen by the owner by two separate connections.

(c) All metal castings or metallic coverings containing or protecting any electric supply line or apparatus shall be connected with earth.

High and Extra-high Voltages

(a) Following provisions shall apply to the connections with earth of 3-phase system for the use of high and extra high voltages:

In case of star-connected system with earthen neutral or in delta-connected system with earthen artificial neutral points, the neutral points shall be earthen by not less than 2 separate connections having its own electrode at the generating station and at the sub-station.

(b) Single-phase high or extra-high voltage system shall be earthen in manner approved by the inspector.

Overhead Lines

(a) **Material and Strength**

- All conductors of overhead lines must be of a breaking strength not less than 700 lbs (317.5 kg.).
- For low voltage with span of less than 15.24 metres (50 ft) and is on the owner's or customer's premises of not less than 300 lbs (136 kg.).

(b) **Clearance above the Ground**

(i) Overhead lines should not be erected across a street at a height less than:
   - For low and medium voltage lines—5.8 m (19 ft)
   - For high voltage line—6.1 m (20 ft)

(ii) Overhead lines should not be erected any street at a height less than :
   - For low medium voltage lines—5.5 m (18ft)
   - For high voltage line—5.8 m (19ft)

(iii) Overhead lines should not be erected elsewhere than along or across any street at a height less than:
   - For low, medium, and high voltage lines (up to 1100 volts),
     If bare—4.6 m (15 ft) and If insulated—4.0 m (13 ft)
   - For high voltage lines above 11,000 volts—5.2 m (17 ft)

(iv) For extra-high voltage lines, the clearance above ground shall not be less than 5.2 m (17 ft) plus 0.3 m (1 ft) for every 33,000 volts or parts thereof by which the voltage of the lines exceeds 33,000 volts.

But the minimum clearance along or across any street shall not be less than 6.1 m (20 ft).

(c) **Clearance for Buildings**

- For low or medium voltage overhead lines, maximum vertical clearance should be 2.44 m (8ft) from the highest point of the building and a minimum horizontal clearance should be 1.22 m (4 ft) from the nearest point from where the lines pass.
- For high voltage overhead lines, minimum vertical clearance should be 3.66 m (12 ft) plus 0.3m (1ft) for energy additional 33,000 volts or part thereof and minimum horizontal clearance should be 1.83 m (6 ft) plus 0.3 m (1 ft) for every additional 33,000 volts or part thereof.

(d) **Protection against Lightning**

- Owner of every overhead line which is exposed as to be liable to injury from lightning shall adopt efficient means for diverting to earth any electrical surge due to lightning.
- The earthing lead for any lightning arrestor shall be directly taken from the lightening arrestor to a separate earth electrode and not through the iron or steel pipe.

15.8 PROVISIONS UNDER WORKMEN'S COMPENSATION ACT

It is an Act to provide compensation by employers to their workmen against injury or loss of life due to accident. Thus, this Act protects the workers as far as possible from hardship arising from accidents. The objective of awarding compensation is to replace the actual loss suffered by the workmen. It came into force from 1 July 1924.

15.8.1 Definitions

Some definitions given under this Act are as follows:

(a) **Dependent:** Dependent means any of the following relatives of a deceased workman namely—

(i) A wife, a minor legitimate son, and unmarried legitimate daughters, or a widowed mother.

(ii) If wholly or in part dependent on the earnings of the workman at the time of his death, a husband, a parent other than a widowed mother, a minor illegitimate son, and an unmarried illegitimate daughter, a daughter illegitimate or illegitimate of married and a minor or if widowed, a minor brother, an unmarried or widowed sister, a widowed daughter-in-law, a minor child of deceased son.

(b) **Partial Disablement:** It is the disablement of temporary nature and the earning capacity is reduced due to disablement of permanent nature.

The list of injuries resulting permanent partial disablement is given in *Schedule I**.

(c) **Total Disablement:** It is such disablement either temporary or permanent in nature, which incapacitates an employee from all work which he/she was capable of performing at the time of accident resulting in such disablement.

Permanent total disablement shall be deemed to result from injuries where the aggregate percentage loss of earning capacity (as specified in *Schedule I*)* amounts to one hundred percent or more.

(d) **Wages:** Wages are the privileges or benefits or remuneration estimated in terms of money other than a travelling allowance or any other contribution paid towards pension, etc.

(e) **Workman:** Workman means any person (other than a person whose employment is of a casual nature and who is employed otherwise than for the purposes of employer's trade or business) who is employed in any such capacity specified in *Schedule II** on monthly wages.

15.8.2 Employer's Liability for Compensation

If an injury is caused to a workman by accident arising out of and in the course of his employment, the employer shall be liable to pay compensation in accordance with the provisions of this Act.

However, the employer shall not be liable in respect of any injury:

(a) That disables the workman for a period less than *three* days.

(b) Not resulting in death caused by an accident which is directly attributed to:

  (i) The workman was under the influence of drink or drugs.

  (ii) Wilful disobedience of the worker to an order issued for the purpose of securing safety of the worker.

  (iii) Wilful removal or disregard to any safety guard.

  But if the worker dies of an accident due to the above reasons, the employers have to pay compensation.

15.8.3 Amount of Compensation

Amount of compensation depends upon the following factors:

(a) The pay of the worker concerned

(b) Extent of injury

(c) The Act divides the injuries under the following four headings:

- causing death
- permanent total disablement
- permanent partial disablement
- temporary disablement

The amount of compensation for each of the above categories is provided in Schedule IV* in the Act.

[**The original Act consists of the Schedules under which the above contents are given. However, the reader if required may refer the book on labour legislation for details of these schedules.*]

15.8.4 Notice of Accident

The injury occurred to worker, how small may it be, should be brought to the notice of employer either personally or through some responsible person.

Every such notice shall give the details such as name and address of the injured person, the cause of the injury, the date and time, place, etc., and should be served to the employer or representative of the employer.

15.8.5 Medical Examination

The worker, who has given notice of an accident, must present himself/herself for the medical examination, if employer offers such examination free of cost by a qualified medical practitioner.

If the worker refuses, fails to be present or disregards for medical examination and other medical aid (free of cost and by qualified medical practitioner) and that such refusal, failure or disregards found to be unreasonable in the circumstances of the case and the injury resulted disablement, then he shall forego the right for compensation of the disablement.

15.8.6 Distribution of Compensation

The payment of compensation in respect of workman whose injury has resulted in death and the payment of lump sum as compensation to a workman or a person under a legal disability shall be made through the commissioner only. If any employer has made the payment directly, it shall not be considered to be a payment of compensation.

However, in the case of diseased workman, an employer may pay an advance up to ₹ 100 to the dependent of the diseased worker for meeting the immediate expenses. This advance shall be deducted by the commissioner from the amount of compensation and repaid to the employer.

If the employer has made the expenditure for funeral of the worker, then this amount (maximum of ₹ 50) shall be deducted by the commissioner from the amount of compensation to be paid thereafter and this amount is repaid to the employer.

15.8.7 Occupational Diseases

If a worker suffers any of the specified occupational disease, (the list of which is given in Schedule III*) and if he/she has worked in factory for at least six months, the employer has to pay compensation. Some of the occupational diseases are illness due to compressed air, poisoning by lead, phosphorus, mercury, benzene, arsenic, radium, and other radioactive substances, X-ray, etc.

[*The original Act consists of the Schedules under which the above contents are given. However, the reader if required may refer the book on labour legislation for details of these schedules.*]

BUSTLE - 15.4

1. *Describe the procedure for treatments from electric shock.*
2. *How would you act or react when a person is subjected to snake bite?*

Test Your Understanding 15.4

1. Explain various safety provisions under electricity rules.
2. Explain various provisions under Workmen's Compensation Act regarding accident of workman.
3. Comment on the statement, 'Safety promotes productivity'.

15.9 CONCLUSION

Prevention is better than cure. Having understood the adverse effects of accidents, one should definitely pay minimum attention to this aspect and take all the necessary steps to prevent and forestall the occurrence of the accidents. It is, therefore, not just plant engineer's but everybody's responsibility in the industry to be safe, work safely, and observe safety.

15.10 RECAP

This chapter focuses on the importance of safety in industry and the role of safety manager during discharging the duties of safety management entrusted with three-fold responsibilities, i.e., before, during, and after the accident. It also spotlights on first aid. The causes and effects of accidents are discussed exhaustively and the precautions are listed out. The efforts of Government and various legislations relevant to industrial safety and compensations are discussed in this chapter. Though this chapter is given at the end of this book, the topic in fact should be foremost to any employee of any industry. Moreover, it is very important to the Plant Engineering employees because, they are closely involved in repairs of small equipment to huge machinery. Further, in many industries, the plant manager is given an additional responsibility of safety management.

15.11 KEYWORDS

1. **First Aid:** The very first and preliminary action to be performed on an injured or who met with an accident.
2. **Accident:** Any sudden or unusual occurrence that interferes with the orderly progress of activity.
3. **Machinery Accidents:** Accidents caused due to inadequate safeguards, bad machine condition, overloading, improper operating, or handling of machines.
4. **Non-machinery Accidents:** Caused due to personal reasons such as age, physical weakness, inexperience and carelessness or from the plant conditions such as poor ventilation and illumination, etc.
5. **Crèches:** Room with required facilities maintained to take care of children under the age of 6 years.
6. **Overtime Wages:** Wages of a worker for the work, worked for more than 9 hours in a day or for more than 48 hours in a week.
7. **Young Person:** A child, who has not completed 14 years.
8. **Adolescent:** A person whose age is between 14 and 17 years.
9. **Dangerous Operations:** If any operation carried out in the factory exposes any employee to a serious type of bodily injury, poisoning or disease, the State Government may declare that operation to be dangerous or hazardous.
10. **Penalties/Punishments:** The charges to be paid for breach of legislative acts.
11. **Dependent:** Dependent means any of the following relatives of a deceased workman namely—

    (i) A wife, a minor legitimate son, and unmarried legitimate daughters, or a widowed mother.

    (ii) If wholly or in part dependent on the earnings of the workman at the time of his death, a husband, a parent other than a widowed mother, a minor illegitimate son, and an unmarried illegitimate daughter, a daughter illegitimate or illegitimate of married and a minor or if widowed, a minor brother, an unmarried or widowed sister, a widowed daughter-in-law, a minor child of deceased son.

12. **Partial Disablement:** It is the disablement of temporary nature and the earning capacity is reduced due to disablement of permanent nature.
13. **Total Disablement:** It is such disablement either temporary or permanent in nature, which incapacitates an employee from all work which he/she was capable of performing at the time of accident resulting in such disablement.

    Permanent total disablement shall be deemed to result from injuries where the aggregate percentage loss of earning capacity (as specified in Schedule I) amounts to one hundred percent or more.
14. **Wages:** Wages are the privileges or benefits or remuneration estimated in terms of money other than a travelling allowance or any other contribution paid towards pension, etc.
15. **Workman:** Workman means any person (other than a person whose employment is of a casual nature and who is employed otherwise than for the purposes of employer's trade or business) who is employed in any such capacity specified in Schedule II on monthly wages.
16. **Compensation:** The returns that a person gets in lieu of the loss incurred due to accidents or any unusual occurrence.
17. **Occupational Disease:** Specified disease, (the list of which is given in Schedule III) due to his occupation or employment and if he/she has worked in factory for at least six months, the employer has to pay compensation. Some of the occupational diseases are illness due to compressed air, poisoning by lead, phosphorus, mercury, benzene, arsenic, radium and other radioactive substances, X-ray, etc.

REVIEW QUESTIONS

1. Discuss the role of a plant manager as safety manager in maintaining safety of the equipment, humans, assets, and environment.
2. What are the functions and responsibilities of safety manager?
3. Do you think attaching 'safety management' to plant engineering department is right? Justify your answer.
4. What are the various legislations connected to industrial safety? Briefly describe the provisions under these acts.
5. Describe the provisions under Workman's Compensation Act. How are they connected to plant engineer?
6. Describe the provisions under Factories Act, 1948. How are they connected to plant engineer?
7. List the dos and don'ts to be observed at hazardous areas of an industry.
8. Explain how the employer and employees both are affected by the accidents.
9. What are the causes of industrial accidents? Suggest practical measures to minimise them.

Bibliography

Aggarwal, S., C., *Maintenance Management*, Prabhu Book Service, New Delhi.

Adam Jr., E., E., and R. J. Ebert., *Production and Operations Management*, Prentice-Hall Inc., Englewood-Cliffs.

Anthony, Kelly, *Maintenance and its Management*, Butterworth-Heinnemann.

Apte, S., *Maintenance Management*, National Productivity Council, New Delhi.

Banwet, D., K., 1981, *Some Studies in Facilities Location Planning*, Ph.D. Diss. IIT, Delhi (Unpublished).

Buffa, E., S., 1976., *Operations Management: The Management of Productive Systems*, John Wiley, New York.

Corder, A., S., *Maintenance Management Techniques*, McGraw-Hill, New York.

Ebeling, E., Charles, 1997, *An Introduction to Reliability and Maintainability Engineering*, McGraw-Hill International Editions, Singapore.

Francis, R., L., and J. A. White., 1974, *Facility Layout and Location: An Analytical Approach*, Prentice-Hall Inc., Englewood-Cliff.

Gardon, F., *Maintenance Engineering*, Applied Science Publishers Limited, London.

Greenhut, M., L., 1956, *Plant Location in Theory and Practice*, University of North Carolina Press, Durham.

Krishnaiah, J., Mohan, 1993, Total Productive Maintenance—Rapid Review Assessment Maintenance.

Lewis, E., E., 1989, *Introduction to Reliability Engineering*, John Wiley.

Majumdar, Nanda, 1998, T.P.M—The Zero Sum Technique, Business Today.

Menipaz, E., 1984, *Essentials of Production and Operations Management*, Prentice-Hall Inc., Englewood-Cliffs.

Mitchell, J., S., *Machinery Analysis and Monitoring*, Pennwell Publishing Co., Oklahoma, U.S.A.

M.J. Neale and Associates, *A Guide to the Condition Monitoring of Machinery*, HMSO, London.

Mohanty, R., P., 1984, *Works Management*, AIMA-Vikas Management Service, Vikas Publication, Delhi

Monks, J., G., 1977, *Operations Management: Theory and Problems*, McGraw-Hill, New York.

Morrow, *Maintenance Engineering Hand Book*, McGraw-Hill, New York.

Raju, N.V.S., *Introducing Condition Based Monitoring to Watch Case Unit Machinery*, M.Tech., Project Report.

Raju, N.V.S., 2004, Ph.D., Thesis titled *Reliability Centered and Condition Based Maintenance Decision system Models for Equipment Reliability Analysis* submitted to Osmania University, Hyderabad.

Raju, N.V.S., and Prof. K. Narayana Rao, 10–11 February, 2001, *Trend Analysis – A Tool For Reliability Centered Maintenance,* paper published at National Seminar on Modern Methods For Economic and Quality Production, conducted at Gramin Polytechnic, Vishnupuri, Nanded, Maharashtra State.

Raju, N.V.S., and Prof. K. Narayana Rao, 27–29 January 2003, *Statistical Tools for Equipment Reliability Analysis in Open Pit Mining,* paper published at National Conference on Modern Trends in Automobiles at Adhiyaman College of Engineering, Hosur, Tamilnadu, India.

Raju, N.V.S., and Prof. K. Narayana Rao, 27–29 January 2003, *Estimation of Equipment Reliability in Open Pit Mining Using PEXE,* paper published at National Conference on Modern Trends in Automobiles at Adhiyaman College of Engineering, Hosur, Tamilnadu, India.

Raju, N. V. S., Prof. K. Narayana Rao, and A. Mruthyunjayam, 7–9 May 2003, *Trend Analysis – A Tool for Achieving TQM,* paper published at National Conference on World Class Manufacturing, conducted by Amritha Institute of Technology, Coimbathore, Tamilnadu, India.

Raju, N.V.S., Prof. K. Narayana Rao, and N. Lakshminarayana, 29–30 June 2004, *Identifying Failure Distribution by Goodness of Fit Tests for Reliability Oriented Maintenance,* paper published at National Conference on State of the Art of Technologies in Mechanical Engineering (NCSAME) conducted by J.N.T.U, Hyderabad.

Raju, N.V.S., Prof. K. Narayana Rao, and N. Lakshminarayana, 29–30 June 2004, *Procedural Steps for Application of Reliability Centered Maintenance Models,* paper published at National Conference on State of the Art of Technologies in Mechanical Engineering (NCSAME) conducted by J.N.T.U, Hyderabad.

Raju, N.V.S., 2004, Ph.D. Thesis titled *Reliability Centered and Condition Based Maintenance Decision system Models for Equipment Reliability Analysis* submitted to Osmania University, Hyderabad.

Raju, N.V.S., N. Lakshmi Narayana, and Prof. P. Rami Reddy, 10 January–10 June, *Statistical Analysis and Inference for Reliability Oriented Maintenance*, published in International Journal of Emerging Technologies and Applications In Engineering, Technology and Sciences (Ij-Eta-Ets), Volume 3 : Issue 1, Page: 153.

Raju, N.V.S., and Prof. K. Narayana Rao, 29–30 September 2002, *Reliability Enhancement by Reliability Centered Maintenance for Overall Equipment Efficiency (OEE),* published at National Conference on Quality Control in Metallurgical Industries, at P.S.G. College, Coimbathore.

Raju, N.V.S., and Prof. K. Narayana Rao, 27–29 January 2002, *Statistical Tools for Equipment Reliability Analysis in Open Pit Mining,* paper presented and published at National Conference on Modern Trends in Automobiles at Adhiyaman College of Engineering, Hosur, Tamilnadu, India.

Rao, K., R., M., P. V. N. Prasad, 1997, *Effectiveness in Maintenance through Reliability Centred Maintenance—Models for Replacement and Overhand*, Proceedings of National Conference on Quality and Reliability (NCQR-97), S. J. College and Engineering Mysore.

Ramireddy, D., V., 1996, Total Productive Maintenance Seminar Report, Department of Reliability Engineering IIT/Bombay.

Reed, R., Jr., *Plant Location Layout and Maintenance*, Homewood, III, Irwin.

Rodriguz, J., R., 1998, *Total Productive Maintenance*, ESCI Hyderabad Training Programme on Production and Maintenance notes circulated.

Seiichi, Nakajima, 1988, *Introduction to TPM*, Productivity Press (UK).

Stewart, H., V., M., *Guide to Efficient Maintenance Management*, London.

Tompkins, J., A., and J. A. White., 1984, *Facilities Planning*, John Wiley, New York.

Turner, W., C., Mize, J. H., and K. E. Case, 1978, *Introduction to Industrial and Systems Engineering*, Prentice-Hall, Englewood-Cliffs.

Index

D

N

O

P

Q

R

S

T

www.ingramcontent.com/pod-product-compliance
Ingram Content Group UK Ltd.
Pitfield, Milton Keynes, MK11 3LW, UK
UKHW061534190726
13845UKWH00005B/96

9 789349 562738